Accounting, Finance, Sustainability, Governance & Fraud: Theory and Application

Series Editor

Kıymet Tunca Çalıyurt, Iktisadi ve Idari Bilimler Fakultes, Trakya University
Balkan Yerleskesi, Edirne, Türkiye

This Scopus indexed series acts as a forum for book publications on current research arising from debates about key topics that have emerged from global economic crises during the past several years. The importance of governance and the will to deal with corruption, fraud, and bad practice, are themes featured in volumes published in the series. These topics are not only of concern to businesses and their investors, but also to governments and supranational organizations, such as the United Nations and the European Union. Accounting, Finance, Sustainability, Governance & Fraud: Theory and Application takes on a distinctive perspective to explore crucial issues that currently have little or no coverage. Thus the series integrates both theoretical developments and practical experiences to feature themes that are topical, or are deemed to become topical within a short time. The series welcomes interdisciplinary research covering the topics of accounting, auditing, governance, and fraud.

Eyüp Çetin · Hilal Özen
Editors

Healthcare Policy, Innovation and Digitalization

Contemporary Strategy and Approaches

Foreword by Nobel Laureate Aziz Sancar

Springer

Editors
Eyüp Çetin
Department of Quantitative Methods
School of Business, Istanbul University
Istanbul, Türkiye

Hilal Özen
Department of Business Administration
Faculty of Business Administration
and Economics
Trakya University
Edirne, Türkiye

ISSN 2509-7873 ISSN 2509-7881 (electronic)
Accounting, Finance, Sustainability, Governance & Fraud: Theory and Application
ISBN 978-981-99-5963-1 ISBN 978-981-99-5964-8 (eBook)
https://doi.org/10.1007/978-981-99-5964-8

This Springer imprint is published by the registered company Springer Nature Singapore Pte Ltd.
The registered company address is: 152 Beach Road, #21-01/04 Gateway East, Singapore 189721, Singapore

Paper in this product is recyclable.

Foreword by Prof. Dr. Aziz Sancar, The Nobel Prize Laureate in Chemistry 2015

Scientists possess a natural inclination to share the knowledge they have painstakingly acquired. This instinct serves as a driving force behind countless scientific research endeavors, including the creation of valuable edited works like this one. Our world is evolving rapidly, bringing forth new challenges in the form of various diseases on a daily basis. Simultaneously, digital technologies are increasingly integrated into medicine and healthcare, influencing every facet of these critical fields. As a result, healthcare has emerged as one of the foremost concerns of our century, encompassing vital topics such as government healthcare policies, healthcare innovation, pandemics, digital healthcare transformation, health informatics, data science, telemedicine, and health economics. It is imperative that these subjects are thoroughly explored through research, and their conclusions are disseminated to professionals, scientists, and the general public.

It is with great pleasure that I pen this foreword for *Healthcare Policy, Innovation, and Digitalization: Contemporary Strategy and Approaches*, published by Springer. The significance of healthcare permeates all aspects of our lives, and this book makes a comprehensive effort to address healthcare from a multifaceted perspective, assembling numerous valuable contributions from the contemporary world. I firmly believe that this book will serve as an invaluable reference for healthcare researchers, professionals, students, and anyone with an interest in these critical topics. On a personal note, I am delighted to extend my heartfelt congratulations and gratitude to all the contributing authors for their invaluable contributions to this edited volume. I reserve special commendation and gratitude for the editors, Professors Eyüp Çetin and Hilal Özen, for their outstanding contribution to the literature.

2023

Prof. Dr. Aziz Sancar, M.D., Ph.D.
Department of Biochemistry and Biophysics
and Department of Biology
The University of North Carolina at Chapel Hill
Chapel Hill, NC, USA

Prof. Dr. Aziz Sancar is a distinguished Turkish Molecular Biologist renowned for his expertise in DNA repair and circadian clock research. In 2015, he was honored with the Nobel Prize in Chemistry, sharing this prestigious recognition with Tomas Lindahl and Paul L. Modrich for their pioneering mechanistic studies of DNA repair. His groundbreaking work has led to significant advancements in our understanding of photolyase and nucleotide excision repair, fundamentally reshaping the landscape of his field.

Currently holding the esteemed position of Sarah Graham Kenan Professor of Biochemistry and Biophysics at the University of North Carolina School of Medicine, he also serves as a valued Member of the UNC Lineberger Comprehensive Cancer Center. Beyond his academic accomplishments, Prof. Dr. Sancar is Co-founder of the Aziz and Gwen Sancar Foundation, a non-profit organization dedicated to promoting Turkish culture and providing support to Turkish students in the United States.

Acknowledgements

We wish to extend our heartfelt gratitude for the inspiring Foreword provided by Nobel Laureate Prof. Aziz Sancar of the University of North Carolina, Chapel Hill, NC, USA. Our sincere appreciation goes out to all the contributors and authors, as well as Nobel Laureate Prof. Aziz Sancar, whose collective efforts have brought this edited book to life.

We also wish to express our deep appreciation to Series Editor Prof. Kıymet Çalıyurt and the entire editorial team at Springer Nature for their unwavering support in bringing this outstanding volume to publication.

Our aspiration is that readers, researchers, and practitioners will find the chapters published in this book both insightful and valuable as they cover a wide range of areas in healthcare policy, innovation, and digitalization.

Prof. Dr. Eyüp Çetin
Prof. Dr. Hilal Özen

Contents

Editors and Contributors

About the Editors

Prof. Dr. Eyüp Çetin is tenured Professor of Operations Research and Statistics in the Department of Quantitative Methods in School of Business at Istanbul University and Affiliate Distinguished Professor at York University, Canada. He is also affiliated with New York Business Global, USA. He has held some managerial positions. He is Former and Founding Dean of the Faculty of Transportation and Logistics and Former Vice Dean of the School of Business at Iştanbul University. He received a Ph.D. in quantitative methods from Istanbul University in 2004. He has taught courses in management science, statistics, business analytics, stochastic models, game theory, and health analytics to all level of business, engineering, and healthcare management students: undergraduates, MBAs, and Ph.D. students. He has been Consultant for numerous companies, and he has taught statistical and decision modeling and business analytics in even global companies such as Vodafone. His current research focuses on business analytics, data science, digital business and health, healthcare management science, mathematical medicine, operations research, and mathematical and statistical modeling. He is also Founder and Editor-in-Chief of prestigious international academic journals such as *European Journal of Pure and Applied Mathematics.*

Prof. Dr. Hilal Özen holds an honors degree in Business Administration (B.S.) from Hacettepe University (2004). She earned her M.S. degree in 2007 and her Ph.D. in 2011, both in Marketing, from Istanbul University School of Business. Currently, she serves as a Professor of Marketing at Trakya University. Her current research interests are focused with great enthusiasm on various areas, including digital marketing, e-health, social media marketing, tourism marketing, sustainability, and consumer decision-making styles. She has published numerous articles on these topics in reputable peer-reviewed journals and presented her research at international conferences. Throughout her academic career, she has been actively involved in teaching a wide range of courses, including principles of marketing, marketing management, research methodology, marketing strategies, current issues in marketing, digital marketing, and healthcare marketing. Her teaching experience spans across all levels of business students, including undergraduate, MS, MBA, and executive MBA students. In addition to her academic pursuits, Hilal is a Co-founder of Coreborn Software and Digital Marketing Company, where she provides valuable consultancy in the field of digital marketing to clients.

Contributors

Fredrick D. Ashbury Department of Oncology, University of Calgary, Calgary, AB, Canada;
Dalla Lana School of Public Health, University of Toronto, Toronto, Canada;
VieCure, Denver, CO, USA

Kostas Athanasakis Department of Public Health Policy, School of Public Health, University of West Attica, Athens, Greece

Armağan Bayram Industrial and Manufacturing Systems Engineering, University of Michigan at Dearborn, Dearborn, USA

Eyüp Çetin Department of Quantitative Methods, School of Business, Istanbul University, Istanbul, Türkiye

John Tuhao Chen Department of Mathematics and Statistics, Bowling Green State University, Bowling Green, OH, US

Lincy Y. Chen School of Industrial and Labor Relations, Cornell University, Ithaca, NY, US

Adele Diederich Department of Psychology, Carl Von Ossietzky University of Oldenburg, Oldenburg, Germany

Yuchi Guo Industrial and Manufacturing Systems Engineering, University of Michigan at Dearborn, Dearborn, USA

Gökçe Kahvecioğlu Supply Chain Optimization Technologies, Amazon, Seattle, WA, USA

Eva K. Lee Center for Operations Research in Medicine and HealthCare, The Data and Analytics Innovation Institute, Tbilisi, Georgia;
USA NSF I/UCRC Center for Health Organization Transformation, Washington, USA;
Industrial and Systems Engineering, Georgia Institute of Technology, Atlanta, Georgia

Lohuwa Mamudu Department of Public Health, California State University, Fullerton, CA, USA

Hilal Özen Department of Business Administration, Faculty of Business Administration and Economics, Trakya University, Edirne, Türkiye

Stefan Pickl Institut für Theoretische Informatik, Mathematik und Operations Research, Universität der Bundeswehr München, Neubiberg, Germany

Ramalingam Shanmugam Honorary Professor of International Studies and Statistics, School of Health Administration, Texas State University San Marcos, San Marcos, TX, USA

Alan Shaw Chair of the Retail Institute Special Interest Research Group and Health Research Lead for the Business School, Leeds Business School, Leeds Beckett University, Leeds, UK

Karan P. Singh Department of Epidemiology and Biostatistics, School of Medicine, The University of Texas at Tyler, Tyler, TX, USA

Kyriakos Souliotis Department of Social and Education Policy, University of Peloponnese, Corinth, Greece;
Health Policy Institute, Athens, Greece

Keith Thompson VieCure, Denver, CO, USA;
Faculty of Medicine, University of Alabama, Birmingham, AL, USA;
Montgomery Cancer Center, Montgomery, AL, USA

Chris P. Tsokos Distinguished University Professor Department of Mathematics and Statistics, University of South Florida, Tampa, FL, USA

Ali Ünlü School of Social Sciences and Technology, Technical University of Munich, Munich, Germany

Marc Wyszynski Department of Mathematics and Computer Science, University of Bremen, Bremen, Germany

Xiao Yu Industrial and Manufacturing Systems Engineering, University of Michigan at Dearborn, Dearborn, USA

Maryna Zharikova Institut für Theoretische Informatik, Mathematik und Operations Research, Universität der Bundeswehr München, Neubiberg, Germany;
Program Tools and Technologies Department, Kherson National Technical University, Kherson, Ukraine

Konstantinos Zisis Department of Public Health Policy, School of Public Health, University of West Attica, Athens, Greece;
Institute for Health Economics, Athens, Greece

List of Figures

List of Tables

Part I
Introduction

Chapter 1
Contemporary Strategy and Approaches in Healthcare Policy, Innovation and Digitalization

Hilal Özen and Eyüp Çetin

Abstract In today's world, people are more aware and conscious about health issues than before. The main reason behind this can be both the developments in health issues and the process we have experienced during the COVID-19 pandemics. We all are more used to digital health applications and innovative health solutions. These developments trigger the changes and also arrangements in healthcare policy issues. All the topics have both theoretical and practical implications in the industry and academic world. This chapter tries to summarize the importance of the new strategies and approaches in healthcare policy, innovation, and digitalization by making an extensive introduction to those three concepts.

Keywords Healthcare policy · Innovation · Digitalization

1.1 Introduction

Especially after COVID-19 pandemic we must accept that the whole world experienced a sharp shift in healthcare industry and was thrown to a different direction. Digitalization was the main issue, most of the countries realized that they should check their policies, and also they had to develop their innovative solutions for healthcare issues. So, health care as a top concept at the last three years had strong connections with innovation, digitalization, and policies which built the topic of this edited book. It is obvious that the need for innovation, digitalization, and enhanced

H. Özen (✉)
Department of Business Administration, Faculty of Business Administration and Economics, Trakya University, Edirne, Türkiye
e-mail: hilalozen@trakya.edu.tr

E. Çetin
Department of Quantitative Methods, School of Business, Istanbul University, Istanbul, Türkiye
e-mail: eyup.cetin@istanbul.edu.tr

E. Çetin and H. Özen (eds.), *Healthcare Policy, Innovation and Digitalization*, Accounting, Finance, Sustainability, Governance & Fraud: Theory and Application,
https://doi.org/10.1007/978-981-99-5964-8_1

policy issues in health care has never been greater than it is today. This book offers a multidisciplinary approach to health care, in both managerial and clinical views. People reading this book will gain a deep understanding of an integrated approach for healthcare strategy in various fields, especially through the lens of new policies, innovation, and digitalization.

With the help of digital technologies in health care, the performance and innovation in the field increased. Machine learning, AI technologies, IoT have all modernized the healthcare performance for predicting the future policy analysis, care settings, and both challenges and opportunities. All those issues were in focus of many studies in the last years (Gopal et al. 2019; Fagherazzi et al. 2020; Kapoor et al. 2020; Matheny et al. 2020; Ziadlou 2020; Javaid and Khan 2021; Lyng et al. 2021; Cannavale et al. 2022), but most of them handled them separately, but in this edited book we aimed to look from a broader approach.

The spread of COVID-19 pandemic boosted digitalization and innovation in digital health technologies both in theory and application. So, the policy part has gained importance as the other issues. There were some papers investigating and criticizing the policies applied by countries and their results. Atkinson et al. (2020) in their study investigated the pandemic policy of UK government and criticized politicians about their decisions. Maor and Howlett (2020) handled the issue from a broader perspective and investigated the governments' policies to fight against COVID-19 by examining the combination of psychological, institutional, and strategic factors. Before and after pandemic the healthcare policy throughout the world is an important issue that should also be investigated. In this chapter, we summarize these three important concepts which are directly related to health care.

In this edited book, we include some remarkable contemporary strategy and approaches in healthcare policy, innovation, and digitalization analyses from micro to macro levels by well-known experts and researchers.

1.2 Healthcare Policy

Healthcare policy is one of the central issues of healthcare service that healthcare providers deliver from micro level to macro frame. In addition to policy, healthcare strategy which may be considered as a roadmap for the adopted policy journey is also equivalently important. The healthcare policies and strategies including treatment methods have been recently become more significant than ever especially during the COVID-19 pandemic.

The literature has been huge to develop healthcare policy and strategies. Beyond, healthcare researchers focusing on policy and strategy have discovered many up to date information and data sources to make research like web search engines (mostly Google), various SNS, and many health-related platforms. The COVID-19 pandemic has accelerated the progress of the literature as expected in such global devastating calamity. To sample a few, Çetin et al. (2020) developed a mathematical framework to fight pandemics in general and applied their models to COVID-19 combat in the United States with real data. Also, we have globally seen the differentiation of COVID-19 vaccine development race; for instance, Pfizer-BioNTech and Moderna employs mRNA-based approach when Johnson & Johnson uses Ad26, and some others adopt inactivated techniques (Çetin et al. 2021).

On the other hand, some local researches were executed about healthcare strategy and policy. AlQutob et al. (2020) mentioned in their study to the importance of policy and strategy during the COVID-19 pandemic lockdown in Jordan and the UAE. They especially highlight the importance of executing appropriate public health strategies and policies. Raoofi et al. (2020) criticized the efficiency of health policies during COVID-19 pandemic in Iran. Similarly, Nah et al. (2020) developed a new optimal policy approach to inform the re-opening plan followed by a postpandemic lockdown by using stochastic optimization. The authors apply their policy proposal to Ontario's re-opening plans in Canada.

There are also some nice coincidences with the origins of diseases and quantitative methodological approaches. For instance, Çetin et al. (2021) ranked the COVID-19 vaccines using grey relational analysis. It is remarkable that they suggest the use of a Wuhan-originated tool, grey systems theory which has spread like a very-useful-pandemic for almost 40 years, to compare, rate, and rank the vaccines to fight another Wuhan-originated agent, coronavirus pandemic.

1.3 Innovation

Innovation in health care is one of the up to date challenge topics which is also directly related to digitalization and policy. Vast developments in technology and also the unexpected changes happening around the world forces both private companies and governments to create innovative solutions for health care. This topic's importance has increased in the last decades because of the challenges happening in health care like the COVID-19, rising costs, and the aging demographic in most of the countires around (Proksch et al. 2019). Innovation in health care is a key term for creating solutions to the stated problems.

Since the term innovation comes from the Latin noun "innŏvātĭo" which refers to "the act or process of introducing new ideas, devices, or methods or to the new ideas, devices, or methods themselves" (Orianaa et al. 2016, p. 47); this topic is a multifacet issue both for healthcare organizations, companies, and policy makers. Their steps will be a lifeblood for improving the healthcare system which has a direct impact on patient care. Thus, innovation in health care gains importance for both practitioners and academics.

Innovation in healthcare industry has an extensive role in many topics which can be summarized as information technology, neuro technology, communication technology, digitalization, biotechnlogy, nanotechnology, etc. So, this topic has a role to close the gap between healthcare providers and patients by improving the service given to them.

1.4 Digitalization

Digitalization in health care can be argued as the most important topic especially in the last three years. Because it can be argued that the pandemic period accelerated the use of digital technologies and the digitalization process of orgnizations. The concept has an incredible broad content from many perspectives like communicaton in both conventional and digital channels, patient care, hospital visits, or sustainability. Especially after COVID-19 pandemic telemedicine and mobile health technologies were the leading topics in healthcare digitalization processes. People needed to change the way they were used to while getting healthcare services. They could not be able to visit their doctors, so they had to adapt to digital technologies that were used in place of physical visits. On the other hand, hospitals and clinics needed to adapt to digital technologies more rapidly. Besides when the sustainable development goals (SDG) of UN checked, digitalization in health care has some effects on SDG 3, SDG 10, and SDG 13 (Özen 2021).

But all of them have "data" in common. Different technologies that are integrated into health care are generating huge amounts of data which should be both processed and evaluated deeply before they can be used effectively (Vaagan et al. 2021) to make decisions and improve the management processes of many firms. Data is the most important variable for all healthcare digitalization concepts. Healthcare data is priceless in today's world. Since machine learning approaches, capacity settings, patient treatment procedures, all have a correlation with data.

Data or with the popular term "big data" is used for getting insights from life. As Agrawal and Prabakaran (2020, p. 525) argued in their article "Big Data will be an integral part of the next generation of technological developments". As in other fields big data's application to health care drives a significant potential for everyone from patients to healthcare professionals by enabling them to gain better insights.

It can be argued that digitalization in health care helps to improve the quality of service and patient care by offering a huge range of possibilities (Butcher and Hussain 2022). With the help of digitalization, most of the barriers and problems may be overcome, and this book will shed a light on those new opportunities.

References

Agrawal R, Prabakaran S (2020) Big data in digital healthcare: lessons learnt and recommendations for general practice. Heredity 124:525–534. https://doi.org/10.1038/s41437-020-0303-2

AlQutob R, Moonesar IA, Tarawneh MR, Al Nsour M, Khader Y (2020) Public health strategies for the gradual lifting of the public sector lockdown in Jordan and the United Arab Emirates during the COVID-19 crisis. JMIR Pub Heal Surveil, 6(3). https://doi.org/10.2196/20478

Atkinson P, Gobat N, Lant S, Mableson H, Pilbeam C, Solomon T, Tonkin-Crine S, Sheard S (2020) Understanding the policy dynamics of COVID-19 in the UK: early findings from interviews with policy makers and health care professionals. Soc Sci Med, 266. https://doi.org/10.1016/j.socscimed.2020.113423

Butcher CJT, Hussain W (2022) Digital healthcare: the future. Future Healthcare J 9(2):113–117. https://doi.org/10.7861/fhj.2022-0046

Çetin E, Kiremitci S, Kiremitci B (2020) Developing optimal policies to fight pandemics and COVID-19 combat in the United States. European J Pure Appl Math 13(2):369–389. https://doi.org/10.29020/nybg.ejpam.v13i2.3700

Çetin E, Özen H, Özen Ö (2021) Primus inter pares: a comparison and ranking of COVID-19 vaccines. J Anal 1(1):1–19. https://doi.org/10.29020/nybg.ja.v1i1.6

Cannavale C, Tammaro AE, Leone D, Schiavone F (2022) Innovation adoption in inter-organizational healthcare networks—the role of artificial intelligence. European J Inno Manag 25(6):758–774

Fagherazzi G, Goetzinger C, Rashid MA, Aguayo GA, Huiart L (2020) Digital health strategies to fight COVID-19 worldwide: challenges, recommendations, and a call for papers. J Med Internet Res 22(6). https://doi.org/10.2196/19284

Gopal G, Suter-Crazzolara C, Toldo L, Eberhardt W (2019) Digital transformation in healthcare—architectures of present and future information technologies. Clinical Chem Lab Med (CCLM) 57(3):328–335. https://doi.org/10.1515/cclm-2018-0658

Javaid M, Khan IH (2021) Internet of Things (IoT) enabled healthcare helps to take the challenges of COVID-19 pandemic. J Oral Biol Craniofacial Res 11(2):209–214. https://doi.org/10.1016/j.jobcr.2021.01.015

Kapoor A, Guha S, Das MK, Goswami KC, Yadav R (2020) Digital healthcare: the only solution for better healthcare during COVID-19 pandemic? Indian Heart J 72(2):61–64. https://doi.org/10.1016/j.ihj.2020.04.001

Lyng HB, Macrae C, Guise V, Haraldseid-Driftland C, Fagerdal B, Schibevaag L, Alsvik JG, Wiig S (2021) Balancing adaptation and innovation for resilience in healthcare—a metasynthesis of narratives. BMC Health Serv Res 21:759. https://doi.org/10.1186/s12913-021-06592-0

Matheny ME, Whicher D, Israni ST (2020) Artificial intelligence in health care a report from the National Academy of Medicine. JAMA 323(6):509–510. https://doi.org/10.1001/jama.2019.21579

Maor M, Howlett M (2020) Explaining variations in state COVID-19 responses: psychological, institutional, and strategic factors in governance and public policy-making. Pol Design Pract 3(3):228–241. https://doi.org/10.1080/25741292.2020.1824379

Nah K, Chen S, Xiao Y, Tang B, Bragazzi N, Heffernan J, Asgary A, Ogden N, Wu J (2020) Scenario tree and adaptive decision making on optimal type and timing for intervention and social-economic activity changes to manage the COVID-19 pandemic. European J Pure Appl Math 13(3):710–729. https://doi.org/10.29020/nybg.ejpam.v13i3.3792

Orianaa C, Patrizioa A, Robertaa BP, Mariannaa C, Claudioa J, Rosannaa T (2016) De innovatione: the concept of innovation for medical technologies and its implications for healthcare policy-making. Health Policy and Technology 5(1):47–64. https://doi.org/10.1016/j.hlpt.2015.10.005

Özen H (2021) Dünya sağlık hizmetlerinin sürdürülebilir kalkınma hedefleri açısından değerlendirilmesi. OPUS–Uluslararası Toplum Araştırmaları Dergisi 17(38):5440–5472. https://doi.org/10.26466/opus.92718789

Proksch D, Busch-Casler J, Haberstroh MM, Pinkwart A (2019) National health innovation systems: clustering the OECD countries by innovative output in healthcare using a multi-indicator approach. Res Policy 48:169–179. https://doi.org/10.1016/j.respol.2018.08.004

Raoofi A, Takian A, Sari AA, Olyaeemanesh A, Haghighi H, Aarabi M (2020) COVID-19 pandemic and comparative health policy learning in Iran. Arch Iran Med 23(4):220–234. https://doi.org/10.34172/aim.2020.02

Vaagan RW, Torkkola S, Sendra A, Farré J, Lovari A (2021) A critical analysis of the digitization of healthcare communication in the EU: a comparison of Italy, Finland, Norway, and Spain. Int J Commun 15:1718–1740

Ziadlou D (2020) Managerial innovation for digital healthcare transformation. In: Leadership, management, and adoption techniques for digital service innovation. IGI Global Publishing. https://doi.org/10.4018/978-1-7998-2799-3.ch008

Prof. Dr. Hilal Özen holds an honors degree in Business Administration (B.S.) from Hacettepe University (2004). She earned her M.S. degree in 2007 and her Ph.D. in 2011, both in Marketing, from Istanbul University School of Business. Currently, she serves as a Professor of Marketing at Trakya University. Her current research interests are focused with great enthusiasm on various areas, including digital marketing, e-health, social media marketing, tourism marketing, sustainability, and consumer decision-making styles. She has published numerous articles on these topics in reputable peer-reviewed journals and presented her research at international conferences. Throughout her academic career, she has been actively involved in teaching a wide range of courses, including principles of marketing, marketing management, research methodology, marketing strategies, current issues in marketing, digital marketing, and healthcare marketing. Her teaching experience spans across all levels of business students, including undergraduate, MS, MBA, and executive MBA students. In addition to her academic pursuits, Hilal is a Co-founder of Coreborn Software and Digital Marketing Company, where she provides valuable consultancy in the field of digital marketing to clients.

Prof. Dr. Eyüp Çetin is tenured Professor of Operations Research and Statistics in the Department of Quantitative Methods in School of Business at Istanbul University and Affiliate Distinguished Professor at York University, Canada. He is also affiliated with New York Business Global, USA. He has held some managerial positions. He is Former and Founding Dean of the Faculty of Transportation and Logistics and Former Vice Dean of the School of Business at Istanbul University. He received a Ph.D. in quantitative methods from Istanbul University in 2004. He has taught courses in management science, statistics, business analytics, stochastic models, game theory, and health analytics to all level of business, engineering, and healthcare management students: undergraduates, MBAs, and Ph.D. students. He has been Consultant for numerous companies, and he has taught statistical and decision modeling and business analytics in even global companies such as Vodafone. His current research focuses on business analytics, data science, digital business and health, healthcare management science, mathematical medicine, operations research, and mathematical and statistical modeling. He is also Founder and Editor-in-Chief of prestigious international academic journals such as *European Journal of Pure and Applied Mathematics*.

Part II
Healthcare Policy

Chapter 2
Prioritization in Health Care: The Influence of Frames on Accepting Prioritization Criteria

Adele Diederich and Marc Wyszynski

Abstract Global pandemics, social and scientific developments such as growing and aging populations, novel and expensive health technologies, and improved medical treatments increase the demand for healthcare services and challenge healthcare systems worldwide. Prioritizing healthcare services according to some pre-defined criteria such as age, health behavior, and social responsibility has been proposed for distributing limited resources. A major concern in healthcare policies is establishing fair and legitimized procedures for distributing scarce healthcare resources. Ways of legitimizing prioritization criteria in terms of fairness and justice are, for instance, drawing on theoretical considerations such as (normative) distributive justice principles, like equity, equality, and need, mostly relying on experts' opinions; or using more practical approaches such as direct-democracy-like elements, i.e., including the general public in decisions on prioritizing scarce healthcare resources. For the latter, opinions and preferences of those are typically elicited by offering them survey questionnaires. However, past research has shown that participants' preferences may be influenced by how the questionnaire is constructed and items are phrased and framed. This chapter discusses the impact of framing on attitudes toward prioritization criteria in healthcare service. In particular, questions are framed in terms of providing and withholding services. By combining psychological theory on judgment and decision-making with recent empirical findings and previous research, we show that the effect of framing on people's preferences for healthcare prioritization criteria is an essential factor that healthcare policymakers may want to consider when using elements of direct democracy.

Keywords Prioritization · Healthcare · Healthcare policymaker · Democracy

---

A. Diederich (✉)
Department of Psychology, Carl Von Ossietzky University of Oldenburg, Oldenburg, Germany
e-mail: adele.diederich@uni-oldenburg.de

M. Wyszynski
Department of Mathematics and Computer Science, University of Bremen, Bremen, Germany
e-mail: m.wyszynski@uni-bremen.de

E. Çetin and H. Özen (eds.), *Healthcare Policy, Innovation and Digitalization*,
Accounting, Finance, Sustainability, Governance & Fraud: Theory and Application,
https://doi.org/10.1007/978-981-99-5964-8_2

2.1 Introduction

Aging populations, new clinical pictures, epidemiological changes, novel and expensive health technologies, and social values, increasing costs, and reductions in tax revenues are working together to deeply stress healthcare systems in most first-world countries. The current global pandemic drastically adds to the overwhelming demand for healthcare services.

The need for controlling or reducing the cost of healthcare delivery to guarantee a stable and sustainable system has long been recognized in both public and privately funded systems (e.g., Hauck et al. 2004). Prioritization of patient care has been proposed as one means to achieve this (e.g., Nagel and Lauerer 2016, for an overview).

Prioritization, aka prioritizing, aka priority setting, has been defined in various ways but typically includes specifying a hierarchy of patients or diseases to be treated, or treatments to be offered. The basis for building a hierarchy or rank order are specified criteria (e.g., Cochrane 2016), some of them to be discussed in the following. Sometimes priority setting refers to resource allocation or rationing, that is, the distribution of limited resources among competing patients or treatments or health programs (e.g., Bowling 1996; Bruni et al. 2008). The opposite of prioritization has sometimes been referred to as posteriorizing (see e.g., Diederich et al. 2012; Diederich and Schreier 2010).

Most stakeholders involved in setting priorities, such as healthcare professionals, pharmaceutical companies, and special interest groups, actively work to make their desires known. A common approach taken by those policymakers is to adopt a utilitarian framework that seeks maximization of societal health benefits through reliance on the cost-effectiveness of health services which has practical implications for physicians and foremost for patients, in particular, when it comes to reforming healthcare systems for controlling or reducing the cost of healthcare delivery (see e.g., Ham 1997; Sabik and Lie 2008). But citizens, financing the healthcare system by taxes or premiums of health insurances and using the services will likely have the greatest difficulty in providing input to these discussions.

Including the perspectives, values, preferences, and expectations of the general public about prioritizing procedures, however, is indispensable for a fair distribution of medical services (e.g., Daniels 2007; Daniels and Sabin 2002). It has been repeatedly argued that including healthy and unhealthy citizens is mandatory to legitimate health policy decisions (e.g., Bowling 1996; Bruni et al. 2008; Diederich 2016b; Diederich and Salzmann 2015; Diederich and Schreier 2009; Mossialos and King 1999). Because the public is the most important stakeholder in the healthcare system, its engagement in priority setting in policy-making is in keeping with the ideals of democracy (Bruni et al. 2008, p. 15). This is even stated in the Ljubljana charter on reforming healthcare in Europe which reads "Healthcare reforms must address citizens' needs, taking into account, through the democratic process, their expectations about health and healthcare" (World Health Organization. Regional Office for Europe 1996, p. 2).

Missing transparency in the decision-making process may be a major reason for the lack of acceptance of healthcare reforms as well as loss of trust and confidence in the healthcare system (e.g., Diederich et al. 2009; Fleck 2016; Ham 1997). If priority-setting decisions are to be accepted, it is important to include the public in the decision-making process (Ham 1997).

This has been tried or achieved in some countries early on. The public has been included in the decision-making process by establishing specific committees or forums (for instance, Norway, Lønning Committee I and II, 1987/1997; Oregon/USA, Oregon Health Service Commission, 1989; The Netherlands, Dunning Committee, 1992/1995; Sweden, Commission of Parliament and Experts, 1993/1995; New Zealand, Core Services Committee/National Health Committee, 1993; Israel, Medical Technical Forum and National Advisory Committee, 1995; Denmark, Danish Counsel of Ethics, 1997; England and Wales, National Institute for Health and Clinical Excellence [NICE], 1999). The format varied from public hearings, discussion forums, and regional conferences, to (telephone-) surveys (for details, see Ham 1997; Sabik and Lie 2008). In Germany, the scientific discourse and public discussions on prioritizing healthcare services started relatively late in comparison with these countries (Diederich et al. 2009). Without going into all different aspects of prioritization (implicit: doctors just do it; explicit: criteria are agreed upon via guidelines or code of medical practice; vertical: within a given disease; horizontal: across different diseases; see e.g., Cochrane 2016), one reason for the cautious discussion in Germany was the confusion with rationing healthcare services. That is, a patient would not get the service at all rather than, for instance, later. When involving the general public, the most realistic and utilized method for large numbers of participants are surveys. There is extensive research on how to construct questionnaires to avoid response biases due to technicalities (see e.g., Dillman et al. 2014; Rossi et al. 2013). Less attention, however, is paid to the framing of the item. Numerous studies in psychology and related fields, however, have shown that participants' preferences may be influenced by how questions are phrased and framed, for instance, in terms of positive or negative statements.

The chapter is organized as follows. To set the stage, we first define frames and framing effects as used in the psychological and related research. Then we briefly summarize some studies on criteria that have been proposed to prioritize healthcare services. The subsequent part reports the results of studies on eliciting attitudes toward prioritization criteria when including frames and shows new data from a recent study. The chapter ends with some concluding remarks.

2.2 Frames and Framing Effects

A decision frame refers to the decision maker's conceptions of choice options among which s/he has to choose, the possible outcomes, and conditional probabilities associated with a particular choice (Tversky and Kahneman 1981, p. 453). In other words, a decision frame indicates how the decision maker perceives, imagines, and interprets the decision situation. Which decision frame a decision maker adopts may depend, on the one hand, on personal characteristics, the decision maker's experiences, and his/her normative values and habits. This type of framing has also been referred to as internal framing (Fischer 1997). On the other hand, perception and interpretation of a frame may be influenced by how the problem is formulated. This type of framing is called external framing. External framings are furthermore used in a "strict" and a "loose" sense (Kühberger 1998). In its strict form, it refers to the description or formulation of one and the same fact but from a different perspective, for instance, by using different wordings like positive and negative expressions. In the current context, an example of a positively framed question about personal responsibility as a prioritization criterion scenario is "Should non-smokers receive partial refunds of their health insurance premiums?" A negatively framed scenario is "Should smokers pay higher copayments to their health insurance?" (Diederich 2016a).

In its loose form, framing includes other manipulations such as presentation formats or different contexts. For example, a person may be described in detail (e.g., sex and occupation) as compared to no details, which are irrelevant for making the decision. The identifiability of the recipient may produce different results. This is known as identifiable victim, identifiable person, or identifiable "others" effect (Jenni and Loewenstein 1997; Kogut and Ritov 2005; Loewenstein et al. 2005, 2006).

Framing effects occur when different formulations of the decision problem or irrelevant aspects in the description of the problem lead to different decisions or evaluations. They are based on preference reversals and preference shifts.

For instance, agreements tend to be higher to positively framed statements as compared to negatively framed statements; donors tend to be more generous to identified victims/persons as compared to unidentified or statistical victims/persons.

In the context of distributive justice, Törnblom (1988) proposes a systematic framework within which the allocation process is explicated in terms of positive and negative outcome allocations. Each of the two outcome allocations includes different forms, which may be accomplished in different ways. Törnblom distinguishes three such modes: delivery, withholding, and withdrawal. Delivery means presenting, transferring, handling, or giving something; withholding is refraining from presenting something; and withdrawal refers to taking away or removing something that the target person possesses (Törnblom 1988, p. 149). Any specific form of outcome allocation may be accomplished in any one of the three accomplishment modes. The consequences, i.e., benefits or harms, depend on specific combinations of outcomes and accomplishment modes. For example, positive consequences or benefits may be obtained by delivering desirable outcomes and by withholding or withdrawing aversive outcomes. Negative consequences or harms can be obtained by delivering aversive outcomes, by withholding positive outcomes, and by withdrawing positively valued resources.

Framing effects have been explored in numerous experiments across many disciplines with a variety of decision scenarios (for reviews see e.g., Gächter et al. 2009; Gilovich et al. 2002; Kahneman et al. 1982; Kühberger 1998; Levin et al. 1998; Steiger and Kühberger 2018). Surprisingly few studies have been conducted to investigate the influence of frames on accepting or rejecting prioritization criteria.

2.3 Prioritization Criteria

Different countries and stakeholders have proposed various criteria for prioritizing healthcare services. Some countries focused on more abstract prioritization principles such as human dignity, need and solidarity, freedom and right of self-determination, necessity of a treatment, and more. Based on these values, prioritization criteria such as quality of the services, cost-effectiveness-efficiency, or personal responsibility were derived. Others postulated concrete allocation guidelines for medical services and new medical technologies as foundation for priority setting. A cost-effectiveness analysis serves as a criterion to include or exclude a treatment (for a review and summary, see Sabik and Lie 2008).

Medical need in terms of severity of illness, the urgency of intervention, or life-threatening situations (e.g., "Rule of Rescue"; Hauck et al. 2004; Jonsen 1986; McKie and Richardson 2003) is generally accepted as a criterion with the highest priority. Across several studies, the majority of respondents agreed that patients suffering a life-threatening disease should be given priority to others (e.g., Diederich et al. 2012; Pinho and Borges 2018; Stumpf et al. 2014). The urgency of treatment (e.g., Diederich et al. 2012) and severity of the disease (e.g., Ryynänen et al. 1999) received broad support as prioritization criteria.

Cost and effectiveness, major topics in health economics (e.g., Krütli et al. 2016; McKie et al. 2009), is often debated as acceptable criterion outside that field (e.g., Busse 1999; Diederich and Salzmann 2015). In the social sciences as well as in philosophy, criteria related to the recipient's personal characteristics attracted most attention, foremost age (e.g., Bognar 2016; Diederich et al. 2011; Fleck 2016; Watters 2016; Williams 1997), and personal responsibility (Buyx 2008; Diederich 2016a; Diederich and Schreier 2010) but also social engagement (e.g., Diederich et al. 2012; Tymstra and Andela 1993), social responsibility (e.g., Nord et al. 1995; Rogge and Kittel 2016; Stumpf et al. 2014; Wegener et al. 2010), and socioeconomic status (e.g., Myllykangas et al. 1996; Pinho and Borges 2018; Ryynänen et al. 1999). In the context of distributive justice, these criteria are often summarized as equity (e.g., Adams 1965; Alwin et al. 2000; Gamliel and Peer 2010; Konow 2001; Lewin-Epstein et al. 2003; Sabbagh 2001).

When rejecting all prioritization criteria despite scarce resources, the equality principle in its simplest form says that every person should have the same chance to receive the resource, for instance, by a lottery (e.g., Lamont and Favor 2017; Sabbagh 2001; Waring 2004). This is the principle that received the most attention in philosophy, strongly influenced by Rawls (1971).

2.4 Prioritization and Frames

While many studies investigated the acceptance of criteria for prioritizing healthcare-related topics and possibly even more to investigate the effects of frames, there are hardly any studies that combine both research areas. In the following, we report the results of them as well as new data we recently collected on this topic.

In an online survey study on priority setting with about 750 female physicians (members of the Ärztinnenbund; see Diederich et al. 2015 for details), one topic was concerned with personal responsibility (Diederich 2016a). Four scenarios included the description of related unhealthy and beneficial behavior: smokers versus non-smokers; adiposity (BMI > 30) versus normal-weight people (BMI 18.5–25); patients with high alcohol consumption versus patients with no alcohol consumption; and patients not exercising versus patients exercising. In the positive frame, participants were asked whether or not they agreed to reward patients for healthy behavior by partially refunding health insurance costs (prioritizing). In the negative frame, they had stated their opinion on whether or not to penalize patients for unhealthy behavior by demanding copayment for healthcare services (posteriorizing). Two preference reversals could be observed for the alcohol and exercise scenarios, but interestingly in opposite directions. For the exercise scenario, 65% of the physicians agreed to refund exercising patients (positive frame) as compared to 43% who agreed to copayments for non-exercising patients (negative frame). This pattern—positively framed statements receive higher agreement scores than negatively framed statements—has been observed in many other studies (see e.g., Gamliel and Peer 2006; Levin et al. 1998; Piñon and Gambara 2005). For the alcohol scenario, 47% of the physicians agreed to refund abstinent patients (positive frame) as compared to 67% who agreed to copayments for drinking patients (negative frame). For the remaining two scenarios (weight and smoking), we could observe preference shifts (i.e., significant differences in agreement proportions but not crossing the 50% line). For the weight scenario, 62% of the physicians agreed to refund patients with a normal weight (positive frame) as compared to 55% who agreed to copayments for obese patients (negative frame). For the smoking scenario, 68% of the physicians agreed to refund non-smokers (positive frame) as compared to 74% who agreed to copayments for smokers (negative frame).

In a representative face-to-face survey of the German public, participants were asked about their opinion on copayments for unhealthy behavior with drug consumption, extreme sports, high alcohol consumption, smoking, sunbathing/solarium, unhealthy diet, and lack of exercise (Diederich et al. 2014). The questions were presented only in a negative frame. Except for unhealthy diet and lack of exercise, the vast majority was in favor of copayments (posteriorizing patients with the respective unhealthy behavior). The same survey included a general statement on preferential treatment of patients with a healthy lifestyle (prioritizing). The vast majority of respondents (89%) did not endorse the statement. This finding is in stark contrast to the results when health-related behavior was specified and presented in a negative frame. This preference reversal may be the result of the framing, but in addition, it may also be an effect related to unpacking a hypothesis, here a statement. Unpacking means partitioning a general description into its components, i.e., health-related lifestyle in concrete health behavior such as smoking, drinking, and so on (see support theory by Tversky and Koehler 1994, accounting for effects caused by unpacking).

Note that many of the studies on prioritizing healthcare services focus on one specific criterion for prioritizing healthcare, e.g., age (see above). Statements centered around it offer binary response options (agree/not agree) or rating scales. Even when several criteria are included in a study, they are often presented in isolation allowing no comparison and trade-offs (Nord et al. 1996). Similar proportions of agreement and similar rating scores for the criteria in question can be interpreted as having received equal importance for distributing the healthcare services, but is less agreeable with the concept of priority setting (building a rank order according to specified criteria) and may leave the decision maker helpless.

Utilizing Törnblom's classification framework, Gamliel and Peer (2010) conducted two framing studies targeting on attitudes to distribution principles in healthcare resource allocation scenarios. Offering questionnaires to undergraduate students, the task in experiment 1 was to rate "perceived fairness" of four positively/negatively framed distribution criteria on a 7-point Likert scale: (1) Merit ("patients who are least/most responsible for their condition would receive/not receive the resource"); (2) tenure ("patients who had been waiting the longest/shortest time would receive/not receive the resource"); (3) need (patients who most/least need the resource would receive/not receive it); and (4) equality ("patients who won/lost a random draw would receive/not receive the resource"). The criteria were embedded in three health scenarios: heart transplants, lung transplants, and AIDS vaccination. Gamliel and Peer (2010) found framing effects for merit, need, and tenure (data collapsed across scenarios). The perceived fairness of these principles was higher in positively framed situations (i.e., to deliver healthcare resources to certain patients) than in negatively framed situations (i.e., not to deliver healthcare resources to other patients). Furthermore, overall fairness ratings were highest for need, followed by equality, then tenure, and last merit.

Experiment 2 included three framed distributive justice criteria: merit, tenure, and equality, and two health scenarios: heart transplantation and AIDS vaccination. Ten statements were constructed to tap on different psychological aspects: cognitive (e.g., "I think the decision is fair"), affective (e.g., "I feel good about this decision"), and behavioral (e.g., "I would donate a small amount of money to an organization aimed at promoting such decisions"). The task was to rate the agreement with the 10 items provided for each of the 12 conditions (2 frames, 2 health scenarios, and 3 distribution criteria) on a 7-point Likert scale. Note that the 12 conditions were presented in a between-subjects design (i.e., each participant received 1 of the 12 conditions with 10 items to be rated on agreement). As for experiment 1, framing effects were observed for merit and tenure but not for equality. Furthermore, on average, equality received the highest fairness ratings and merit the lowest.

Equality may indeed minimize inequality in healthcare service allocation (Cookson and Dolan 1999), but the acceptance rate of selecting patients randomly for treatment is generally low to very low in surveys (Diederich 2016a; Krütli et al. 2016; Mossialos and King 1999; Wegener et al. 2010). Note, however, that this has been practiced in real life. For instance, in 2020, the pharmaceutical corporation Novartis raffled 100 doses of the gene therapy "Zolgensma".[1] Other examples of random distributions of scarce healthcare resources such as dialysis, AIDS, and multiple sclerosis medicines have been reported as well (see Waring 2004 for an overview).

Similar to the study by Gamliel and Peer (2010), we constructed several healthcare scenarios: heart transplant, artificial respiration, treatment against unusual infection, treatment against COVID-19, vaccination against unusual infection, vaccination against COVID-19. Note that the scenarios treatment against unusual infection versus treatment against COVID-19 and vaccination against unusual infection versus vaccination against COVID-19 may be considered as frames in a loose form similar to the identifiable "others" effect (Jenni and Loewenstein 1997; Kogut and Ritov 2005; Loewenstein et al. 2005).

Each scenario described a certain number of patients suffering from a particular disease, and a limited number of resources (e.g., doses of a medicine or a special treatment) only sufficient for a portion of the patients. For example, one scenario described the problem of organ allocation to potential recipients of a heart transplant and read as follows: "Imagine two persons in a hospital suffering from heart disease. Both persons' lives are at risk unless they undergo a heart transplant. All other treatments have been tried and have failed. The hospital has only one matching heart for transplantation". Each scenario was followed by positively (receiving) or negatively (not receiving/withholding) framed statements related to distribution criteria.

To match the respective frame statements, we included antagonistic word pairs like most/least responsible; longest/shortest waiting time; and so on. Table 2.1 lists the distribution criteria and the accordingly framed statements.

Social responsibilityPatients most important to maintain essential public services or facilities (e.g., medical care, childcare, grocery stores, and police) receive the resource. Patients least important to maintain essential public services or facilities (e.g., medical care, childcare, grocery stores, and police) do not receive the resource.

Social security payments Patients who paid most to the social security system (e.g., health insurance premium) in the past year receive the resource. Patients who paid least to the social security system (e.g., health insurance premium) in the past year do not receive the resource.

In an online survey, about 500 participants were asked to rate the fairness of each statement for a given scenario on 5-point Likert scales with response categories unfair, rather unfair, neutral, rather fair, and fair.

We performed contingency analyses using Pearson's Chi-squared test for categorical data and standardized residuals as follow-up tests (Agresti 2019, p. 39). Statistically significant results are marked by "*" ($p < 0.05$), "**" ($p < 0.01$), or "***" ($p < 0.001$). In the following, we show the proportions of combined rather

[1] Zolgensma is a very expensive therapy (costs: roughly 2.000.000 € per dose; Dyer 2020) to treat spinal muscular atrophy (Type I and II), a rare disease with a high mortality rate (the vast majority of infantile patients die before the age of two (see e.g., Birnkrant et al. 1998; Hirtz et al. 2005).

Table 2.1 Allocation criteria framed in terms of receiving and withholding the health resource

Criterion	Receive framing	Withhold framing
Equality	Patients who have won a lottery receive the resource	Patients who have lost a lottery do not receive the resource
Need	Patients who need the resource the most receive it	Patients who need the resource the least do not receive it
Waiting time	Patients who wait the longest time for the resource receive it	Patients who wait the shortest time for the resource do not receive it
Personal responsibility	Patients least responsible for their condition receive the resource	Patients most responsible for their condition do not receive the resource
Social responsibility	Patients most important to maintain essential public services or facilities (e.g., medical care, child care, grocery stores, and police) receive the resource	Patients least important to maintain essential public services or facilities (e.g., medical care, child care, grocery stores, and police) do not receive the resource
Social security payments	Patients who paid most to the social security system (e.g., health insurance premium) in the past year receive the resource	Patients who paid least to the social security system (e.g., health insurance premium) in the past year do not receive the resource

fair and fair ratings and combined rather unfair and unfair ratings. For simplicity, we label them fair and unfair only. The results of the neutral ratings are omitted.

Need as a criterion for prioritizing patients and healthcare services received the highest proportions of fair ratings and the lowest proportions of unfair ratings across all six scenarios in both frames. We observed a statistically significant preference shift: In the receive (positive) frame, more participants (86%) gave fair marks as compared to 82% (*) of the participants in the withhold (negative) frame. Three percent of the participants gave unfair ratings in the receive frame and 4% of them in the withhold frame.

Waiting time as a prioritization criterion was included in two scenarios and received overall the second largest proportion of fair ratings (54%) and the second lowest on unfair ratings (18%) across the scenarios and frames. We observed one preference reversal and two reference shifts within each of the three response categories. In the receive frame, 68% of the participants gave fair ratings, whereas, in the withhold frame, only 39% (***) of the respondents considered waiting time as a fair prioritization criterion. Unfair ratings were given by 10% of the respondents in the receive frame and by 26% (***) in the withhold frame.

Personal responsibility as prioritizing criterion, included in four scenarios, was seen as fair by 43% of the respondents across the respective scenarios and frames and as unfair by 30%. We found a preference shift in one response category: More respondents, i.e., 33%, gave unfair ratings in the receive frame than in the withhold frame, i.e., 27% (*).

Social responsibility as a prioritization criterion, also included in four scenarios, was judged controversial. Overall, the proportion of fair and unfair ratings was 38% in both cases. We could observe preference shifts within each response category but in opposite directions. In the receive frame, 47% of the respondents rated it as fair as compared to 30% in the withhold frame (***). For unfair ratings, the proportions were swapped. In the receive frame, 30% of the respondents rated it as unfair as compared to 47% in the withhold frame (***).

Equality as a criterion for prioritizing healthcare services was included in six scenarios. Across scenarios and frames, 59% of the respondents gave unfair ratings, and 18% saw an equal allocation of healthcare services as fair. Note that this result is in parts different from the Gamliel and Peer (2010) study who observed highest acceptance for equality but in accordance with other survey data (Diederich 2016a; Krütli et al. 2016; Mossialos and King 1999; Wegener et al. 2010). However, like Gamliel and Peer (2010), we found no framing effects in the response categories of the equality criterion.

The payments to the social security system criterion were included in four scenarios and received the lowest fairness ratings. Of the respondents, 84% rated it as unfair to consider it as a prioritization criterion, and only 6% of the respondents gave fair ratings. No framing effect was observed.

Framing effects in a loose form could have occurred when comparing the fairness ratings between the unusual infection scenarios to the COVID-19 scenarios. Preference shifts were observed for several prioritization criteria. In particular, when proposing to distribute the healthcare service according to need, the proportion of unfair ratings was higher in the COVID-19 frame (5%) than in the unusual infection frame (2%**). Social responsibility as distribution criterion was considered as fair in the COVID-19 frame (42%) as compared to the unusual infection frame (35%*). Distributing healthcare services by chance (equality criterion) showed more extremes between the identified disease and the unspecific one. In particular, in the COVID-19 frame, 15% of respondents gave fair ratings and 62% unfair ratings. In the infectious disease framing, 23% (**) gave fair ratings and 54% (*) unfair ratings.

To summarize, both framing forms, that is framing in a strict and framing in a loose sense, influenced the acceptance of criteria for prioritizing healthcare services. In particular, attitudes seem to depend on whether the allocation of a specific health resource is framed as either receiving or not receiving (withholding) it. For instance, people evaluate the need, waiting time, and social responsibility criteria as more favorable in the receive frame than in the withhold frame. A more ambiguous picture emerges for the personal responsibility criterion. Whether or not a framing effect occurs and also the direction of the effect seem to depend on further factors such as item construction, analysis methods, and effects related to "unpacking" (Tversky and Koehler 1994) a statement. Attitudes toward other criteria (i.e., equality and payments made to the social system), however, were not affected by the framing of receiving or withholding the resource. Another framing, rather in a loose sense, is the way how the disease to be treated is called. Giving a disease different label (e.g. COVID-19 or just infectious disease) seems to further affect the acceptance of certain prioritization criteria.

2.5 Concluding Remarks

In times of decreasing election turnouts and rising political apathy, the legitimization of democratic institutions is challenged in many European counties (Leininger 2015). One popular way to re-establish legitimacy is including elements of direct democracy such as public involvement by eliciting attitudes using surveys and questionnaires. Prioritizing healthcare services has been discussed on and off for many years. While in some countries discussions and public involvement is included in standard procedures, it is avoided in other countries, particularly in Germany, where the topic seems to get more attention when the budgets of the statuary healthcare insurance get tighter. Previously confused with rationing, prioritizing is now, thanks to the regulations during the COVID-19 pandemic, recognized by the general public as ranking of patients or treatments according to some criteria such as age, vulnerability, social responsibility, and possibly more (but see other definitions of prioritizing discussed earlier; Cochrane 2016).

Contrary to some objections to include the public in the decision-making process for not being objective or not knowledgeable and self-serving (Bruni et al. 2008; Diederich et al. 2009), the public may provide an important perspective about opinions on fair distributions and values of the citizens who are the main funders and users of the healthcare system. It may increase the quality and acceptance of decisions on prioritization criteria (see e.g., Bruni et al. 2008; Diederich et al. 2009, 2015; Ham 1997; Stumpf and Raspe 2012).

But who decides on the criteria for prioritizing healthcare services to be discussed (publicly)? This is less obvious. For instance, the criteria for distributing the COVID-19 vaccine to the citizens were recommended by the Standing Committee on Vaccination (STIKO; consisting of researchers and physicians) and finally determined by the government. The criteria include age (oldest first, and youngest last), system importance (e.g., healthcare professionals first), or medical urgency (e.g., people with pre-existing health conditions first). This was neither discussed nor based on evidence at that stage.

Research in the social sciences and philosophy often includes personal responsibility and age as criteria, sometimes connected to and guided by principles derived from distributive justice theory. Other times criteria are postulated ad hoc. For instance, some major German stakeholders in public health, i.e., physicians, health insurances, and scattered health politicians, recently proposed that patients who are unwilling to take a COVID-19 vaccine should pay (part of the) costs of medical treatment in case of contracting it (see e.g., Bruche 2021; Menke 2021). However, the proposal did not receive ample political and public support, and it was not pursued further. Obviously, one could have framed it differently by paying the costs of COVID-19 treatment only for those who took the vaccine. The acceptance may have been different.

Framing effects have been observed in health-related topics besides prioritization. Take, for instance, the willingness of deceased organ donation. Depending on a country's defaults, donation agreement rates differ drastically. In countries where citizens are donors by default but can opt-out, the rate is close to one hundred percent agreement. In countries where citizens are non-donors by default but can opt-in the agreement rates vary but are less than 30% (see e.g., Johnson and Goldstein 2003; Rieu 2010; Rithalia et al. 2009; Smith et al. 2013). Other examples of framing effects can be observed in promoting preventive healthcare. Various studies have demonstrated that gain-framed messages (i.e., statements emphasizing the benefits of performing a concrete health behavior) increase the willingness to use measures for prevention, such as using sunscreen and plaque-fighting mouth rinse (e.g., Detweiler et al. 1999; Rothman et al. 1999); loss-framed messages (i.e., statements emphasizing the costs of not performing a concrete health behavior), however, increase the willingness to use early detection measures such as mammography screening or plaque-detecting disclosing rinse (e.g., Abood et al. 2005; Banks et al. 1995; Rothman et al. 1999, 2006). These are impressive real-life results showing that wording and rewording, i.e. framing, can have substantial consequences.

The current chapter reviewed and reported findings that may help to improve public involvement in political decisions, such as prioritizing healthcare resources and political communication of decisions on resource allocation. While we advocate the involvement of the citizens in the decision-making process on prioritizing healthcare, we also want to make them (and the policymakers) aware of how simple frames make influence their preferences. The reported results may further raise awareness that opinion surveys (e.g., market research and opinion polls) can sometimes be manipulated to influence the outcome (e.g., social attitudes or election projections).

References

Abood DA, Black DR, Coster DC (2005) Loss-framed minimal intervention increases mammography use. Women's Heal Iss 15(6):258–264

Adams JS (1965) Inequity in social exchange. In: Berkowitz L (ed) Advances in experimental social psychology, vol 2. Academic Press, pp 267–299

Agresti A (2019) An introduction to categorical data analysis, 3rd edn. Wiley

Alwin DF, Borgatta EF, Montgomery RJV (2000) Social justice. In: Encyclopedia of sociology, 2nd edn, vol 4. Macmillan Reference USA, pp. 2695–2711

Banks SM, Salovey P, Greener S, Rothman AJ, Moyer A, Beauvais J, Epel E (1995) The effects of message framing on mammography utilization. Health Psychol 14(2):178–184

Birnkrant DJ, Pope JF, Martin JE, Repucci AH, Eiben RM (1998) Treatment of type I spinal muscular atrophy with noninvasive ventilation and gastrostomy feeding. Pediatr Neurol 18(5):407–410

Bognar G (2016) Priority setting and age. In: Nagel E, Lauerer M (eds) Prioritization in medicine: an international dialogue. Springer International Publishing, pp 163–177

Bowling A (1996) Health care rationing: the public's debate. BMJ 312(7032):670–674

Bruche J (2021) Corona: Müssen Ungeimpfte höhere Krankenkassenbeiträge zahlen? Welle der ungeimpften CoViD-19-Patienten schwappt in Krankenhäuser. Südwestrundfunk. https://www.swr.de/swraktuell/baden-wuerttemberg/braucht-es-krankenkassentarife-fuer-ungeimpfte-100.html. Accessed on March 5, 2022

Bruni RA, Laupacis A, Martin DK (2008) Public engagement in setting priorities in health care. CMAJ 179(1):15–18

Busse R (1999) Priority-setting and rationing in German health care. Health Policy 50(1–2):71–90

Buyx AM (2008) Personal responsibility for health as a rationing criterion: why we don't like it and why maybe we should. J Med Ethics 34(12):871–874
Cochrane JR (2016) Fundamental evaluation criteria in the medicine of the twenty-first century. In: Nagel E, Lauerer M (eds) Prioritization in medicine: an international dialogue. Springer International Publishing, pp 11–37
Cookson R, Dolan P (1999) Public views on health care rationing: a group discussion study. Health Policy 49(1–2):63–74
Daniels N (2007) Just health: meeting health needs fairly. Cambridge University Press
Daniels N, Sabin J (2002) Setting limits fairly: can we learn to share medical resources? Oxford University Press
Detweiler JB, Bedell BT, Salovey P, Pronin E, Rothman AJ (1999) Message framing and sunscreen use: gain-framed messages motivate beach-goers. Health Psychol 18(2):189–196
Diederich A (2016a) Age and personal responsibility as prioritization criteria? The view of the public and of physicians. In: Nagel E, Lauerer M (eds) Prioritization in medicine: an international dialogue. Springer International Publishing, pp 241–257
Diederich A (2016b) What health care services does the public want and who should decide? ask them! Israel J Heal Policy Res 5:47
Diederich A, Salzmann D (2015) Public preferences regarding therapeutic benefit, costs of a medical treatment and evidence-based medicine as prioritization criteria. J Public Health 23(3):137–148
Diederich A, Schreier M (2009) Kriterien der Priorisierung aus gesellschaftlicher Sicht. Z Evid Fortbild Qual Gesundhwes 103(2):111–116
Diederich A, Schreier M (2010) Zur Akzeptanz von Eigenverantwortung als Posteriorisierungskriterium. Bundesgesundheitsblatt-Gesundheitsforschung-Gesundheitsschutz 53(9):896–902
Diederich A, Winkelhage J, Schnoor M, Margrit S (2009) Priorisierung. Öffentlicher Diskurs erforderlich. Deutsches Ärzteblatt 106(14):A 654–656
Diederich A, Winkelhage J, Wirsik N (2011) Age as a criterion for setting priorities in health care? A survey of the German public view. PLoS ONE 6(8):e23930
Diederich A, Swait J, Wirsik N (2012) Citizen participation in patient prioritization policy decisions: an empirical and experimental study on patients' characteristics. PLoS ONE 7(5):e36824
Diederich A, Schwettmann L, Winkelhage J (2014) Does lifestyle matter when deciding on copayment for health care? a survey of the general public. J Public Health 22(5):443–453
Diederich A, du Bois G, Dörr D (2015) Einstellungen zu Priorisierungen in der medizinischen Versorgung: Ergebnisse einer Befragung des Deutschen Ärztinnenbundes (DÄB) (Working Paper 40 / 2015; Priorisierung in Der Medizin). Jacobs University, FOR 655
Dillman DA, Smyth JD, Christian LM (2014) Internet, phone, mail, and mixed-mode surveys: the tailored design method. Wiley
Dyer O (2020) Health ministers condemn Novartis lottery for Zolgensma, the world's most expensive drug. BMJ 368:m580
Fischer K (1997) Tun oder Lassen? Die Rolle von framing-Prozessen bei der Wahl von Handlung oder Unterlassung in Entscheidungssituationen. Peter Lang
Fleck L (2016) Just caring: fair innings and priority setting. In: Nagel E, Lauerer M (eds) Prioritization in medicine: an international dialogue. Springer International Publishing, pp 197–216
Gächter S, Orzen H, Renner E, Starmer C (2009) Are experimental economists prone to framing effects? a natural field experiment. J Econ Behav Organ 70(3):443–446
Gamliel E, Peer E (2006) Positive versus negative framing affects justice judgments. Soc Just Res 19(3):307–322
Gamliel E, Peer E (2010) Attribute framing affects the perceived fairness of health care allocation principles. Judgm Decis Mak 5(1):11–20
Gilovich T, Griffin D, Kahneman D (eds) (2002) Heuristics and biases: the psychology of intuitive judgment. Cambridge University Press
Ham C (1997) Priority setting in health care: learning from international experience. Health Policy 42(1):49–66

Hauck K, Smith PC, Goddard M (2004) The economics of priority setting for health care: a literature review, vol 28878. The International Bank for Reconstruction and Development/The World Bank
Hirtz D, Iannaccone S, Heemskerk J, Gwinn-Hardy K, Moxley R, Rowland LP (2005) Challenges and opportunities in clinical trials for spinal muscular atrophy. Neurology 65(9):1352–1357
Jenni K, Loewenstein G (1997) Explaining the identifiable victim effect. J Risk Uncertain 14(3):235–257
Johnson EJ, Goldstein D (2003) Do defaults save lives? Science 302(5649):1338–1339
Jonsen AR (1986) Bentham in a box: technology assessment and health care allocation. J Law Med Ethics 14(3–4):172–174
Kahneman D, Slovic P, Tversky A (eds) (1982) Judgment under uncertainty: heuristics and biases. Cambridge University Press.
Kogut T, Ritov I (2005) The "identified victim" effect: an identified group, or just a single individual? J Behav Decis Mak 18(3):157–167
Konow J (2001) Fair and square: the four sides of distributive justice. J Econ Behav Organ 46(2):137–164
Krütli P, Rosemann T, Törnblom KY, Smieszek T (2016) How to fairly allocate scarce medical resources: ethical argumentation under scrutiny by health professionals and lay people. PLoS ONE 11(7):e0159086
Kühberger A (1998) The influence of framing on risky decisions: a meta-analysis. Organ Behav Hum Decis Process 75(1):23–55
Lamont J, Favor C (2017) Distributive justice. In: Zalta EN (ed) The Stanford encyclopedia of philosophy (Winter 2017). Metaphysics Research Lab, Stanford University. https://plato.stanford.edu/archives/win2017/entries/justice-distributive/. Accessed on March 5, 2022
Leininger A (2015) Direct democracy in Europe: potentials and pitfalls. Global Pol 6(S1):17–27
Levin IP, Schneider SL, Gaeth GJ (1998) All frames are not created equal: a typology and critical analysis of framing effects. Organ Behav Hum Decis Process 76(2):149–188
Lewin-Epstein N, Kaplan A, Levanon A (2003) Distributive justice and attitudes toward the welfare state. Soc Just Res 16(1):1–27
Loewenstein G, Small D, Strnad J (2005) Statistical, identifiable and iconic victims and perpetrators
Loewenstein G, Small DA, Strnad J (2006) Statistical, identifiable, and iconic victims. In: McGaffery E, Slemrod J (eds) Behavioral public finance: toward a new agenda. Russel Sage Foundation, pp 32–46
McKie J, Richardson J (2003) The rule of rescue. Soc Sci Med 56(12):2407–2419
McKie J, Hurworth R, Richardson J, Shrimpton B (2009) Treatment costs and priority setting in health care: a qualitative study. Australia New Zealand Heal Pol 6(1)
Menke F (2021) Provokanter Corona-Vorschlag: Ungeimpfte zur Kasse bitten. Westdeutscher Rundfunk. https://www1.wdr.de/nachrichten/themen/coronavirus/ungeimpfte-corona-kosten-102.html. Accessed on March 5, 2022
Mossialos E, King D (1999) Citizens and rationing: analysis of a European survey. Heal Pol 49(1):75–135
Myllykangas M, Ryynänen O-P, Kinnunen J, Takala J (1996) Comparison of doctors', nurses', politicians' and public attitudes to health care priorities. J Heal Serv Res Pol 1(4)
Nagel E, Lauerer M (eds) (2016) Prioritization in medicine: an international dialogue. Springer International Publishing
Nord E, Richardson J, Street A, Kuhse H, Singer P (1995) Maximizing health benefits vs egalitarianism: an Australian survey of health issues. Soc Sci Med 41(10):1429–1437
Nord E, Street A, Richardson J, Kuhse H, Singer P (1996) The significance of age and duration of effect in social evaluation of health care. Health Care Anal 4(2):103–111
Pinho M, Borges AP (2018) A three-country survey of public attitudes towards the use of rationing criteria to set healthcare priorities between patients. Inter J Ethics Syst 34(4):472–492
Piñon A, Gambara H (2005) A meta-analytic review of framming effect: risky, attribute and goal framing. Psicothema 17(2):325–331
Rawls J (1971) A theory of justice, vol 971. Harvard University Press

Rieu R (2010) The potential impact of an opt-out system for organ donation in the UK. J Med Ethics 36(9):534–538
Rithalia A, McDaid C, Suekarran S, Myers L, Sowden A (2009) Impact of presumed consent for organ donation on donation rates: a systematic review. BMJ 338:a3162
Rogge J, Kittel B (2016) Who shall not be treated: Public attitudes on setting health care priorities by person-based criteria in 28 nations. PLoS ONE 11(6):e0157018
Rossi PH, Wright JD, Anderson AB (2013) Handbook of survey research. Academic Press.
Rothman AJ, Martino SC, Bedell BT, Detweiler JB, Salovey P (1999) The systematic influence of gain-and loss-framed messages on interest in and use of different types of health behavior. Pers Soc Psychol Bull 25(11):1355–1369
Rothman AJ, Bartels RD, Wlaschin J, Salovey P (2006) The strategic use of gain- and loss-framed messages to promote healthy behavior: how theory can inform practice. J Commun 56(s1):S202–S220
Ryynänen O-P, Myllykangas M, Kinnunen J, Takala J (1999) Attitudes to health care prioritisation methods and criteria among nurses, doctors, politicians and the general public. Soc Sci Med 49(11):1529–1539
Sabbagh C (2001) A taxonomy of normative and empirically oriented theories of distributive justice. Social Justice Research 14(3):237–263
Sabik LM, Lie RK (2008) Priority setting in health care: Lessons from the experiences of eight countries. Inter J Equity Heal 7:4
Smith NC, Goldstein DG, Johnson EJ (2013) Choice without awareness: ethical and policy implications of defaults. J Public Policy Mark 32(2):159–172
Steiger A, Kühberger A (2018) A meta-analytic re-appraisal of the framing effect. Zeitschrift Für Psychologie 226(1):45–55
Stumpf S, Raspe H (2012) Deliberative Bürgerbeteiligung in der Priorisierungsdebatte: Welchen Beitrag können Bürger leisten? Z Evid Fortbild Qual Gesundhwes 106(6):418–425
Stumpf S, Hecker S, Raspe H (2014) Kriterien für die Priorisierung medizinischer Leistungen im Licht eines regionalen Surveys—Ergebnisse und methodologische Fragen. Das Gesundheitswesen 76(4):221–231
Törnblom KY (1988) Positive and negative allocations: a typology and a model for conflicting justice principles. Adv Group Processes 5:141–168
Tversky A, Kahneman D (1981) The framing of decisions and the psychology of choice. Science 211(4481):453–458
Tversky A, Koehler DJ (1994) Support theory: a nonextensional representation of subjective probability. Psychol Rev 101(4):547–567
Tymstra T, Andela M (1993) Opinions of Dutch physicians, nurses, and citizens on health care policy, rationing, and technology. J Am Med Assoc 270(24):2995–2999
Waring DR (2004) Medical benefit and the human lottery: an egalitarian approach to patient selection (Weisstub DN, ed). Springer
Watters SM (2016) Fair innings as a basis for prioritization: an empirical perspective. In: Nagel E, Lauerer M (eds) Prioritization in medicine: an international dialogue. Springer International Publishing, pp 179–196
Wegener B, Mason D, International Social Justice Project (ISJP) (2010) International social justice project, 1991 and 1996. Inter-university Consortium for Political and Social Research [distributor]
Williams A (1997) Intergenerational equity: an exploration of the 'Fair Innings' argument. Health Econ 6(2):117–132
World Health Organization. Regional Office for Europe (1996) The Ljubljana Charter on reforming health care in Europe [Technical documents]. World Health Organization. Regional Office for Europe. https://apps.who.int/iris/handle/10665/347612. Accessed on March 5, 2022

Adele Diederich is Cognitive Psychologist with a research focus on perception and decision-making. She develops theories and formal models and tests them by experiments and empirical observations. In decision-making, she applies dynamic-stochastic models for binary choice options with multiple attributes; Bayesian approaches to aggregate experts' opinions, or conjoint analysis (discrete choice model) to elicit preferences in survey data (e.g., for prioritizing medical services). She was Chair of the German Research Foundation (DFG) research group on Setting Priorities in Medicine and appointed Member of the Committee Prioritizing in the Health System at the German Medical Association (Bundesärztekammer). She was Member of the DFG research group on Need-Based Justice and Distribution Procedures where she investigates the effects of frames on when allocating resources. She serves as Editor-in-Chief for Journal of Mathematical Psychology.

Marc Wyszynski is Multidisciplinary Researcher with expertise in psychology and political science. He earned his PhD in Psychology from Jacobs University Bremen and has a master's degree in professional public decision-making from the University of Bremen. He has also obtained a bachelor's degree in political science from the University of Hannover. Currently, he is Postdoctoral Researcher at the Department of Mathematics and Computer Science at the University of Bremen. His research focuses on cognitive biases and the role of individual differences in decision-making. His multidisciplinary background enables him to bring a unique perspective to his research, incorporating theory and research methodology from various fields. His research has significant implications for decision-making processes in various settings, including individual decision-making and public policy.

Chapter 3
Using Pharmacoepidemiologic Studies to Inform Drug Policy and Spending: A Health Economics Perspective

Konstantinos Zisis, Kostas Athanasakis, and Kyriakos Souliotis

Abstract Healthcare resource allocation today is more difficult than ever. The asymmetry between available resources and health needs requires methodologically sound and transparent decision-making processes for resource allocation. Clinical effectiveness and cost-effectiveness of health technologies are major determinants of pricing and reimbursement decisions—however, the above must be demonstrated both in the controlled and in the real-world setting. Pharmacoepidemiology can play a major role in evidence-informed pharmaceutical policy decisions, determining what "actually works" in real-life conditions. We discuss the role of pharmacoepidemiology in the decision-making context of pharmaceutical policy and, in specific, its role in the resource allocation framework, focusing on its impact on Health Technology Assessment, the efficiency of spending, and handling of uncertainty in coverage decisions, the latter achieved through risk-sharing schemes.

Keywords Pharmacoepidemiology · Pharmacoeconomics · Health technology assessment · Real world evidence · Pharmaceutical policy

K. Zisis (✉) · K. Athanasakis
Department of Public Health Policy, School of Public Health, University of West Attica, Athens, Greece
e-mail: kzisis@uniwa.gr

K. Athanasakis
e-mail: kathanasakis@uniwa.gr

K. Zisis
Institute for Health Economics, Athens, Greece

K. Souliotis
Department of Social and Education Policy, University of Peloponnese, Corinth, Greece

Health Policy Institute, Athens, Greece

K. Souliotis
e-mail: ksouliotis@uop.gr

E. Çetin and H. Özen (eds.), *Healthcare Policy, Innovation and Digitalization*, Accounting, Finance, Sustainability, Governance & Fraud: Theory and Application,
https://doi.org/10.1007/978-981-99-5964-8_3

3.1 Introduction

Over the last decades, the continuous increases in life expectancy, in conjunction with the epidemiological transition and, primarily, with the introduction of innovative health technologies, have contributed to an increase in demand for health services and, subsequently, a growing trend in health spending at the international level (Robinson et al. 2012). This change in the pattern of needs, demands, and spending, when viewed against the limited financial resources available to cover them, has led to the introduction of decision-making "systems" or "contexts" for healthcare resource allocation that needs to be both methodologically rigorous as well as be able to promote transparency and accountability (Baltussen and Niessen 2006).

This evolution in decision-making specifically applies to health technologies—primarily, and at least for the time being, to pharmaceuticals. Pharmaceutical expenditures are the healthcare cost components that grow at a more rapid pace compared to other functions of the system and have now come to account for over 20% of overall healthcare spending (Organisation for Economic Co-operation and Development 2020). Given the constant introduction of new pharmaceutical technologies in the system and the subsequent claims for their insurance coverage (i.e. an allocation of a proportion of the limited resources of the system to the new technology), priority-setting frameworks for the allocation of resources are needed—and, given that resource allocation decisions are generally complex and have competing trade-offs (Thokala et al. 2016), they must be based on multiple dimensions and criteria (Tromp and Baltussen 2012; Bloom et al. 2006). In particular, safety, quality, clinical effectiveness, and at the same the cost-effectiveness are among the key prerequisites for decision on resource allocation toward technology, to assure affordability, fiscal sustainability, patient access, and coverage of unmet needs. Nowadays, it is necessary to demonstrate the above outcomes in the clinical research environment but it's also particularly desirable for their establishment in the actual real-world practice setting. Real-world evidence (RWE), which is the clinical evidence derived by real-world data (RWD) concerning the use of a medicinal product, has an increased and emerging role in the decision-making process (FDA). Demonstration of real-world value for money is in continuous need which means that every Dollar/Euro should be spent efficiently to achieve an incremental health effect for populations. Pharmaceutical policies should be based on evidence-based approaches. Evidence-based methods help policymakers to make well-informed decisions in health care, and this has been a very successful milestone in our new era, initiated by the pioneer Professor Archibald Cochrane (Stavrou et al. 2014). The science of pharmacoepidemiology can have a substantial contribution in this area by providing very useful data regarding the safety, effectiveness, and drug utilization behaviors of human populations.

Taking the above into account, the objective of this chapter is to highlight the role and describe the current setting and impact of pharmacoepidemiology in evidence-based pharmaceutical policy decisions. As a second objective, we will outline the potential and future opportunities to consider pharmacoepidemiology studies in current and future pharmaceutical practice setting such as clinical research, health technology assessment (HTA), coverage and reimbursement policies, through the need for continuous monitoring of medicinal products in different subgroups.

3.2 Pharmacoepidemiology: Definition, History, and Applications

According to the FDA (FDA 2018), pharmacoepidemiology studies are observational studies that provide an overview and assess how drugs are used by human populations and what effects they produce. In particular, the science of pharmacoepidemiology provides an approach to the interactions between drugs and populations regarding the health effects and risks of the utilization of drugs (Montastruc et al. 2019). Essentially, it is the science that connects pharmacology and epidemiology. The main goals of pharmacoepidemiology are to study and assess the most important issues of drugs in a real-world setting, actual practice, and routine care environment: (i) effectiveness, (ii) utilization, (iii) safety, and (iv) cost related to care and cost-effectiveness of drugs (Bégaud 2019; Wettermárk 2013; Takahashi et al. 2012).

The science of pharmacoepidemiology is based on observational studies, of which there are various types: drug safety and drug utilization are the most well-known and most usual relevant studies. The main and secondary sources of data used in pharmacoepidemiology are surveys, registries, administrative health databases, claims databases (billing and pharmacy), and electronic medical records (EMR) (Nishtala and Bala 2019). In more detail, main data might be data collected primarily by investigators concerning subject participants, prospective observational studies assessing drug use, and patient registries which are mainly used for post-authorization monitoring of drugs while can provide a unique opportunity for post-approval collection data regarding safety (Alarkawi et al. 2018; Jonker et al. 2017). These sources have both advantages and disadvantages. For instance, disease registries can provide valuable insights into routine care costs and outcomes (Larsson et al. 2012; Gliklich et al. 2014). In regard to pharmacy claims, despite their valuable source for identification of data related to costs and outcomes, such as non-adherence (Comer et al. 2015), several limitations related to them such as missing information on drugs not reimbursable might not be available in these databases while generalizability issues might also be presented (Sinnott et al. 2017). As a result, both benefits and disadvantages are presented by these sources.

FDA was the pioneer of reviewing and assessing side effects. Following a period until the 1950s when there was limited interest in the side effects of medicines, the agency initiated to collect and review reports on side effects for drug surveillance programs, especially from the 1960s onwards (Balçık et al. 2016). At about the same time, in the European setting, drug utilization research studies were initiated to assess differences between countries in drug use and prescription factors of physicians. The first meeting was organized in 1969 by the World Health Organization (WHO) in Oslo, where an unanimous call of scientists requested an organized and common drug classification system. After this period, rapid development of the international dialogue led to the establishment of critical mass research groups, which subsequently led to the development of sets of methodologies and guidelines (Wettermark et al. 2016). Pharmacoepidemiology further advanced in the mid-1980s through the establishment of the International Society of Pharmacoepidemiology in conjunction with the introduction of large databases and new methods.

3.2.1 Pharmacoepidemiology and Pharmacoeconomics: The Impact on Pharmaceutical Policies

Many examples highlighting the positive impact of pharmacoepidemiology on pharmaceutical policy, particularly on both coverage and reimbursement decisions. In a study carried out in British Columbia and Ontario of Canada (Paterson et al. 2006), it was found that pharmacoepidemiology research through administrative data was valuable for informing drug decision-makers regarding drug coverage issues. In particular, through the cooperation between university-based pharmacoepidemiologists with access to administrative health data and prescription drug utilization studies and drug policy planners, higher rates of evidence in terms of real-world data were provided for effective drug coverage policies. Furthermore, pharmacoepidemiology research might have additional positive implications in cost savings in conjunction with drug pricing policies. An illustrative example is the study by Spillane (Spillane et al. 2015) who found that the contribution of pharmacoepidemiological research through monitoring and assessment of cost-related data could produce savings of €35 million for the third-party payer in Ireland due to the introduction of reference pricing and generic substitution policies. The demonstration of these savings was the result of analysis and cost data assessment on national pharmacy claims data, in the context of the Irish Health Service Executive-Primary Care Reimbursement Service community pharmacy claims (HSE-PCRS). So how post-marketing pharmacoepidemiologic studies might contribute and provide insights for both pricing and coverage decisions? Many examples are highlighting the added value of a post-market review of drugs to impact pricing and reimbursement. A recent report by OECD (Organisation for Economic Co-operation and Development 2019) outlines such cases.

Pharmacoeconomic analyses are the ideal type of analyses for presenting economic evidence. More specifically, pharmacoeconomic analyses focus on the optimal allocation of resources needed to gain maximum health effects by the usage of medicines—and this is done by comparing alternative drugs in terms of costs and benefits. This is where pharmacoepidemiology research, through drug utilization studies, can be added in conjunction with pharmacoeconomic analyses to lead to potential meaningful influence in reimbursement decisions, since they provide data on real-world practice (Sacristán and Soto 1994; Levine and Lelorier 2012). For instance, patient compliance and adherence in a real-world setting, which is also an aim of pharmacoepidemiologic research, could be a part of pharmacoeconomics to assess the effectiveness and cost-effectiveness of health technologies. The importance of adherence has been also noticed by several articles in the literature (Souliotis et al. 2022, Hiligsmann et al. 2012; Hughes et al 2007). Thus, the combination of both sciences may demonstrate both cost-effectiveness and safety life-changing medicinal products and bring the desired value for money.

The use and importance of RWD and RWE in health care have increased in recent years while health technology assessment (HTA) agencies (Lou et al. 2020) and regulatory authorities show an interest in reviewing these data to monitor and evaluate safety and effectiveness in the post-market environment, either as primary or complementary evidence, since they provide the main advantage of presenting benefits and risks in day-to-day practice (Pulini et al. 2021). RWD are data obtained outside randomized-controlled trials (RCTs') and particularly from observational (retrospective or prospective) studies and registries (Berger et al. 2017). Pharmacoepidemiology also has the advantage of applying the use of drugs in the real-world environment as it leads to the emergence of RWE concerning to safety and effectiveness of medicinal products in human populations, through the application of epidemiological methods. It provides the knowledge to policymakers on what is safe and more effective in clinical practice and distinguishes which subgroups will gain maximum benefit from the treatment under study with the use of several techniques. Generally, in observational studies this is not a usual practice; however, a promising study (Qi et al. 2021) noted that by introducing a proposed modeling approach in a large observational study, they identified subgroups with different responses to antibiotics treatment for acute rhinosinusitis. Nonetheless, pharmacoepidemiologic studies are subject to significant limitations associated with bias and confounding, mainly due to the non-randomization of subjects in their design that leads to arduous causability. This is a significant difference compared to RCTs and may be resolved through interventions either in the design process of studies or in the analysis (Caparrotta et al. 2019). Another limitation of RWE in regulatory processes is that despite its acceptance by regulatory authorities as satisfactory clinical evidence on post-approval drug effectiveness and safety, it is not largely integrated into the regulatory decision-making process (Baumfeld Andre et al. 2020). However, it seems that RWE value is growing and higher rates of acceptance by authorities are necessary to highlight the drug efficacy and safety in a real-world environment.

Another main feature of studies related to pharmacoepidemiology is that they highlight comparative effectiveness research (CER). Although RCTs are the gold evidence for the assessment of comparative risks and benefits of medicinal products, they have their limitations associated with external validity, inability to capture all adverse events due to lower timeframe which might operate than needed, exclusion of vulnerable populations, and in most drug approval cases a comparison of an active substance against no treatment (Babar 2019; van Luijn et al. 2007). According to the Institute of Medicine (Institute of Medicine 2009), CER is defined as "the generation and synthesis of evidence that compares the benefits and harms of alternative methods to prevent, diagnose, treat, and monitor a clinical condition or to improve the delivery of care". Both drug utilization and safety studies aim at generating evidence from real-world environments regarding the effectiveness and risks of alternative drugs. But what are the advantages of using these studies to highlight CER in a post-marketing environment? Healthcare utilization studies have a lower cost, represent actual clinical practice, providing the base for real-world drug effectiveness, and do not exclude vulnerable and older populations (Schneeweiss and Avorn 2005). CER may better guide and improve the clinical decision-making process while it should be noted that is generally taken into consideration for both coverage and pricing decisions by agencies in six European Union countries and Great Britain (Sorenson 2010). Furthermore, an important aspect of CER that has emerged recently is the conduct of indirect comparisons which can add significant value when direct comparison data are missing. Both payers and HTA agencies have shown interest to consider such data. However, and despite the dynamism CER might provide, it is still not accounted for in decision-making and to an extent that it should be (Frois and Grueger 2017). As a result, evidence and data derived by day-to-day CER may provide additional information on both safety and clinical effectiveness outcomes between alternative drugs for well-informed decisions but more efforts and policies are needed to establish CER as a meaningful measure for the decision-making of drugs. Such a policy might be the establishment of CER value framework as a certain measure across countries to adopt a well-defined set of benefits and risks concerning real-world drug effectiveness and to serve as a beginning point of comparative drug assessment data, valuable for payers.

3.2.2 Incorporating Pharmacoepidemiology in the Health Technology Assessment Process

The Health Technology Assessment (HTA) is crucial in the decision-making process and for resource allocation. HTA has been one of the most important processes in pharmaceutical policy that promotes value, prioritization, and ethical and fair allocation of resources (Kanavos et al. 2019; Mitton et al. 2019; Specchia et al. 2015). Science has many similar definitions which converge on a specific general phrase: approach based on value for money. The new definition of HTA, constructed through a convergence process among key HTA international institutions and published in May 2020, outlines HTA as the *multidisciplinary process of determining the value of health technology at different points of its lifecycle by using explicit methods, to inform decision-making to contribute to an equitable, efficient, and high-quality health system* (O'Rourke et al. 2020). According to World Health Organization (WHO), it is the systematic evaluation of the characteristics and effects of health technologies and interventions, including their direct and indirect outcomes while this transparent process determines the worth of a health technology and provide guidance on its utilization in healthcare systems worldwide. (WHO) The process is conducted by interdisciplinary groups using analytical frameworks and various methods, including clinical, epidemiological, health economics, and other methodologies. Considering the above views, HTA, as a decision-making platform, inherently includes both "positive" as well as "normative" aspects, captured in the activities of the assessment and appraisal of data, respectively.

The HTA process includes a comparative exercise of incoming interventions compared on the current "standard of care" in the clinical and economic dimensions, alongside other aspects of the availability of the technology (economic, legal, ethical, and societal). The weighting of clinical evidence plays an important role in the assessment of health technologies (Vreman et al. 2019). Pharmacoepidemiology has a crucial role in assessing cost-effectiveness and comparative effectiveness between pharmaceutical products, while it also contributes to the appropriate clinical evidence needed for the overall value of therapies. Both clinical and economic evidence collection is one of the most important factors influencing decision-making in the HTA process (Yuasa et al. 2021; Allen et al. 2017). Although randomized-controlled trials are the primary sources of clinical evidence needed by HTA agencies (Zisis et al. 2020) under optimal conditions, pharmacoepidemiology studies assess how these medicines are applied and contribute to overall RWE. Literature shows that incorporation of RWD, derived from pharmacoepidemiology research, increases over time for drug appraisal in the HTA process. In England, the National Institute for Health and Care Excellence (NICE) has received an increase of approximately 38% submissions for rare-disease medicines which involved RWD between 2015 and 2019 compared to the period of 2010–2014 (Harwood et al. 2019). In addition, RWE has been primarily included for the assessment of oncology medicines by HTA agencies and orphan drugs for efficacy, drug usage, economic modeling, and long-term efficacy reasons (Jao et al. 2018). It should be also noted that HTA agencies in Europe are quite interested in using RWD for relative effectiveness assessment (REA) of medicinal products in terms of pharmacoeconomic analyses, while differences in how they use these data for REA have emerged (Makady et al. 2018).

There is no doubt that RWE deriving by pharmacoepidemiology studies, either through drug utilization or safety databases, should be considered as a complementary source of clinical evidence for medicinal products in the HTA appraisal process by filling evidence gaps in drug usage, safety, and long-term efficacy outcomes. RWE derived by such studies provides reliable data based on longer duration than RCTs, and its value is expected to increase over the next years. Such an example highlighting valuable complementary data has been described in another study (Izano et al. 2020). The authors attempted to investigate the association of bisphosphonate therapy beyond five years, with a lower risk of hip fracture by reviewing women (retrospective cohort study) treated with the drug for five years. Although clinical trials demonstrated the efficacy of this drug for the first three to five years of therapy, there seems to be uncertainty on the efficacy beyond this timeframe. Their findings outlined that the continuation of drug therapy beyond five years would not provide hip fracture benefits. An additional pharmacoepidemiologic study compared associations, timing, and prognosis of renal adverse events following different immune checkpoint inhibitors (ICIs) in real-world clinical practice. Despite knowledge of the association between kidney impairments with ICIs, this study (Chen et al. 2020) reviewed approximately

15-years data (2005–2019) from FDA's Adverse Event Reporting System (FAERS) and concluded that all ICIs included in the study were associated with renal adverse effects in a real-world setting, with a median presentation time for these effects of 48 days.

3.2.3 How Can Pharmacoepidemiology Contribute to the Efficiency and Management of Health Spending? A Health Economics Perspective

Nowadays, achieving maximum efficiency in healthcare is one of the most important issues for health systems due to the increased demand for health services, innovative technologies, and demographic changes. Efficiency shows the way for optimal healthcare delivery and thus better health policies, while the cost savings associated with efficiency improvement are higher (Martin et al. 2009).

Pharmacoepidemiologic studies have a unique advantage of supporting decision-makers regarding efficient and transparent resource allocations. This advantage stems from the fact that drug utilization and safety studies can identify side effects associated with evaluated drugs that may lead to additional treatments to manage them and distinguish behaviors and drug adherence rates by populations and subgroups. For instance, a longitudinal study on a German database (Holt et al. 2010) highlighted that, on average, an increase in total healthcare cost by € 137 was associated with one additional prescribed potentially inappropriate medication (PIM) for the elderly. PIMs are "medications that should be avoided due to their risk which outweighs their benefit and when there are equally or more effective but lower risk alternatives are available" (Page et al. 2010). By clarifying and communicating to the public side effects of medicines, the cost of treatment from both societal and healthcare system perspective is expected to decline (Storm 2000).

It is therefore understandable that the early insights of the potential side effects of the assessed medicinal products and the continuous monitoring of behaviors concerning medication adherence provide the opportunity for well-informed decision-making to produce evidence-based decisions, improving the efficiency of the overall healthcare delivery.

3.2.4 Handling Uncertainty Through Risk-Sharing Schemes in Pharmaceutical Policy. Is There Any Room for Pharmacoepidemiology Research?

Pharmaceutical policies are surrounded by uncertainty. Uncertainty is inherent in the effectiveness of an innovative drug (for instance an advanced medicinal product) in the actual clinical setting, compared to the results of an RCT; uncertainties surround a decision by a regulatory or HTA agency, when the process is not based on complete evidence (Vreman et al. 2020), such as the case of accelerated approvals; uncertainty exists in the macro level, regarding the actual impact of a policy. The existence of uncertainty inevitably leads to costs: societal costs to patients, potential costs to the healthcare system, and direct cost to pharmaceutical companies. After the outbreak of COVID-19, substantial efforts have been made by regulatory authorities to expedite pathways such as rapid reviews and approvals (especially for COVID-19-related products), implement flexible processes to facilitate clinical trials, and enhance collaboration and harmonization efforts for trials on COVID-19 related products and vaccines (Bolislis et al. 2021).

Uncertainty around clinical evidence is also evident in risk-sharing schemes or risk-sharing agreements (RSAs'), which are pre-planned agreements between pharmaceutical companies and payers based on various criteria (Adamski et al. 2010), including cost, effectiveness, and safety. Such agreements are mutually beneficial schemes, between the third-party payers and the marketing authorization holders (MAHs), in which the continuation or non-continuation of reimbursement or even the existing rate of reimbursement of medicines depends on the economic and health outcomes of the treatment (Garrison et al. 2013). Pricing pressures (value-based pricing), costs reduction due to the emergence of the global financial crisis, and other concerns relevant to clinical effectiveness in the real-world environment have impacted on the uptake of such agreements (Piatkiewicz et al. 2018). An example was the agreement between United Kingdom (UK) agencies and a pharmaceutical manufacturer about coverage of bortezomib proteasome inhibitor for the treatment of multiple myeloma, in which refunds are being offered for each defined outcome that was not achieved (Garber and McClellan 2007).

Pharmacoepidemiology research, through drug safety and utilization studies, may improve the rate of evidence regarding the safety and clinical effectiveness of medicinal products which are already used in health systems and thus contribute to reducing uncertainty. This is of high importance for many medicinal products, particularly for oncology drugs, where in some cases a low rate of value for money has been demonstrated (Espín et al. 2011; Salas-Vega et al. 2017) and where the division of target populations into subgroups leads to limited real-world data. These types of studies might enhance the efficacy and safety of medicinal products in conjunction with the acceptability of cost, while drug utilization studies can provide insights to payers concerning to which patients might not gain substantial benefits, expressed either by outcomes or endpoints, either intermediate or surrogate.

3.3 Conclusion

This review aimed to highlight the importance and necessity of integrating pharmacoepidemiology research into health policy and in particular pharmaceutical policy, by bringing forward the use of evidence-based on the real-world value of a technology. It is important to recognize the role of drug utilization and safety studies in addressing issues not captured by RCTs and their supplementary real-world evidence-practice. Therefore, these types of studies might address gaps arising from other sources of data and contribute to the minimization of uncertainty which accompanies decisions relating to HTA process and risk-sharing agreements, while providing additional comparative effectiveness outcomes.

References

Adamski J, Godman B, Ofierska-Sujkowska G, Osińska B, Herholz H, Wendykowska K, Laius O, Jan S, Sermet C, Zara C, Kalaba M, Gustafsson R, Garuolienè K, Haycox A, Garattini S, Gustafsson LL (2010) Risk sharing arrangements for pharmaceuticals: potential considerations and recommendations for European payers. BMC Health Serv Res 7(10):153. https://doi.org/10.1186/1472-6963-10-153

Alarkawi D, Ali MS, Bliuc D et al (2018) The challenges and opportunities of pharmacoepidemiology in bone diseases. JBMR Plus 2(4):187–194

Allen N, Walker SR, Liberti L, Salek S (2017) Health technology assessment (HTA) case studies: factors influencing divergent HTA reimbursement recommendations in Australia, Canada, England, and Scotland. Value Heal 20(3):320–328. https://doi.org/10.1016/j.jval.2016.10.014

Babar Z (2019) Encyclopedia of pharmacy practice and clinical pharmacy, 1st edn. Elsevier Academic Press

Balçık YP, Şantaş G (2016) Pharmacoepidemiology. IOSR J Pharm 6:2250-3013

Baltussen R, Niessen L (2006) Priority setting of health interventions: the need for multi-criteria decision analysis. Cost Eff Resour Alloc 21(4):14. https://doi.org/10.1186/1478-7547-4-14

Baumfeld Andre E, Reynolds R, Caubel P, Azoulay L, Dreyer NA (2020) Trial designs using real-world data: the changing landscape of the regulatory approval process. Pharmacoepidemiol Drug Saf 29(10):1201–1212. https://doi.org/10.1002/pds.4932

Bégaud B (2019) A history of pharmacoepidemiology. Therapie 74(2):175–179. https://doi.org/10.1016/j.therap.2018.10.003

Berger ML, Sox H, Willke RJ, Brixner DL, Eichler HG, Goettsch W, Madigan D, Makady A, Schneeweiss S, Tarricone R, Wang SV, Watkins J, Daniel MC (2017) Good practices for real-world data studies of treatment and/or comparative effectiveness: recommendations from the joint ISPOR-ISPE Special Task Force on real-world evidence in health care decision making. Pharmacoepidemiol Drug Saf 26(9):1033–1039. https://doi.org/10.1002/pds.4297

Bloom BR, Michaud, CM, La Montagne JR, Simonsen L (2006) Priorities for global research and development of interventions. In: Jamison DT, Breman JG, Measham AR et al (eds) Disease control priorities in developing countries, 2nd edn. World Bank, Washington, DC. ISBN 978-0-8213-6179-5. PMID 21250329. Retrieved 2016-06-30

Bolislis WR, de Lucia ML, Dolz F, Mo R, Nagaoka M, Rodriguez H, Woon ML, Yu W, Kühler TC (2021) Regulatory agilities in the time of COVID-19: overview, trends, and opportunities. Clin Ther 43(1):124–139. https://doi.org/10.1016/j.clinthera.2020.11.015

Caparrotta TM, Dear JW, Colhoun HM, Webb DJ (2019) Pharmacoepidemiology: using randomised control trials and observational studies in clinical decision-making. Br J Clin Pharmacol 85(9):1907–1924. https://doi.org/10.1111/bcp.14024

Chen G, Qin Y, Fan QQ, Zhao B, Mei D, Li XM (2020) Renal adverse effects following the use of different immune checkpoint inhibitor regimens: a real-world pharmacoepidemiology study of post-marketing surveillance data. Cancer Med 9(18):6576–6585. https://doi.org/10.1002/cam4.3198

Comer D, Couto J, Aguiar R, Wu P, Elliott D (2015) Using aggregated pharmacy claims to identify primary nonadherence. Am J Manag Care 21(12):e655–e660

Espín J, Rovira J, Gracia L, Andalusian School of Public Health (2011) Experiences and impact of European risk–sharing schemes focusing on oncology medicines experiences and impact of European risk-sharing schemes focusing on oncology medicines, available from: http://ec.europa.eu/DocsRoom/documents/7608/attachments/1/translations/en/renditions/pdf

FDA (2018) Real-world evidence, available from: https://www.fda.gov/science-research/science-and-research-special-topics/real-world-evidence. Accessed on May 3, 2021

Frois C, Grueger J (2017) Pricing of pharmaceuticals: current trends and outlook and the role of comparative effectiveness research. In: Birnbaum H, Greenberg P (eds) Decision making in a world of comparative effectiveness research. Adis, Singapore. https://doi.org/10.1007/978-981-10-3262-2_7

Garber AM, McClellan MB (2007) Satisfaction guaranteed—"payment by results" for biologic agents. N Engl J Med 357(16):1575–1577. https://doi.org/10.1056/NEJMp078204

Garrison LP, Towse A, Briggs A, de Pouvourville G, Grueger J, Mohr PE, Severens JL, Siviero P, Sleeper M (2013) Performance-based risk-sharing arrangements-good practices for design, implementation, and evaluation: report of the ISPOR good practices for performance-based risk-sharing arrangements task force. Value Heal 16(5):703–719. https://doi.org/10.1016/j.jval.2013.04.011

Gliklich RE, Dreyer NA, Leavy MB (eds) (2014) Registries for evaluating patient outcomes: a user's guide [Internet], 3rd edn. Agency for Healthcare Research and Quality (US), Rockville, MD. Report No.: 13(14)-EHC111

Harwood M, Deighton A, Mickle A, et al (2019) The use of real-world data in health technology assessment of medications for rare diseases. Value Heal 22(Suppl 3):S863. Poster presentation at the ISPOR Europe

Hiligsmann M, Boonen A, Rabenda V, Reginster JY (2012) The importance of integrating medication adherence into pharmacoeconomic analyses: the example of osteoporosis. Expert Rev Pharmacoecon Outcomes Res 12(2):159–166. https://doi.org/10.1586/erp.12.8

Holt S, Schmiedl S, Thürmann PA (2010) Potentially inappropriate medications in the elderly: the PRISCUS list. Dtsch Arztebl Int 107(31–32):543–551. https://doi.org/10.3238/arztebl.2010.0543

Hughes D, Cowell W, Koncz T, et al (2007) International society for pharmacoeconomics and outcomes research economics of medication compliance working group. Methods for integrating medication compliance and persistence in pharmacoeconomic evaluations. Value Heal 10(6):498–509

Institute of Medicine (IOM) Initial national priorities for comparative effectiveness research. The National Academies Press, Washington, DC

Izano MA, Lo JC, Adams AL, Ettinger B, Ott SM, Chandra M, Hui RL, Niu F, Li BH, Neugebauer RS (2020) Bisphosphonate treatment beyond 5 years and hip fracture risk in older Wwmen. JAMA Netw Open 3(12):e2025190. https://doi.org/10.1001/jamanetworkopen.2020.25190

Jao R, Jaksa A, Pontynen A, et al (2018) Health technology assessment (HTA) agencies consideration of real world evidence (RWE). Value Heal 21(21):S 7. https://doi.org/10.1016/j.jval.2018.04.032

Jonker CJ, van den Berg HM, Kwa MSG, Hoes AW, Mol PGM (2017) Registries supporting new drug applications. Pharmacoepidemiol Drug Saf 26(12):1451–1457. https://doi.org/10.1002/pds.4332

Kanavos P, Tzouma V, Fontrier AM, Souliotis K (2019) Implementing health technology assessment (HTA) in Greece. Myths, Reality Cautionary Tales, Archiv Hellen Med 36(4):444–451

Larsson S, Lawyer P, Garellick G, Lindahl B, Lundström M (2012) Use of 13 disease registries in 5 countries demonstrates the potential to use outcome data to improve health care's value. Health Aff (millwood) 31(1):220–227. https://doi.org/10.1377/hlthaff.2011.0762

Levine M, Lelorier J (2012) Pharmacoepidemiology and pharmaceutical reimbursement policy. In: Brian L, Strom MD, Kimmel SE, Hennessy S (eds) Pharmacoepidemiology, 5th edn. Wiley

Lou J, Kc S, Toh KY, Dabak S, Adler A, Ahn J, Bayani DBS, Chan K, Choiphel D, Chua B, Genuino AJ, Guerrero AM, Kearney B, Lin LW, Liu Y, Nakamura R, Pearce F, Prinja S, Pwu

RF, Shafie AA, Sui B, Suwantika A, Teerawattananon Y, Tunis S, Wu HM, Zalcberg J, Zhao K, Isaranuwatchai W, Wee HL (2020) Real-world data for health technology assessment for reimbursement decisions in Asia: current landscape and a way forward. Int J Technol Assess Health Care 36(5):474–480. https://doi.org/10.1017/S0266462320000628

Makady A, van Veelen A, Jonsson P, Moseley O, D'Andon A, de Boer A, Hillege H, Klungel O, Goettsch W (2018) Using real-world data in health technology assessment (HTA) practice: a comparative study of five HTA agencies. Pharmacoeconomics 36(3):359–368. https://doi.org/10.1007/s40273-017-0596-z

Martin LA, Neumann CW, Mountford J et al (2009) Increasing efficiency and enhancing value in health care: ways to achieve savings in operating costs per year. In: *IHI* innovation series white paper. Institute for Healthcare Improvement, Cambridge, MA

Mitton C, Seixas BV, Peacock S, Burgess M, Bryan S (2019) Health technology assessment as part of a broader process for priority setting and resource allocation. Appl Health Econ Health Policy 17(5):573–576. https://doi.org/10.1007/s40258-019-00488-1

Montastruc JL, Benevent J, Montastruc F, Bagheri H, Despas F, Lapeyre-Mestre M, Sommet A (2019) What is pharmacoepidemiology? definition, methods, interest and clinical applications. Therapie 74(2):169–174. https://doi.org/10.1016/j.therap.2018.08.001

Nishtala P, Bala S (2019) Sources of data used in pharmacoepidemiology and pharmacovigilance. In: Encyclopedia of pharmacy practice and clinical pharmacy, 1st edn. Elsevier, Amsterdam, The Netherlands, pp 391–399

OECD (2019) Using routinely collected data to inform pharmaceutical policies. Analytical Report for OECD and European Countries, available from: https://www.oecd.org/health/health-systems/Using-Routinely-Collected-Data-to-Inform-Pharmaceutical-Policies-Analytical-Report-2019.pdf. Accessed on May 10, 2021

OECD (2020) European Union, health at a glance: Europe 2020: state of health in the EU cycle. OECD Publishing, Paris

O'Rourke B, Oortwijn W, Schuller T, International Joint Task Group (2020) The new definition of health technology assessment: a milestone in international collaboration. Int J Technol Assess Health Care 36(3):187–190. https://doi.org/10.1017/S0266462320000215

Page RL 2nd, Linnebur SA, Bryant LL, Ruscin JM (2010) Inappropriate prescribing in the hospitalized elderly patient: defining the problem, evaluation tools, and possible solutions. Clin Interv Aging 7(5):75–87. https://doi.org/10.2147/cia.s9564

Paterson JM, Laupacis A, Bassett K, Anderson GM (2006) BC-Ontario pharmacosurveillance for decision-making collaborative. using pharmacoepidemiology to inform drug coverage policy: initial lessons from a two-province collaborative. Health Aff (Millwood) 25(5):1436–1443. https://doi.org/10.1377/hlthaff.25.5.1436

Piatkiewicz TJ, Traulsen JM, Holm-Larsen T (2018) Risk-sharing agreements in the EU: a systematic review of major trends. Pharmacoecon Open 2(2):109–123. https://doi.org/10.1007/s41669-017-0044-1

Pulini AA, Caetano GM, Clautiaux H, Vergeron L, Pitts PJ, Katz G (2021) Impact of real-world data on market authorization, reimbursement decision & price negotiation. Ther Innov Regul Sci. 55(1):228–238. https://doi.org/10.1007/s43441-020-00208-1

Qi W, Abu-Hanna A, van Esch TEM, de Beurs D, Liu Y, Flinterman LE, Schut MC (2021) Explaining heterogeneity of individual treatment causal effects by subgroup discovery: an observational case study in antibiotics treatment of acute rhino-sinusitis. Artif Intell Med 116:102080. https://doi.org/10.1016/j.artmed.2021.102080

Robinson S, Williams I, Dickinson H, Freeman T, Rumbold B (2012) Priority-setting and rationing in healthcare: evidence from the English experience. Soc Sci Med 75(12):2386–2393. https://doi.org/10.1016/j.socscimed.2012.09.014

Sacristán JA, Soto J (1994) Drug utilisation studies as tools in health economics. Pharmacoeconomics 5(4):299–312. https://doi.org/10.2165/00019053-199405040-00005

Salas-Vega S, Iliopoulos O, Mossialos E (2017) Assessment of overall survival, quality of life, and safety benefits associated with new cancer medicines. JAMA Oncol 3(3):382–390. https://doi.org/10.1001/jamaoncol.2016.4166

Schneeweiss S, Avorn J (2005) A review of uses of health care utilization databases for epidemiologic research on therapeutics. J Clin Epidemiol 58(4):323–337. https://doi.org/10.1016/j.jclinepi.2004.10.012

Sinnott SJ, Bennett K, Cahir C (2017) Pharmacoepidemiology resources in Ireland-an introduction to pharmacy claims data. Eur J Clin Pharmacol 73(11):1449–1455. https://doi.org/10.1007/s00228-017-2310-7. Epub Aug 17 2017. Erratum in: Eur J Clin Pharmacol. 2017 73(11):1457

Sorenson C (2010) Use of comparative effectiveness research in drug coverage and pricing decisions: a six-country comparison. Issue Brief (commonw Fund) 91:1–14

Souliotis K, Giannouchos TV, Golna C et al (2022) Assessing forgetfulness and polypharmacy and their impact on health-related quality of life among patients with hypertension and dyslipidemia in Greece during the COVID-19 pandemic. Qual Life Res 31:193–204. https://doi.org/10.1007/s11136-021-02917-y

Specchia ML, Favale M, Di Nardo F et al (2015) HTA working group of the Italian society of hygiene, preventive medicine and public health (SItI). How to choose health technologies to be assessed by HTA? a review of criteria for priority setting. Epidemiol Prev 39(4 Suppl 1):39–44

Spillane S, Usher C, Bennett K, Adams R, Barry M (2015) Introduction of generic substitution and reference pricing in Ireland: early effects on state pharmaceutical expenditure and generic penetration, and associated success factors. J Pharm Pol Pract 8(Suppl 1):O10

Stavrou A, Challoumas D, Dimitrakakis G (2014) Archibald Cochrane (1909–1988): the father of evidence-based medicine. Interact Cardiovasc Thorac Surg 18(1):121–124. https://doi.org/10.1093/icvts/ivt451

Storm BL (2000) Pharmacoepidimiology, 3rd edn. Wiley

Takahashi Y, Nishida Y, Asai S (2012) Utilization of health care databases for pharmacoepidemiology. Eur J Clin Pharmacol 68(2):123–129. https://doi.org/10.1007/s00228-011-1088-2

Thokala P, Devlin N, Marsh K et al (2016) Multiple criteria decision analysis for health care decision making—an introduction: report 1 of the ISPOR MCDA emerging good practices task force. Value Heal 19(1):1–13

Tromp N, Baltussen R (2012) Mapping of multiple criteria for priority setting of health interventions: an aid for decision makers. BMC Health Serv Res 13(12):454. https://doi.org/10.1186/1472-6963-12-454

U.S. Food & Drug Administration (2018) Framework for FDA's real-world evidence program, December, available from: https://www.fda.gov/media/120060/download. Accessed on May 4, 2021

van Luijn JC, Gribnau FW, Leufkens HG (2007) Availability of comparative trials for the assessment of new medicines in the European Union at the moment of market authorization. Br J Clin Pharmacol 63(2):159–162. https://doi.org/10.1111/j.1365-2125.2006.02812.x

Vreman RA, Bouvy JC, Bloem LT, Hövels AM, Mantel-Teeuwisse AK, Leufkens HGM, Goettsch WG (2019) Weighing of evidence by health technology assessment bodies: retrospective study of reimbursement recommendations for conditionally Approved Drugs. Clin Pharmacol Ther 105(3):684–691. https://doi.org/10.1002/cpt.1251

Vreman RA, Naci H, Goettsch WG, Mantel-Teeuwisse AK, Schneeweiss SG, Leufkens HGM, Kesselheim AS (2020) Decision making under uncertainty: comparing regulatory and health technology assessment reviews of medicines in the United States and Europe. Clin Pharmacol Ther 108(2):350–357. https://doi.org/10.1002/cpt.1835

Wettermark B (2013) The intriguing future of pharmacoepidemiology. Eur J Clin Pharmacol 69(Suppl 1):43–51. https://doi.org/10.1007/s00228-013-1496-6. Epub 2013 May 3

Wettermark B, Elseviers M, Almarsdóttir AB et al (2016) Introduction to drug utilization research. In: Methods and Application. Wiley

World Health Organization, Health technology assessment, retrieved from: Health technology assessment-Global (who.int). Accessed on Apr 23, 2021

Yuasa A, Yonemoto N, Demiya S, Foellscher C, Ikeda S (2021) Investigation of factors considered by health technology assessment agencies in eight countries. Pharmacoecon Open 5(1):57–69. https://doi.org/10.1007/s41669-020-00235-6

Zisis K, Naoum P, Athanasakis K (2020) Qualitative comparative analysis of health economic evaluation guidelines for health technology assessment in European countries. Int J Technol Assess Health Care 10(37):e2. https://doi.org/10.1017/S0266462320002081

Konstantinos Zisis is Health Economist and Policy Analyst by training. He studied Health Management & Health Policy and Planning. Currently, he is PhD Candidate in health technology assessment (HTA) at the Department of Public Health Policy, School of Public Health, University of West Attica. He is also working as Country Approval Specialist in the clinical trials industry. He focuses his research in the fields of economic evaluation in health care, evidence-based health care, and HTA having published several papers and abstracts in health policy and economics, while he is Regular Reviewer for international journals.

Kostas Athanasakis Ph.D., is Assistant Professor in Health Economics and Health Technology Assessment at the Department of Public Health Policy of the University of West Attica (Greece) and Member of the Laboratory for Health Technology Assessment (LabHTA). His research and teaching efforts mainly focus on health technology assessment, economic evaluation of healthcare interventions, the economics of pharmaceutical markets, and pharmaceutical policy. Other areas of interest include health systems sustainability and resilience, the economic analysis of healthcare markets, and public finance. He has authored/co-authored over 250 full text papers and abstracts as well as 14 books in the field of health services research and economics and policy during the last years.

Kyriakos Souliotis is Professor of Health Policy and Dean at the School of Social and Political Sciences at the University of Peloponnese. He was Vice Rector of Social and Regional Development (2016–2017). He is currently President of the Scientific Board of the Hellenic Association of Political Scientists and President of the Scientific Committee of the Greek Patient Association. From May 2010 to December 2011, he served as President of OPAD (Health Care Organization for Public Servants), and from May 2011 to September 2012, as Vice-President of EOPYY (National Organization for Health Care Services Provision). From June 2010 to August 2013, he was Member of the National Ethics Committee for Clinical Trials and the Steering Committee for Rare Diseases. He has published 32 books and more than 350 chapters in books and papers in peer-reviewed journals on health policy and economics, organization and administration of health services, economic inequalities, etc.

Chapter 4
Accelerating Personalized Medicine Adoption in Oncology: Challenges and Opportunities

Fredrick D. Ashbury and Keith Thompson

Abstract Cancer therapy has been shifting away from "one-size-fits-all" approaches to treatment decisions that are predicated on the molecular profile of the patient's cancer. Personalized medicine, precision medicine, precision oncology, or "omics-guided" therapy are expressions used for this paradigm shift often interchangeably. The uptake of precision medicine, while advancing in cancer care, has faced several adoption challenges, including education, policy, and practical factors. Facilitating the transformation toward personalized medicine that will improve patient outcomes in the oncology setting requires a coordinated effort among policymakers, cancer agencies, health systems, and industry. To implement precision medicine effectively in cancer practice requires an informatics solution beyond the legacy electronic medical record platforms currently available to clinical teams.

Keywords Cancer care · Oncology · Artificial intelligence · Health system

F. D. Ashbury (✉)
Department of Oncology, University of Calgary, Calgary, AB, Canada
e-mail: fred.ashbury@viecure.com

Dalla Lana School of Public Health, University of Toronto, Toronto, Canada

F. D. Ashbury · K. Thompson
VieCure, Denver, CO, USA

K. Thompson
Faculty of Medicine, University of Alabama, Birmingham, AL, USA

Montgomery Cancer Center, Montgomery, AL, USA

K. Thompson
e-mail: keith.thompson@viecure.com

E. Çetin and H. Özen (eds.), *Healthcare Policy, Innovation and Digitalization*,
Accounting, Finance, Sustainability, Governance & Fraud: Theory and Application,
https://doi.org/10.1007/978-981-99-5964-8_4

4.1 Introduction

The International Agency for Research on Cancer (IARC) has estimated that approximately 1 in 8 men and 1 in 11 women will die from cancer. In 2020, more than 10 million people worldwide died of cancer. The global number of cancer deaths each year is anticipated to nearly double by 2030 (Sung et al. 2021) with more than 27 million new cancer cases estimated to occur.

Mitigating premature deaths due to cancer is a strategic priority in most countries, including strategic investments being made by governments, not-for-profit organizations and private industry in research to identify novel treatments—and regrettably to a much lesser extent investments—in new technologies to identify cancers as early as possible to optimize treatment (Raza 2019).

The contribution of science to map the human genome has greatly advanced our understanding of the biology of cancer, and the work in this field is moving rapidly. Cancer therapy has been materially shifting away from "one-size-fits-all" treatment options to approaches that are predicated on the molecular profile of the patient's cancer. Personalized medicine, precision medicine, precision oncology, or "omics-guided" therapy are expressions commonly used for this paradigm shift, and these terms are often used interchangeably. The uptake of precision medicine, while advancing in cancer care, has faced several adoption challenges, including personal, organizational, systems-level, policy, and practical factors (Ashbury et al. 2021). Overcoming these challenges will require a coordinated effort at the national, system, and practice levels, but these efforts cannot be understood without considering the larger context of key pressures impacting cancer care delivery. These exogenous factors include workforce burnout and shortages, increasing incidence of cancer and cancer complexity, an explosion of new knowledge generated by science, and a dearth of informatics solutions to help the clinical team cope with these pressures.

This chapter will build the case for including effective artificial intelligence solutions to augment clinicians' natural intelligence at the point-of-care. By outlining exogenous influences and pressures on cancer practice, the argument for the benefit of such a solution will be evident.

The response required to effectively manage these pressures will depend on political will, ease of implementation, ease of use, and establishment of novel models of care to ensure the proper application of omics-guided solutions to improve patient outcomes.

4.2 Oncology Workforce Challenges

Cancer care practitioners worldwide have become busier over the past 10 years than ever before. More patients are waiting in reception areas to be checked in, and others are swelling the queues for labs and imaging studies. More patients are occupying radiation clinics, infusion suites, and exam rooms. Statistics from GLOBECAN in

2012 reported that some 14 million new cancer cases were diagnosed and over 8 million people died of cancer worldwide (Torre et al. 2015). The same data estimates that incidence of cancer will double by 2030 (Sung et al. 2021). To address this growing need, more providers, clinicians, and staff will be sorely needed to help diagnose, treat, and support an increasing cancer patient population. In addition to human resource capacity increases, more efficient and effective models of care should be implemented to manage the demand for cancer care.

The magnitude of increasing numbers of new and continuing patients on cancer care delivery is even more pronounced in rural and resource-limited areas of the world, where current oncology workforce estimates show that the demand for oncology care professionals is already far outstripping the current supply of available clinical experts, thereby jeopardizing cancer care delivery and patient outcomes (Mathew 2018; Stefan 2015) exacerbating disparities.

These workforce challenges are not limited to a specific oncology discipline. For example, in the United States, reports go back some three decades describing the issues and challenges of training, recruiting, and retaining professionals in the radiation oncology field (Shah and Royce 2021). Some jurisdictions are reporting improvements in recruiting professionals to fill radiation oncology positions, such as Canada (Wu et al. 2022), but the experience in low- and middle-income countries fares much worse (Nefreteri et al. 2019). Similar reports exist describing shortages of professionals in medical oncology (Lwin et al. 2018; Shih et al. 2021), oncology pharmacists (Shih et al. 2021), surgical oncology (Clarke 2022), and nursing (Challinor et al. 2020). Moreover, cross cutting all these oncology disciplines is the urgent need to have a diverse workforce that reflects the dynamic and diverse patient population.

Workforce burnout is a major challenge exacerbating care delivery capacity (Yates and Samuel 2019). A combination of increases in the numbers of patients who require treatment and follow up care, growing workplace demands, and inadequate electronic health record platforms have created a perfect storm of stress on personnel to the point that many are leaving the workforce for early retirement or other employment options. Moreover, the pandemic has created even greater stress on oncologists and oncology professionals, where some have been drawn into primary care practice to help treat and care for cancer and non-cancer patients with COVID-19 (Sing et al. 2022; Ashbury 2021).

4.3 Cancer Care—Knowledge Explosion and Increasing Complexity

Cancer is not a single disease, and care often requires a complex, multidisciplinary approach. Advances in the molecular profiling of cancers to facilitate treatment decisions have been unprecedented; it has been a veritable knowledge explosion (Kaspers 2022). More than ever before, practitioners can provide hope to cancer patients for improved quality of life, cure, or prolong life. Omics-focused research has expanded dramatically thanks to investments from government, not-for-profit foundations, and the private sector. Initiatives such as the Cancer Genome Atlas and the Worldwide Innovation Network, comprising some 40 institutions encompassing nearly every region of the globe, have been researching and applying knowledge of the biology of cancers to identify even more treatment options predicated on the results of next-generation sequencing labs.

The substantial growth in biomarkers in many cancers, including urologic, breast, lung, colon, melanoma, to name a few, and improvements in germline and somatic sequencing technologies (tissue and liquid biopsy) is yielding even greater opportunities for better therapeutic outcomes. Nevertheless, ordering of sequencing tests for eligible patients does not occur in all cases. Research indicates that the proportion of eligible patients that are sequenced to identify potentially actionable therapeutic targets has increased (Sturgill et al. 2021), particularly in the community oncology setting. However, substantial improvement in ensuring genomic testing is done in eligible patients is still needed. The reasons for this limited uptake of genomic sequencing include cost to the patient, organizational factors (absence of guidelines on which test to order, lack of institutional policies, lab resources, and providing adequate tissue), individual factors (beliefs in the efficacy of molecular testing, knowledge, and experience), and practical issues (lack of resources to order tests, time required to order tests, and delays in obtaining test results) (3), and geography (rural/remote physicians report more barriers to test ordering than those practicing in suburban and urban settings) (Roberts et al. 2021). Gray and colleagues have reported that both oncologists and patients have modest or poor levels of knowledge regarding genetic testing and genomic sequencing, and physicians lack experience with large-scale sequencing platforms (Gray et al. 2016).

In addition, in the past 18 months, the Food and Drug Administration in the United States has approved more than 60 new immunotherapy and targeted therapy agents that oncologists can add to the treatment arsenal. Lack of awareness of these new drugs is one factor that can limit adoption, high cost is another. On top of the challenges of selecting novel agents for treatment is the knowledge and tools required to administer the drugs (alone or in combination with other agents or treatment modalities) and to monitor patients during treatment, considering that the side effects of immunotherapies and targeted treatments are less well known until wider application in the real-world setting occurs (Bustillos et al 2022; Rapoport et al. 2021; Coschi and Juergens 2021).

The challenge, therefore, is to be able to sift, sort, and select through the wealth of research and data, to identify and determine the best "fit" for an individual patient's circumstances, and to do so—in near real time—to offer effective solutions. Keeping track of the new test and drug approvals, management of the drugs and their associated toxicities is a daunting, even overwhelming, task for any person. Acknowledging that the flood gates of knowledge have opened, while exciting, is also overpowering when thinking about how to apply what we know when the patient is waiting in an exam room to discuss treatment options. Nobody can process the myriad data elements alone.

4.4 Implementing Solutions at Point-of-Care—Key Issues

We are seeing new models of cancer care delivery emerging throughout the world in response to the pandemic and reduction in the oncology workforce, including incorporating family practice physicians into care delivery (Druel et al. 2020) welcomed by patients to support shared decision-making (Noteboom et al. 2021), greater use of telemedicine/telehealth/digital health technologies to facilitate remote patient monitoring (Cannon 2018; Chan et al. 2020; Patt et al. 2021), in-home care delivery (Mooney et al 2021) including supportive care (Nipp et al. 2022), and patient self-management strategies with or without integrated digital technologies (Luo et al. 2022). Middle- and low-income countries, and even practices in high-income countries, provide care to patients that cannot access these technologies due to income constraints, in which cases the patients are further disadvantaged.

Moreover, cancer practices in high- and some middle-income countries (more so the former) generally comprise a complex ecosystem of information technologies. These technologies can include an EHR (or multiple EHRs), patient portals, scheduling platforms, a web of interfaces to support the integration of laboratory tests, and the like. In low- and many middle-income countries, however, the informatics ecosystem is much less diverse, due to cost and other factors (Ngusie et al. 2022). The challenge, when multiple technologies co-exist, is to integrate these disparate platforms in a coherent and cohesive way to allow for ease of transfer of the material data required to manage the appropriate diagnosis, treatment, supportive care, survivorship, and end-of-life care of cancer patients while simultaneously managing the practice. These complex challenges are exacerbated by the lack of standardization of clinical data undermining the interpretation and use of these data (Bernstam et al. 2022). The result: an increase in the risk that the treatment decision is not optimal for the patient's situation.

4.5 The Role of Artificial Intelligence to Complement Natural Intelligence

A significant challenge today remains the ability to process the massive amounts of patient-specific data, including demographic, social determinants of health, clinical data (including next-generation sequencing results) and insurance data. Most oncology practices, and especially those in low- and middle-income countries, have inadequate informatics tools to enable them to process these data easily to make treatment decisions and must rely on manual record keeping and/or outdated electronic health records technologies ("EHRs"), which does not allow the ability to maintain patient privacy standards. Moreover, these legacy EHRs often can only accept an electronic transmission of sequencing results in report format or by scanning the documents received into a folder to be read by the oncology team when the results are returned several weeks after the test order, thereby causing additional delay in the start of potentially life-saving therapy. The burden on the individual provider is simply too great to manage these complex and prodigious data points, and being unable to do so, a reliance on outdated systems becomes the only course of action.

The application of artificial intelligence (AI), or the ability of computers to do certain tasks done typically by humans that require human intelligence, represents a recent and ground-breaking trend in the oncology setting. The field of AI is broad, and terminology for the types of data that can be processed by AI is equally diverse, including machine learning, natural language processing, deep learning, unsupervised learning, and so on. As Chua and colleagues have stated, "although these complexities create substantial barriers for machines, they also create large hurdles for humans and represent opportunities, where AI algorithms could have significant impact on the quality and value of cancer care, especially in communities that lack oncology subspecialists" (Chua et al. 2021, p. 4140). AI is being used in pathology, radiography, to improve diagnostic accuracy in gastrointestinal cancers, treatment response, and other outcomes (Chua et al. 2021). In precision oncology, AI is also being used to analyze genomic data to help identify potential therapeutic targets to facilitate decision-making (Penson et al. 2020).

Unlike humans, computers do not suffer from fatigue; more data only requires more processing time. However, central to the application of AI is the processing of discrete data to generate considerations at point-of-care in near or real time to facilitate clinical decision-making. As such, efforts need to be made to ensure data sources (e.g., sequencing labs, EHRs, etc.) have well-defined transfer protocols to pass their data discretely to the AI engine. Unfortunately, healthcare data are not typically stored in a structured format, and as such, AI effectiveness will be much more limited (Elkhader and Elemento 2022). Moreover, drawing data from multiple sources places a greater burden on oncology practices to maintain and generally exacerbate implementation challenges (Parikh et al. 2019). Therefore, AI integrated with EHRs emphasizing discrete data is the path forward to enhance the power of AI to assist clinical decision-making. Unfortunately, the majority of EHR platforms have been sorely lacking in their evolution to facilitate processing of genomic data, which is surprising in the light of the accelerated growth in sequencing tests and increase in their uptake, thereby challenging oncologists to apply these results when making treatment decisions (Ohno-Machado et al. 2018). The key components will be integration of next-generation sequencing and genetic testing results as discrete data, discrete capture of demographics, vitals, diagnosis, staging, histology, tumor markers, tumor dimensions, co-morbidities, medications, treatments, symptoms, supportive care interventions, patient-reported outcomes, and so on. AI can, therefore, act as the "second opinion" to assist clinicians in busy practices, particularly in the community oncology setting, by processing patient-specific data supported by evidence-based content as "rules" to consider. Furthermore, given oncology patients' situations are not static, AI can assist with the interpretation of these dynamic and complex changes.

4.6 Concluding Remarks

Transitioning to personalized medicine to improve patient outcomes in the oncology setting requires a coordinated effort among policymakers, cancer agencies, health systems, and industry. To implement precision medicine appropriately in cancer practice requires an informatics solution beyond the legacy electronic medical record platforms currently available to clinical teams. AI platforms provide an important step forward if these can be systematically architected and "fed" with discrete data to allow for processing of these data against the latest evidence and guidelines to facilitate clinical decision-making. The day is approaching quickly when physicians, clinicians, and patients will have comprehensive data and information available at their fingertips, including data on genetics, genomics, proteomics, metabolomics, and so on. AI will support physicians' natural intelligence and skills to easily order appropriate diagnostic tests, receive, process, and interpret results in a timely manner to promote quality decision-making for the overall benefit of the patient.

Conflict of Interest Disclosure I, Fredrick Ashbury, disclose that my affiliation with VieCure, a company I have co-founded and invested in that operates within the oncology space. Aside from this, I have no personal, financial, or professional conflicts of interest. I certify that all my current and foreseeable conflicts of interest, or their absence, have been disclosed here.

I, Keith Thompson, disclose that my affiliation with VieCure, is a company I have invested in that operates with the oncology space. Aside from this, I have no personal, financial, or professional conflicts of interest. I certify that all my current and foreseeable conflicts of interest, or their absence, have been disclosed here.

References

Ashbury FD (2021) Covid-19 and supportive cancer care: key issues and opportunities. Curr Opin Oncol 33(4):295–300

Ashbury FD, Thompson K, Williams C, Williams K (2021) Challenges adopting next-generation sequencing in community oncology practice. Curr Opin Oncol 33(5):507–512

Bernstam EV, Warner JL, Krauss JC, Ambinder E, Rubinstein WS, Komatsoulis G, Miller RS, Chen JL (2022) Quantitating and assessing interoperability between electronic health records. JAMIA 29(5):753–760

Bronfenbrenner U (1977) Experimental toward an ecology of human development. Am Psychol 32:513–531. https://doi.org/10.1037/0003-066X.32.7.513

Bustillos H, Indorf A, Alwan L, Thompson J, Jung L (2022) Xerostomia: an immunotherapy-related adverse effect in cancer patients. Supp Care Cancer 30:1681–1687

Cannon C (2018) Telehealth, mobile applications, and wearable devices are expanding cancer care beyond walls. Sem Oncol Nursing 34(2):118–125

Challinor JM, Alqudimat MR, Teixeira TOA, Oldenmenger WH (2020) Oncology nursing workforce: challenges, solutions, and future strategies. Lancet Oncol 21(12):e564–e574

Chan A, Ashbury FD, Fitch MI, Koczwara B, Chan RJ (2020) Cancer survivorship care during COVID-19—perspectives and recommendations from the MASCC Survivorship Study Group. Support Care Cancer 28(8):3485–3488

Chua IS, Gaziel-Yablowitz M, Korach AT, Kehl KL, Levitan NA, Arriaga YE, Jackson GP, Bates DW, Hassett (2021) Artificial intelligence in oncology: path to implementation. Cancer Med 10(12):4138–4149

Clarke CN (2022) Disparities in creating a diverse surgical oncology physician workforce: just a leaky pipeline? Surg Oncol Clinics 31(1):21–27

Coschi CH, Juergens RA (2021) The price of success: immune-related adverse events from immunotherapy in lung cancer. Curr Oncol 28(6):4392–4407

Druel V, Gimenez L, Paricaud K, Delord J-P, Grosclaude P, Boussier N, Rouge Bugat M-E (2020) Improving communication between the general practitioner and the oncologist: a key role in coordinating care for patients suffering from cancer. BMC Cancer 20:495. https://doi.org/10.1186/s12885-020-06993-0

Elkhader J, Elemento O (2022) Artificial intelligence in oncology: from bench to clinic. Semin Cancer Biol 84:113–128

Gray S, Park E, Najita J et al (2016) Oncologists' and cancer patients' views on whole-exome sequencing and incidental findings: results from the CanSeq study. Genet Med 18:1011–1019

Kaspers GJL (2022) 20 years of expert review of anticancer therapy. Expert Rev Anticancer Ther 22(1):1–2. https://doi.org/10.1080/14737140.2022.2019019

Luo X, Chen YZ, Chen J, Zhang Y, Li MF, Xiong CX, Yan (2022) Effectiveness of mobile health-based self-management interventions in breast cancer patients: a meta-analysis. Supp Care Cancer 30:2853–2876

Lwin Z, Broom A, Sibbritt D, Francis K, Karapetis CS, Karikios D, Harrup R (2018) The Australian medical oncologist workforce survey: the profile and challenges of medical oncology. Sem in Oncol 45(5–6):284–290

Mathew A (2018) Global survey of clinical oncology workforce. JCO Global Oncol 4:1–12

Mooney K, Titchener K, Haaland B, Coombs LA, O'Neil B, Nelson R, McPherson JP, Kirchhoff AC, Beck AC, Ward JH (2021) Evaluation of oncology hospital at home: unplanned health care utilization and costs in the Huntsman at Home real-world trial. J Clin Oncol 39(23):2586–2593

Nefreteri S, Elmore C, Praijogi B, Polo Rubio JA, Zubizarreta E (2019) The global radiation oncology workforce in 2030: estimating physician training needs and proposing solutions to scale up capacity. Appl Rad Oncol, 9–16

Ngusie HS, Kassie SY, Chereka AA, Enyew (2022) Healthcare providers' readiness for electronic health record adoption: a cross-sectional study during pre-implementation phase. BMC Heal Serv Res 22. https://doi.org/10.1186/s12913-022-07688-x

Nipp RD, Shulman E, Smith M, Brown PMC, Johnson PC, Gaufberg E, Vyas C, Qian CL, Neckermann I, Hornstein SB, Reynolds MJ, Green J, Temel JS, Areej E-J (2022) Supportive oncology care at home interventions: protocols for clinical trials to shift the paradigm of care for patients with cancer. BMC Cancer 22:383. https://doi.org/10.1186/s12885-022-09461-z

Noteboom EA, Perfors LAA, May AM, Stegmann ME, Duijts SFA, Visserman EA, Engelen V, Richel C, van der Wall E, de Wit N, Helsper CW (2021) GP involvement after a cancer diagnosis; patients' call to improve decision support. BJGP Open 5(1). https://doi.org/10.3399/bjgpopen20X101124

Ohno-Machado L, Kim J, Gabriel RA, Kuo GM, Hogarth MA (2018) Genomics and electronic health record systems. Hum Mol Genet 27:R48–R55

Parikh RB, Teeple S, Navathe AS (2019) Addressing bias in artificial intelligence in health care. JAMA 322:2377–2378

Patt DA, Wilfong L, Toth S, Broussard S, Kanipe K, Hammonds J (2021) Telemedicine in community cancer care: how technology helps patients with cancer navigate a pandemic. JCO Oncol Pract 17(1):e11–e15

Penson A, Camacho N, Zheng Y et al (2020) Development of genome-derived tumor type prediction to inform clinical cancer care. JAMA Oncol 6:84–91

Rapoport B, Cooksley T, Johnson DB, Anderson R (2021) Supportive care for new cancer therapies. Curr Opin Oncol 33(4):287–294

Raza A (2019) The first cell: and the human costs of pursuing cancer to the last. Basic Books, New York

Roberts MC, Spees LP, Freedman AN, Klein WMP, Prabhu Das I, Butler EN, de Moor JS (2021) Oncologist-reported reasons for not ordering multimarker tumor panels: results from a nationally representative survey. JCO Prec Oncol 5:701–709

Shah C, Royce TJ (2021) Chicken little or goose is cooked.? the state of the US radiation oncology workforce: workforce concerns in US radiation oncology. Int J Rad Oncol Biol Phys 110(2):268–271

Shih Y-CT, Kim B, Halpern MT (2021) State of physician and pharmacist oncology workforce in the United States in 2019. JCO Oncol Pract 17(1):e1–e10

Sing S, Farrelly A, Chan C, Nicholls B, Nazeri-Rad N, Bellicoso D (2022) Prevalence and workplace drivers of burnout in cancer care physicians in Ontario, Canada. JCO Oncol Pract 18(1):e60–e71

Stefan DC (2015) Cancer care in Africa: an overview. JCO Global Oncol 1:30–3a

Sturgill EG, Misch A, Lachs R, Jones CC, Schlauch D, Jones SF, Shastry M, Yardley DA, Burris HA III, Spigel DR, Hamilton EP, McKenzie AJ (2021) Next-generation sequencing of patients with breast cancer in community oncology clinics. JCO Prec Oncol 5:1297–1311

Sung H, Ferlay J, Siegel RL, Laversanne M, Soerjomataram I, Jemal A, Bray F (2021) Global cancer statistics 2020: GLOBOCAN estimates of incidence and mortality worldwide for 36 cancers in 185 countries. CA Cancer J Clin 71(3):209–249

Torre LA, Bray F, Siegel RL, Ferlay J, Lortet-Tieulent J, Jemal A (2015) Global cancer statistics, 2012. CA Cancer J Clin 65:87–108

Wu CHD, Malik N, Kim M, Stuckless T, Halperin R, Archambault J, Thompson R, Ringash J, Brundage M, Loewen SK (2022) Employment outcomes for Canadian radiation oncology graduates: 2020 assessment and longitudinal trends. Adv Rad Oncol 7(3). https://doi.org/10.1016/j.adro.2022.100915

Yates M, Samuel V (2019) Burnout in oncologists and associated factors: a systematic literature review and meta-analysis. Eur J Cancer Care. https://doi.org/10.1111/ecc.13094

Fredrick D. Ashbury Co-founder and Chief Scientific Officer at VieCure. He is responsible for developing and maintaining VieCure's oncology clinical content, genomic-science knowledge identification and codification, and the design and implementation of diagnostic, treatment, post-treatment survivorship, and palliative care/end-of-life rules that underpin the platform. He has published more than 125 peer-reviewed articles and major reports and holds university appointments in the Department of Oncology, University of Calgary, Canada, and the Dalla Lana School of Public Health at the University of Toronto, Canada. He sits on the board of the Multinational Association of Supportive Care in Cancer and is Member of ASCO. He is Editor-in-Chief of Supportive Care in Cancer.

Keith Thompson is Medical Oncologist, Former Member of ASCO's CancerLinq initiative, and Vice-President, Medical Affairs at VieCure. He has published on community oncology health services delivery and uptake of clinical practice guidelines. He holds an academic appointment in the Faculty of Medicine at the University of Alabama. He sits on the editorial board of the Journal of Clinical Oncology.

Part III
Innovation

Chapter 5
Framework for Epidemic Risk Analysis

Maryna Zharikova and Stefan Pickl

Abstract Global urbanization presents numerous health challenges and necessitates a deeper comprehension of how diseases spread in urban settings. The chapter proposes a novel framework for epidemic risk analysis by incorporating human mobility. The proposed framework combines evidence-based risk analysis driven by facts with the advantages of flexible precautionary models taking into account causation-and-effect connections and spatial reference. The precautionary approach is implemented in the form of a set of interrelated models, which, as a result, give estimates of a potential risk, obtained in advance and distributed in space. To obtain these estimates, the following models are built: a spatial model, a model of human mobility, as well as a model of disease spread. The spatial model comprises a set of confined locations of people's concentration. These locations are subdivided into several levels, such as residential buildings (base locations where people live), and workplace locations / locations of study (locations that people attend regularly). Over the spatial model, we build the model of human mobility represented as a network with the nodes being locations of different levels. Each node at every time moment is characterized by a certain number of people. The arcs of the network show human flows within the spatial model, which influence disease spread. The disease spreading model reflects disease movement through the nodes of the network. Each node of the network has its status regarding a certain disease, such as susceptible, infected, or recovered. A transition of the node from one status to another is represented as an event. A spread of the disease through the nodes of the network is represented by an event-tree network. We also consider external factors (meta-factors) that can influence the rate of spread of the disease such as weather conditions (temperature,

M. Zharikova (✉) · S. Pickl
Institut für Theoretische Informatik, Mathematik und Operations Research, Universität der Bundeswehr München, Neubiberg, Germany
e-mail: maryna.zharikova@unibw.de

S. Pickl
e-mail: stefan.pickl@unibw.de

M. Zharikova
Program Tools and Technologies Department, Kherson National Technical University, Beryslavske Shose, 24, Kherson 73040, Ukraine

E. Çetin and H. Özen (eds.), *Healthcare Policy, Innovation and Digitalization*, Accounting, Finance, Sustainability, Governance & Fraud: Theory and Application,
https://doi.org/10.1007/978-981-99-5964-8_5

humidity) and air quality. A distinctive feature of the proposed approach is a spatial reference of network nodes. This chapter addresses the framework for epidemic risk analysis. The proposed framework is based on two main principles such as an evidence-based approach and a precautionary principle. These principles are used in the following way. Using the precautionary principle, scenarios of disease spread and risk assessments are built in advance and can be used for disease prevention and mitigation. With the help of the evidence-based approach, the risk assessments obtained using the precautionary approach are adjusted. Justified risk assessments can be used for making decisions to counter the disease. The proposed framework makes it possible to diagnose the situation within the area of interest and sheds new light on disease-spreading analysis. The most dangerous areas can be identified, and measures can be taken to stop the disease from spreading. The work contributes to the creation and putting into practice the policies that slow down or prevent the spread of illnesses.

Keywords Disease spread · Human mobility · Event tree · Epidemic risk

5.1 Introduction

The world is becoming more urban every day. Emerging infectious diseases have been made more prevalent as a result of rising urbanization, population density, human encroachment, and worldwide interconnection. These diseases can spread more quickly, posing numerous problems for global health.

Population density is one factor that contributes to the spread of disease. Cities are characterized by their dense population, which is characterized by frequent close human contact. The chance of spreading viruses may rise as a result. The rapid spread of infectious diseases and a domino effect can be caused by the increased number of passive encounters between people as a result of high population density. (Neiderud 2014). Epidemics and pandemics are two of the most obvious examples of the relationship between population density and infectious illnesses (Fakhruddin 2020).

Another factor in the spread of diseases in the urban environment is human mobility (Grais et al. 2003; Hufnagel et al. 2004; Brownstein et al. 2006; Colizza et al. 2006). The majority of studies on how mobility affects the transmission of pandemics concentrate on long-distance travel across nations, states, and cities (Aguilar et al. 2022), but the spread of viruses is also caused by the movement of people between cities within the same country, as well as be the movement within a certain city when people go to work, study, etc. In urban areas, the dominant statuses of transport are automobiles, buses, metros, bicycles, and pedestrians. In particular, the mixture of different statuses of transport in residential areas, combined with the socioeconomic activities, leads to the fact that there is constant congestion in the stations, offices, and stores, making personal contact inevitable (Tatem et al. 2012; Finger et al. 2016; Massaro et al. 2019).

Poor housing can also be a factor in the spread of diseases in the urban environment. It may be related to insufficient water supplies as well as waste management and hygiene and can lead to the spreading of insect and rodent vector diseases.

Worldwide urbanization calls for a deeper understanding of the interconnections in societies and the vulnerabilities within them. The cornerstone of finding efficient techniques to stop, control, and manage epidemic spreads is monitoring, evaluating, and predicting the impact of infectious diseases on the wellness of a society.

For researching the spread of disease, creating measures to control and prevent new outbreaks, and reducing their catastrophic impact on a population, the use of models in public health decision-making has become more crucial (Chen 2014; Neiderud 2014). The development of the methods of big data analytics, artificial intelligence, and machine learning has opened significant opportunities to address the above-mentioned issues.

This chapter addresses the framework for epidemic risk analysis. The proposed framework is based on two main principles such as the evidence-based approach and the precautionary principle. Today´s and future challenges require modifications for these principles and the developing new approaches, which are based on them. In the proposed framework, these principles are used according to the following. Using the precautionary principle, scenarios of disease spread and risk assessments are built in advance and can be used for disease prevention and mitigation. With the help of the evidence-based approach, the risk assessments obtained using a precautionary approach are adjusted. Justified risk assessments can be used for making decisions to counter the disease.

5.2 Recent Works

Currently, there are two main approaches for simulating infectious disease transmission and epidemic risks (Ibeas et al. 2013), such as evidence-based and precautionary.

5.3 Evidence-Based Principle

The rise of computational and statistical methodologies gave rise to the development of the systematic use of clinical studies and data. The development of computer sciences in the twentieth century put more weight on statistical analysis and interpretation of medical data. The implementation of evidence-based methodology started playing a big role.

A population's needs are served through evidence-based public health, which combines the best available research with expert opinion and expertise from stakeholders. Data from observational research, surveillance, and modeling are crucial for generating the public health evidence base for infectious diseases (European Centre for Disease Prevention and Control 2011).

Currently, there are many formulations of the term and principles of the evidence-based approach. Eddy formulated principles on how to transpose data-driven evidence to design policies and guidelines and was the first to define the "evidence-based approach" which "explicitly describes the available evidence that pertains to a policy and ties the policy to evidence but does not explicitly estimate the magnitudes or compare the benefits and harms" (Eddy 1990; Pickl et al. 2021).

The research group around David Sackett and Gordon Gyatt described the term "evidence-based medicine" in the following way: "Evidence-based medicine de-emphasizes intuition, unsystematic clinical experience, and pathophysiologic rationale as sufficient grounds for clinical decision-making and stresses the examination of evidence from clinical research"(Sackett et al. 1996; Pickl et al. 2021). The probably most cited definition was given by Sackett et al.: "Evidence-based medicine is the conscientious, explicit, and judicious use of current best evidence in making decisions about the care of individual patients" (Pickl et al. 2021).

The evidence-based approach to epidemic risk analysis usually uses such methods as statistics, case-based reasoning, machine learning, etc.

For example, in (Shi et al. 2016), the authors present a universal model of illness risk assessment based on unstructured clinical medical text analyzing its characteristics. They create the convolutional neural network by training it on a large amount of historical clinical text data to extract disease risk features. Using patient disease-related data from the hospital's big data, a disease risk assessment is created.

5.3.1 Precautionary Principle

When scientific information about an environmental or human health threat is ambiguous and the stakes are high, the precautionary principle permits decision-makers to take preventive steps.

The precautionary principle has become more frequently used by public health advocates around the world as the foundation for preventive measures (Bernard and Goldstein 2001). The precautionary principle is used in the evaluation of the effectiveness and/or cost-effectiveness of interventions designed to prevent future harm.

There are a lot of definitions of this term but for the objectives of this chapter, a definition included in the Rio Declaration of 1992 is one that is helpful (Rio Declaration 1992): "when an activity raises threats of harm to human health or the environment, precautionary measures should be taken even if some cause-and-effect relationships are not fully established scientifically".

The precautionary principle is appropriate when preliminary objective scientific analysis shows that there are reasons to believe that the degree of protection may not be adequate to prevent possibly harmful effects on the environment, human, animal, or plant health. This approach allows taking precautions when there is uncertainty (Raffensperger and Tickner 1999; Kriebel and Tickner 2001).

UNESCO, in collaboration with the World Commission on the Ethics of Scientific Knowledge and Technology (COMEST), published a report on the precautionary principle aiming to properly explain how to apply the principle from an ethical point of view. The conditions for the application of the principle were stated as follows:

- There are numerous scientific uncertainties.
- There are realistic scenarios (or models) of the potential damage that are supported by science (based on some scientifically plausible reasoning).
- It is impossible to lessen uncertainties in the short term without also raising ignorance of other important aspects through increased idealization and abstraction.
- The potential harm is morally intolerable, sufficiently substantial, or even permanent for current or future generations.
- It is imperative to take action now because waiting will make it much more difficult or expensive to mount an effective defense (Pickl et al. 2021).

The precautionary approach to epidemic risk analysis is usually based on modeling the mobility of people in social systems and the spread of diseases (Ibeas et al. 2013). The spread of diseases is associated here with the contact of people with each other, mainly in a closed crowded space. Typically, disease spread models are based on the human mobility model.

There are some approaches for modeling disease spread based on human mobility models using networks. In most such works, the nodes of networks represent people (Azizi et al. 2020; Kim et al. 2021). Such networks can be individual or population based. Each node in population-based networks represents a group of people, as opposed to individual-based networks, where each node represents a specific person (Chen 2014).

There are several individual-based approaches. In (Azizi et al. 2020), a person is categorized as being vulnerable, infected, or recovered from the disease under investigation at any particular time t. The authors use the networks for modeling the pattern of awareness among those who are vulnerable. In (Kim et al. 2021), the authors develop a scale-free network model for epidemic spreading simulation that is performed for a time series event by constructing an epidemic status matrix (z) to represent the status of the nth node at time step t. For each node, the value of the epidemic status matrix at time step t can be 0, 1, or 2, indicating that a node is susceptible, infective, or recovered, respectively. The simulation is performed using the Monte Carlo method. In (Nadini et al. 2020), people are represented as agents. The agents communicate based on proximity criteria, move in a two-dimensional physical world, and spread disease.

Individual-based approaches give a very detailed picture of the movement of people. The limitation of these approaches is that they are suitable for small areas comparable in size to the size of a single residential area. However, for modeling human mobility in a city or other large areas, this approach is difficult to use.

There are some other works proposing population-based approaches where the movement of people is modeled not at the level of each person, but at the level of groups of people moving from one region to another. In (Kim et al. 2019), disease spread is considered as a sequence of disease outbreaks across meta-populations by including human movement as topological pathways across locations in a diverse social system. The movement of people is modeled not at the level of each person, but at the level of the flow of people from one region to another.

The above-mentioned network population-based approaches represent dynamics of infectious diseases based on interconnections between (groups of) people, assume homogeneity in disease transmission, and employ a non-spatial lens to view the spread of disease. These models help predict the number of the affected population, but they do not specifically address the causes of epidemic development. (Chen 2014).

A spatial approach to modeling human mobility is proposed in (Aguilar et al. 2022; Bassolas et al. 2019). The area is subdivided into S2 cells, and an origin–destination matrix reflects the weekly travels (trip flow) between pairs of cells. (Bassolas et al. 2019) imposes hotspots by limiting the total number of trips made from each cell. Based on the variances in out-mobility across cells, each cell is ranked in ascending order and given a hotspot level. Mobility hierarchy is imposed by the cells' ordering in ascending hotspot level order.

The above review leads to the conclusion that much has been done so far in the field of epidemic risk analysis, however, there is a need for new approaches that can combine evidence-based risk analysis driven by facts with the advantages of flexible precautionary models, taking into account cause-and-effect relationships and spatial reference.

Therefore, the chapter aims at providing a framework for epidemic risk analysis in a heterogeneous dynamic social system based on a combination of precautionary and evidence-based approaches. The precautionary approach is used to assess epidemic risk in advance. For this purpose, a population-based approach is used, and the model of disease spread developed in the form of an event-tree network with the nodes being buildings representing the places of people's concentration and the arcs being people flows. The meta-level of the network is also considered, in which the influence of external factors, such as weather conditions, on the arcs is considered. A distinctive feature of the proposed approach is a spatial reference of network nodes. The evidence-based approach is used to adjust epidemic risk assessments obtained using the precautionary approach for making response decisions.

5.4 Combination of Evidence-Based and Precautionary Approaches to Epidemic Risk Analysis

The proposed framework for epidemic risk analysis is based on a combination of evidence-based and precautionary approaches.

The precautionary approach captures the idea of the importance of regulatory intervention, even while the underlying evidence is lacking and the costs of regulation are high economically (Science for Environment Policy 2017).

To implement the precautionary approach, various scientifically based models and methods are used that can be applied to various areas and conditions taking into account different types of uncertainty. As a result, this approach is more general but less precise than evidence based. The precautionary approach can be used in advance to monitor and diagnose the spread of diseases within certain areas and analyze the potential risks.

However, expecting regulatory science to provide completely definitive knowledge on epidemic spread and dangers is impractical because there will always be some degree of uncertainty in any scientific investigation. It might be challenging to decide precisely when and how to respond to prospective threats (Uggla et al. 2012). Such questions do not have conclusive solutions: The precautionary principle is an example of a principle that is not a rule that dictates specific actions to take or results to be achieved. (Science for Environment Policy 2017).

The evidence-based approach is less general than precautionary, but since it is based on the available evidence, the decisions made on its base are likely to be correct.

Dependence on the availability of evidence base leads to the inflexibility of this approach. There is not likely to be a large body of published literature to draw from to support specific control approaches for new and emerging infectious diseases, or even for some known infectious diseases. (European Centre for Disease Prevention and Control 2011). Because of the rarity of the first instances, the urgency of the crisis, and the general public's rejection of control groups or various interventions in diverse populations, trials are particularly challenging to envision in scenarios involving new diseases (European Centre for Disease Prevention and Control 2011).

Considering the advantages and disadvantages of evidence-based and precautionary approaches (Table 5.1), the chapter proposes a framework for epidemic risk analysis based on a combination of these two approaches.

In the proposed framework, the risk is defined as the likelihood of a loss due to a disease. Risk is assessed for some area of interest as a product of the probability that all people inside this area will get sick and severity (effect of disease on populations at risk). This probability depends on people's mobility from one area to another.

Severity is measured by the impact of disease across several domains related to the health of humans, services in the healthcare field, infrastructure of inpatient health, health services in the community, and health services for the public (Peters et al. 2019). There are some quantitative metrics in each domain that can be used to assess how the disease affects a product or service.

The impact is measured against a baseline (e.g., the number of fatalities or ambulance transports during an incident measured against the baseline for the community during a period when no disease occurred) and is graded according to the size of the change. For each disease, a composite risk score is assessed from the weighted averages of scores in all domains (Peters et al. 2019).

Table 5.1 Advantages and disadvantages of the evidence-based and precautionary approaches

	Evidence-based approach	Precautionary approach
Advantages	1. The approach is based on the best available evidence data that are used to make decisions (they are likely to be correct)	1. The approach is applicable when there is a dearth of substantial scientific understanding about the subject, allows addressing different types of uncertainties 2. The approach is flexible: the developed models can be used for different areas 3. Conclusions are logically justified, and it is possible to trace a causal relationship
Disadvantages	1. The approach is not flexible because it depends on available evidence base 2. Finding and gathering evidence data may take a long time 3. The approach is characterized by a lack of flexibility: it may be difficult to extrapolate estimated impacts from real occurrences that have happened elsewhere to a specific area 4. The approach does not directly compare the benefits and harms or estimate the magnitudes 5. It is impossible to logically substantiate conclusions and trace a causal relationship 6. There are numerous situations when gathering evidence is impossible due to logistical and ethical constraints or because the diseases are so rare that studies are not feasible 7. Trials are particularly challenging to envision in scenarios involving developing diseases due to the rarity of the initial instances and the urgency of the required responses 8. Context is crucial for understanding and using the results	1. Big computational complexity 2. The approach is characterized by a complexity of used models 3. Information provided by this approach is not conclusive because uncertainty, ambiguity, and ignorance are unavoidable parts of scientific research 4. It might be challenging to decide precisely when and how to respond to potential threats because the precautionary principle does not provide rules that demand particular actions or results

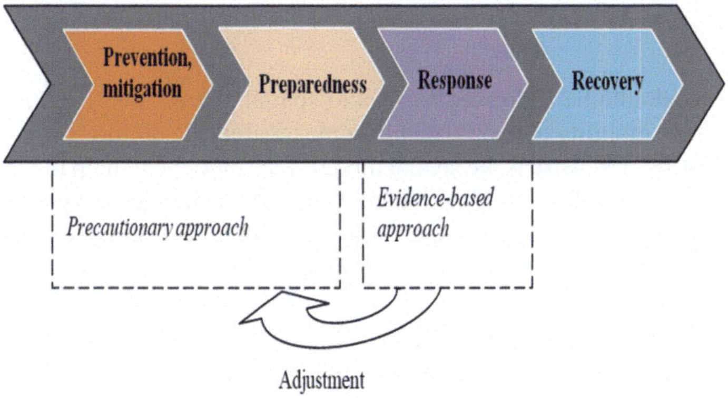

Fig. 5.1 Risk management cycle

The precautionary approach is implemented in the form of a set of interrelated models, which, as a result, give estimates of the potential risk, obtained in advance and distributed in space. To obtain these estimates, the following models are built: a spatial model, a model of people's mobility, as well as a model of the spread of diseases depending on their type.

The evidence-based approach is used to adjust potential risk estimates in real time.

Thereby, using precautionary and evidence-based approaches, the risk is assessed in two stages:

(1) Following the precautionary approach, potential risk assessments are developed;
(2) With the help of the evidence-based approach, the risk assessments obtained at the first stage are refined.

The developed approach to risk assessment covers all stages of the risk management cycle except recovery. Depending on the stage, either a precautionary or evidence-based approach can be used (Fig. 5.1).

5.5 Precautionary Approach to Risk Analysis

The precautionary approach is implemented in the form of a set of interrelated models, which, as a result, give estimates of the potential risk, obtained in advance and distributed in space. To obtain these estimates, the following models are built: a spatial model, a model of people's mobility, as well as a model of the spread of diseases depending on their type.

5.5.1 Spatial Model

Any epidemic should be addressed as a socio-spatial process evolving in space and time through the mobility of people within some territory. Therefore, the basic model of the developed framework is the spatial model. This model will help in representing the spatial–temporal information and in analyzing the dynamic spread of diseases.

The domain specifics require a spatial model with a flexible scale depending on the tasks and the territorial coverage of the disease since some of the diseases cover the territory of an entire country, while others can occur locally requiring a higher level of detail. Thus, the analysis of disease spread must be provided within a certain territory of consideration called the area of interest (AOI).

The main places of disease distribution are crowded confined spaces such as residential buildings, places of work/study, and shops. Therefore, these objects should be represented in the spatial model.

The spatial model of the AOI is described as a multi-level structure representing spatial objects of different levels (Fig. 5.2).

Figure 5.2 shows an example of a spatial model consisting of four layers. The first layer is a level of a geographical coordinate system, which represents the AOI as continuous space and can be implemented as a basic layer of a spatial model. Looking ahead, all the objects located within the AOI can be geo-located using a geographical coordinate system.

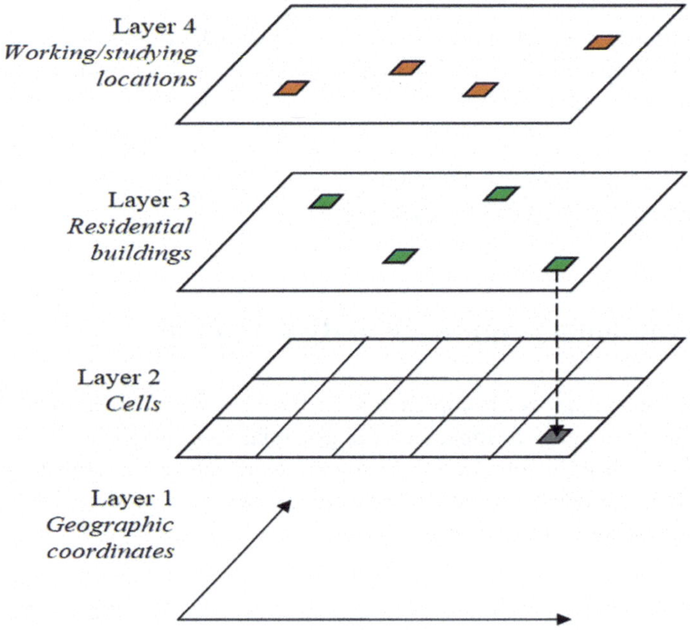

Fig. 5.2 Spatial model

At the second level, we impose a metrical grid D of isometric square cells with the size being $\delta \times \delta$. As a result, the continuous space is discretized by the grid $D = \{d_{ij}\}$ of isometric square cells d_{ij}, where ij represents the coordinates of the cell within the grid D. Dividing the area into discrete cells allows moving from a continuous representation of space to a discrete one, simulate the flow of people from one cell to another, as well as to assess the risk for each cell and describe the situation at the cell level for decision-making.

It is advisable to enable varying the cell size δ to make it possible to change the scale of the discretized AOI.

The additional layers of the spatial model contain the objects representing the places of people's concentration. Such objects can be the buildings or transport where people are concentrated, each of which is a spatial object. Figure 5.2 shows three layers (from the third to the fifth), containing such objects. This set is not exhaustive. In reality, there can be more such layers in the spatial model. Each object is projected onto a certain cell of the base layer (the layers of cells).

5.5.2 Human Mobility Within the Spatial Model

People move from one place of their concentration to another. Among all the places of people's concentration, we will consider buildings, that is, closed static spaces in which the spread of infectious diseases is possible. All places of people's concentration are displayed as spatial objects within the corresponding layers of the spatial model. Therefore, the movement of people in the spatial model will represent the movement between these layers. All spatial objects are projected onto the second layer of cells. Each such object is located inside a certain cell. As a result, the movement of people between objects is reduced to their movement between the cells (Fig. 5.3).

Each spatial object representing a confined place where people are concentrated has a set of parameters such as the average time spent by people in a given building, the most probable period of time when people visit a given building, and the number of people currently in the building. Such parameters can be measured in any measurement unit. Table 5.2 gives the example of a group of parameters with qualitative values for various object groups.

A set of values of parameters for a certain object defines its state $s \in S$, , which changes depending on different influencing factors (weather, season of the year, day of the week, etc.). The object state changes discretely with a certain time step Δt.

Each object can be characterized by a certain number of people, and some of them can be infected. Therefore, depending on the danger of infection, it is proposed to determine the categories of object states by the level of incidence called statuses. Status (incidence level) for each object in the easiest way can be classified as low, medium, high, and critical.

Suppose $W = \{w_0, ..., w_i, ..., w_F\}$ is an ordered set of the object statuses, where w_0 is the initial status, w_F is the final status, and w_i is the transitional status. Suppose

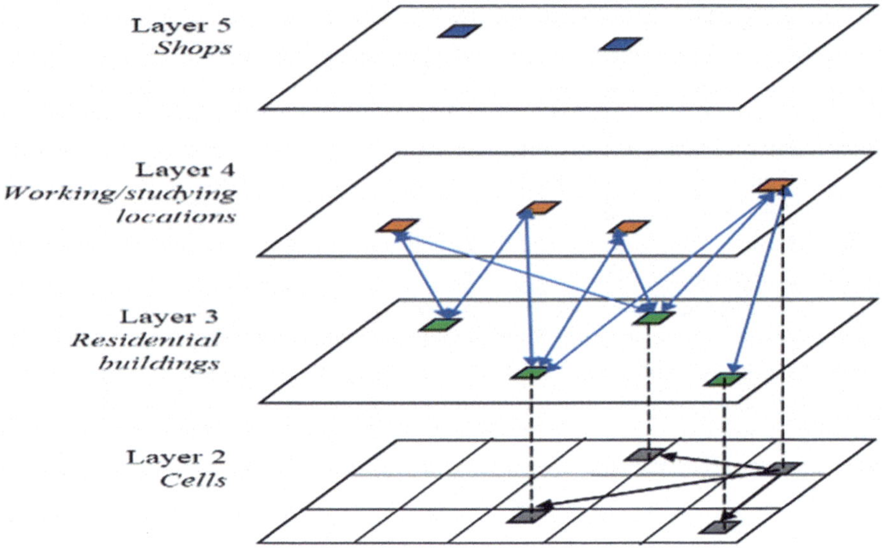

Fig. 5.3 People's mobility

Table 5.2 Parameter values for various objects

Objects	The most probable period of time for visiting		Average time spent by people	
	Working days	Weekends	Working days	Weekends
Residential buildings	Evening, night	All day long	Very big	Very big
Working/studying places	The first half of the day	–	Big	–
Exhibitions, theaters, concerts (cultural events)	–	All day long	Very small	Average
Shops	Evening	All day long	Small	Average
Public transport	Morning, peak hour, evening	All day long	Very small	Big

ϑ is a category function such that $\vartheta : O \times S \rightarrow W$, where O is a set of objects. Each state category is a certain subspace of the n-dimensional attribute values space.

Each random change of values of any subset of object parameters can change the object status. This change is not necessary, but possible.

We consider each significant (perhaps, jump-like) change of the object attribute's value, which forces the object to change its status, as an event, and denote it by y, so that $y : w_i \rightarrow w_j$. The models of the human mobility and disease dynamic can be represented based on a model of changing the statuses of the objects. Assume, that each object goes through a sequence of qualitatively different statuses. The states should be evaluated during continuous observations (monitoring) that allow obtaining time-ordered sequences of events.

We propose to model human mobility as a dynamic network built over the spatial model. Nodes of the network represent confined places of people's concentration in the form of geo-referenced spatial objects. At each moment of time, each node is associated with the status of the object w_i and the likelihood of the transition of the object to another state w_j during the current time interval Δt. The arcs represent the dynamic intensity of people flow N_{ij}, which is the number of individuals traveling from node i to node j during a current time interval Δt.

We also highlight external factors that affect the likelihood of objects transitioning from one status to another called meta-factors. They include weather conditions such as air temperature and air humidity. Meta-factors affect relationships between objects using meta-arcs, speeding up or slowing down the transition of objects from one status to another.

Meta-factors, as well as objects, have some number of parameters that determine their states ω_i. It is also possible to identify qualitatively different statuses of meta-factors that affect the rate of spread of diseases in different ways (Fig. 5.4). For example, for such a meta-factor as air temperature, the following statuses can be distinguished in a simplified form: low, medium, high.

Such a spatially distributed human mobility model is heterogeneous because each object is characterized by the different intensities of people flow, which changes during the day and depending on the day of the week (Table 5.2). The network should allow simulating human mobility with a certain time step Δt.

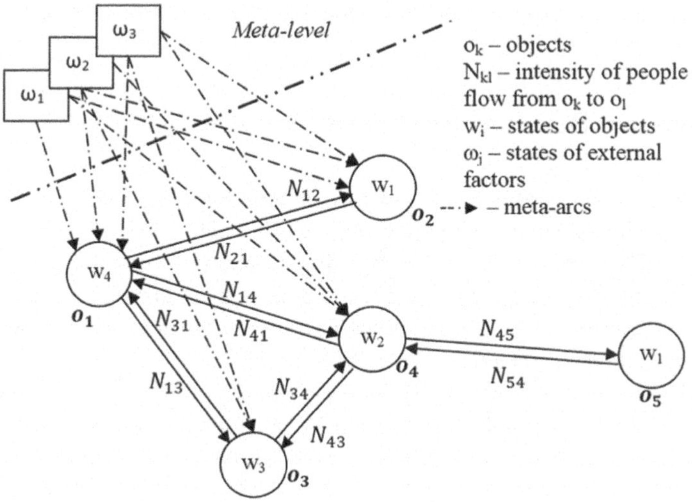

Fig. 5.4 Fragment of the network

5.5.3 *Disease Spread Model*

A precautionary approach involves constant assessment of the situation regarding human mobility and assessment of potential risk in advance to prevent the occurrence and spread of diseases.

To implement this approach, the considered framework proposes the development of templates for the dynamics of infectious diseases that outline every conceivable possible scenario of disease spreading. For each type of disease, a separate template is built, which is a network built based on the human mobility network. Each node of the disease spread network represents an object that at each moment of time is in a certain status w_i in relation to a certain disease. The statuses, e.g., can be the following: susceptible (w_1), unprotected (w_2), infected (w_3), recovered (w_4). As a result of people's flows from one node to another, each node can switch from one status to another. This transition is represented as an event.

At each moment of time, a certain node o_i depending on its state is associated with a certain status w_k, and the set of likelihoods λ_j of transition of this node to another status w_j. These likelihoods of switching the node o_i from one status to another depending on the total intensity of input people flow, as well as on the external factors being in the certain statuses (Fig. 5.4) and can be obtained by simulation or by expert estimates. Thus, the disease spread network contains information about disease dynamics in the form of likelihoods of switching the nodes between statuses at each discrete time moment.

The proposed framework for epidemic risk analysis contains the base of disease-spreading templates, which is open for additions and changes. The result of applying this approach will be an assessment of a spatially distributed dynamic potential risk.

5.6 Evidence-Based Approach to Risk Analysis

The evidence-based approach is used to calculate adjusted risk, which modifies potential risk obtained using the precautionary approach.

Risk is evaluated based on further planning to guarantee that the various vulnerable populations (for instance, children, the elderly, persons who have a cognitive impairment, those who have chronic illnesses, people who are poor, etc.) have access to emergency response resources (Peters et al. 2019).

The potential score and a population impact score, which takes into account the relative size of each vulnerable population in the planning jurisdiction using evidence data with census data and people's health profiles, produce the adjusted risk.

Depending on the modified risk assessment, decisions can be made whether any of the following are required for a group of people for the disease under consideration: special emergency communication protocols, specific preparations for mass sheltering and care, specific evacuation plans, specific plans for human services and medical management (Peters et al. 2019).

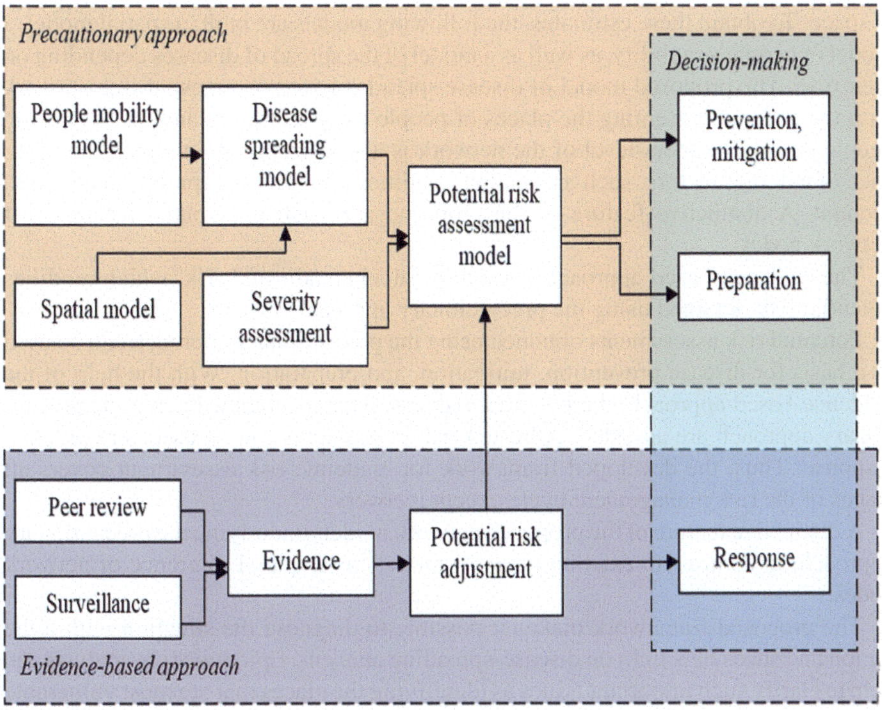

Fig. 5.5 Framework for epidemic risk analysis

5.7 Framework for Epidemic Risk Analysis

The proposed framework is based on two main principles such as the evidence-based approach and the precautionary principle. These principles are used in the following manner. Using the precautionary principle, scenarios of disease spread and potential risk assessments are built in advance. Potential risk assessments can be used as a basis for disease prevention, mitigation, and preparation. With the help of the evidence-based approach, the potential risk assessments obtained using the precautionary approach are adjusted. Adjusted risk assessments can be used for epidemic response (Fig. 5.5).

5.8 Conclusions

The chapter proposes a novel framework for epidemic risk analysis by incorporating human mobility. The proposed framework combines evidence-based risk analysis driven by facts with the advantages of flexible precautionary models, taking into account causation-and-effect connections and spatial reference.

The precautionary approach is implemented as a set of interrelated models, which, as a result, give estimates of the potential risk, obtained in advance and distributed in space. To obtain these estimates, the following models are built: a spatial model, a model of people's mobility, as well as a model of the spread of diseases depending on their type. The proposed model of disease spread is based on an event-tree network with the nodes representing the places of people's concentration and the arcs being people flows. The meta-level of the network is also considered, in which the influence of external factors, such as weather conditions, on disease spread is taken into account. A distinctive feature of the proposed approach is a spatial reference of network nodes.

The evidence-based approach is used to calculate adjusted risk, which modifies potential risk obtained using the precautionary approach.

Potential risk assessments obtained using the precautionary approach can be used as a basis for disease prevention, mitigation, and preparation. With the help of the evidence-based approach, the potential risk assessments obtained using the precautionary approach are adjusted. Adjusted risk assessments can be used for epidemic response. Thus, the developed framework for epidemic risk assessment covers all stages of the risk management cycle except recovery.

A distinctive feature of the proposed network models underlying the precautionary approach, in contrast to existing network models, is a spatial reference of network nodes.

The proposed framework makes it possible to diagnose the situation within the region and sheds new light on disease-spreading analysis. Epidemic risk analysis can help to clarify such important issues as identifying the places that are most vulnerable to epidemic spread and are the points of entry of disseminating the epidemic. The work aids in the formulation and application of regulations intended to slow or stop the spread of illnesses.

There are some potentially important areas for future research.

An interesting direction for future work is modeling multiple diseases. For this purpose, different event-tree networks can be built for different types of infectious diseases. Since the occurrence of one disease can cause the occurrence of another, it will be useful to explore the relationships between these networks.

The evidence-based approach should be elaborated. An interesting approach to implementing the evidence-based principle is case-based reasoning. The case base should contain ready-made scenarios for the spread of the disease and the solutions corresponding to them. Such a case base would serve as a good complement to the precautionary approach.

References

Aguilar J, Bassolas A, Ghoshal G, Hazarie S, Kirkley A, Mazzoli M, Meloni S, Mimar S, Nicosia V, Ramasco JJ, Sadilek A (2022) Impact of urban structure on infectious disease spreading. Nature Port 12:3816. https://doi.org/10.1038/s41598-022-06720-8

Azizi A, Montalvo C, Espinoza B, Kang Y, Castillo-Chavez C (2020) Epidemics on networks: reducing disease transmission using health emergency declarations and peer communication. Infect Disease Modell 5:12–22. https://doi.org/10.1016/j.idm.2019.11.002

Bassolas A, Barbosa-Filho H, Dickinson B, Dotiwalla X, Eastham P, Gallotti R, Ghoshal G, Gipson B, Hazarie SA, Kautz H, Kucuktunc O, Lieber A, Sadilek A, Ramasco JJ (2019) Hierarchical

organization of urban mobility and its connection with city livability. Nat Commun 10:4817. https://doi.org/10.1038/s41467-019-12809-y
Bernard D, Goldstein MD (2001) The precautionary principle also applies to public health actions. Am J Pub Heal 91(9)
Brownstein JS, Wolfe CJ, Mandl KD (2006) Empirical evidence for the effect of airline travel on inter-regional influenza spread in the United States. PLoS Med 3(10). https://doi.org/10.1371/journal.pmed.0030401
Chen D (2014) Modeling the spread of infectious diseases: a review: analyzing and modeling spatial and temporal dynamics of infectious diseases. Wiley. https://doi.org/10.1002/9781118630013.ch2
Colizza V, Barrat A, Barthélemy M, Vespignani A (2006) The role of the airline transportation network in the prediction and predictability of global epidemics. Proc Nat Acad Sci USA 103. https://doi.org/10.1073/pnas.0510525103
Eddy DM (1990) Practice policies: where do they come from? JAMA 263(9):1265–1275. https://doi.org/10.1001/jama.1990.03440090103036
European Centre for Disease Prevention and Control (2011) Evidence-based methodologies for public health—how to assess the best available evidence when time is limited and there is lack of sound evidence. Stockholm
Fakhruddin BSHM, Blanchard K, Ragupathy D (2020) Are we there yet? the transition from response to recovery for the COVID-19 pandemic. Prog Disas Sci 7. https://doi.org/10.1016/j.pdisas.2020.100102
Finger F, Genolet T, Mari L, Constantin de Magny G, Manga NM, Rinaldo A, Bertuzzo E (2016) Mobile phone data highlights the role of mass gatherings in the spreading of cholera outbreaks. Procs Nat Acad Sci USA 113(23). https://doi.org/10.1073/pnas.1522305113
Grais RF, Ellis JH, Glass GE (2003) Assessing the impact of airline travel on the geographic spread of pandemic influenza. Eur J Epidemiol 18(11). https://doi.org/10.1023/A:1026140019146
Hufnagel L, Brockmann D, Geisel T (2004) Forecast and control of epidemics in a globalized world. Procs Nat Acad Sci USA 101(42). https://doi.org/10.1073/pnas.0308344101
Ibeas Á, Cordera R, dell'Olio L, Coppola P (2013) Modelling the spatial interactions between workplace and residential location. Transp Res Part A 49:110–122. https://doi.org/10.1016/j.tra.2013.01.008
Kim M, Paini D, Jurdak R (2019) Modeling stochastic processes in disease spread across a heterogeneous social system. PNAS 116(2):401–406
Kim K, Yoo S, Lee S, Lee D, Lee K-H (2021) Network analysis to identify the risk of epidemic spreading. Appl Sci 11:2997. https://doi.org/10.3390/app11072997
Kriebel D, Tickner J (2001) Reenergizing public health through precaution. Am J Pub Heal 91(9)
Massaro E, Kondor D, Ratti C (2019) Assessing the interplay between human mobility and mosquito borne diseases in urban environments. Sci Rep 9(1):16911. https://doi.org/10.1038/s41598-019-53127-z
Nadini M, Zino L, Rizzo A, Porfiri M (2020) A multi-agent model to study epidemic spreading and vaccination strategies in an urban-like environment. Appl Netw Sci 5:68. https://doi.org/10.1007/s41109-020-00299-7
Neiderud C-J (2014) How urbanization affects the epidemiology of emerging infectious diseases. Infect Ecol Epidemiol 5:27060. https://doi.org/10.3402/iee.v5.27060
Peters R, Hipper TJ, Kricun H, Chernak E (2019) A quantitative public health risk assessment tool for planning for at-risk population. Publ Heal Pract 109(S4)
Pickl S, Budde D, Buchetmann B, Ferstl A, Klimek M, Thienel P (2021) WHO project. Integrating a general fast-track, agile process into the WHO framework for the development and quality assurance of normative products. Core Competence Center for Operations Research Management Tenacity Experience Safety & Security Alliance
Raffensperger C, Tickner J (eds) (1999) Protecting public health and the environment. Implementing the precautionary principle. Island Press, New York, NY

Rio Declaration on Environment and Development (1992) United Nations; 1992. Publication E.73.II.A.14, Stockholm, Sweden

Sackett DL, Rosenberg WM, Gray JA, Haynes RB, Richardson WS (1996) Evidence based medicine: what it is and what it isn't. BMJ (Clinical Research Ed.) 312(7023):71–72. https://doi.org/10.1136/bmj.312.7023.71

Science for Environment Policy (2017) Future brief: the precautionary principle: decision-making under uncertainty, İssue 18. ISBN 978-92-79-67015-2, ISSN 2363-278X. https://doi.org/10.2779/709033

Shi X, Hu Y, Zhang Y, Li W, Hao Y, Alelaiwi A, Rahman SMM, Hossain MS (2016) Multiple disease risk assessment with uniform model based on medical clinical notes. Special Sect Heal Big Data 4:7075–7083

Tatem A, Huang Z, Das A, Qi Q, Roth J, Qiu Y (2012) Air travel and vector-borne disease movement. Parasitology 139(14). https://doi.org/10.1017/S0031182012000352

Uggla Y, Forsberg M, Larsson S (2012) Dissimilar framings of forest biodiversity preservation: uncertainty and legal ambiguity as contributing factors. Forest Policy Econ 62:36–42. https://doi.org/10.1016/j.forpol.2015.07.007

Maryna Zharikova graduated in 1999 from Kherson National Technical University, Kherson, Ukraine, with a specialist degree in programming. In 2004, she received her PhD degree in Mathematical Modeling and Computational Methods from Kherson National Technical University. In 2018, she received her Doctoral degree in Information Technologies from Academy of Printing, Lviv, Ukraine. She is Professor in the Software Tools and Technologies Department at Kherson National Technical University, Ukraine, and in the Faculty of Computer Science, Institute for Theoretical Computer Science, Mathematics and Operations Research, at Bundeswehr University Munich, Germany. She has gained valuable experience in using information and computational technologies for compound risk analysis of natural hazards and infectious disease outbreaks. She has a particular wealth of experience and skills in event-tree multi-hazard modeling and spatially distributed multi-hazard risk analysis based on geoinformation technologies, big data technologies, and machine learning methods.

Stefan Pickl was born in Darmstadt, Germany, on September 29, 1967. He studied mathematics, electrical engineering, and philosophy at the Technical University of Darmstadt (Diploma in 1993; ERASMUS-scholarship at the EPFL Lausanne); doctor's degree at the TU Darmstadt in 1998 followed by his habilitation at the University of Cologne in 2005. From the years 2000 to 2005, he was Scientific Assistant and Project Manager at the Center for applied Computer Sciences in Cologne (ZAIK). In 2000, he received the PhD thesis award by the German Society for Operations Research, followed by international "best-paper awards" in the years 2003, 2005, 2007, and 2015. He published more than 250 articles. He is Vice-President of the German Committee for Disaster Reduction (DKKV). There, he is mainly engaged in the development of resilient systems and integrated assessment models. During the years 2004 to 2022, he was invited for assignments as Visiting Professor in the States, Asia, and Europe. As Guest Scientist, he was "Visiting Scientist" in Los Alamos National Labs, in the SANDIA Laboratory, at the MIT as well as at the Santa Fe Institute for complex systems. He is Founding Director of COMTESSA (Core Competence Center for Operations Research, Management-Tenacity-Excperience, Safety and Security ALLIANCE). From 2013 to 2022, he was International Coordinator of the innovative security projects RIKOV and Resilience of the Franco-German High Speed Train Network (REHSTRAIN) and REVEARS. In 2022, he was Chair of CRITIS2022—International Conference on Critical Information Infrastructures Security. He is Chair of the section "Bodensee-dreieck" of DWT Germany. He is Honorary Chair at The University of Nottingham Malaysia Campus.

Chapter 6
Transmissibility and Death Index from Pandemic COVID-19 Among Nations Across Continents

Ramalingam Shanmugam and Karan P. Singh

Abstract In this article, stochastic models are introduced to address, capture, and illustrate how much of the pandemic COVID-19 transmissibility occurred in regular as well as anomalous patterns in nations within continents: Africa, America, Asia, Europe, and Oceania. The results are interesting, not only drastic differences in the neighboring nations but also the existence of spiral pulling up of the COVID-19 death rates. In this process, a new death index is developed to portray the uniqueness to fight against the COVID-19 cases versus deaths. The index offers an insight for the health administrators to formulate preventive strategies for a successful tackling of this unprecedented uneven pandemic.

Keywords Bivariate inverse binomial models · Geometric models · Reproduction number · Rho value · Infectious disease · Periodic fluctuations · Tracking the pandemics · Africa · America · Asia · Europe · Oceania

6.1 Introduction and Motivation

Creating indices to compare several groups in a study project is quite aged old. For example, Agrawal et al. (2020) introduced a practical model and indices for comparing collaborative databases. The novel Coronavirus Disease 2019 (COVID-19) has spread to at least 184 countries worldwide, with over one hundred seventeen million confirmed cases. As of February 13, 2022, the number of deaths due to

---

R. Shanmugam
International Studies and Statistics, School of Health Administration, Texas State University San Marcos, San Marcos, TX, USA
e-mail: rs25@txstate.edu

K. P. Singh (✉)
Department of Epidemiology and Biostatistics, School of Medicine, The University of Texas at Tyler, Tyler, TX, USA
e-mail: karan.singh@uthct.edu

E. Çetin and H. Özen (eds.), *Healthcare Policy, Innovation and Digitalization*, Accounting, Finance, Sustainability, Governance & Fraud: Theory and Application,
https://doi.org/10.1007/978-981-99-5964-8_6

COVID-19 is over 5.81 million (Worldometer 2023). The details are given in Shanmugam et al. (2021, 2022) for a few recent models on the spread of COVID-19, Shanmugam (2014, 2021), Shanmugam and Singh (2021), and Shanmugam and Chattamvelli (2015) for details on stochastic modeling of healthcare systems.

The pandemic of viral infection of COVID-19 originated in the Republic of China. The virus is named severe acute respiratory syndrome coronavirus 2, or SARS-CoV-2. These viruses spread in close proximity, and hence, a rapid viral transmission in the human population occurs. The respiratory droplets of the COVID-19 infected individual are caught by another healthy person being in proximity. The Intelligence Community in USA assessed that the virus of COVID-19, probably emerged initially as small exposure that occurred no later than November 2019 with the first known cluster of COVID-19 cases arising in Wuhan, China in December 2019, Umakanthan et al. (2020) for details.

Since the diagnosis of the virus causing COVID-19 in year 2019, the governments around the world have been investing time and efforts to contain it. In spite of the collective and individualistic attempts to slow or end the spread of COVID-19, the number of infected cases is328,072,948 (765,222,168) and the number of deaths among the cases had risen to 5,539,567 (6,921,901) (*Global COVID-19 Tracker—Updated as of January 17 | KFF*) (KFF 2023). Such a tragic state of this infectious disease has triggered a great deal of concerns worldwide. The public health researchers have been trying to assess the level of infectivity due to COVID-19 virus including its original and variants called Delta or Omicron.

An approach to rate its virality in a nation at a given time, the researchers have been estimating what is known as *reproductive number* based on differential-difference equations of the measurable variables of the stochastic public health mechanism. The initial reproductive rate is R_0 (pronounced as *R nought* according to Milligan and Barrett 2015). The population is assumed to be **susceptible** to infection. The number (R_0) is the number of cases directly caused by an infected individual throughout his infectious period. R_0 is used to determine the ability of a disease to spread within a given population. The reproduction number represents the transmissibility of a disease. "R nought" is also referred to as the reproduction number. R tells you the average number of people who will contract a contagious disease from one person with that disease. It specifically applies to a population of people who were previously free of infection and haven't been vaccinated. Interpretations of "R nought" are done as follows, Agrawal et al. (2020), Barratt et al. (2018) and Milligan and Barrett (2015) for details.

If R is less than 1, then infection causes less than one new infection, and in which case, the disease will decline and eventually die out. If R equals 1, each existing infection causes one new infection and in which case, the disease will stay alive and stable, but there won't be an outbreak or an epidemic. If R is more than 1, each existing infection causes more than one new infection and the disease will be transmitted between people, and there may be an outbreak or epidemic. Importantly, a disease's R value only applies when everyone in a population is completely vulnerable to the disease. At later time (t) is denoted by R_t.

When $R_0 > 1$, the health researchers interpret that the infection would spread across the population and necessary preventions need to be applied. Otherwise (that is, $R_0 < 1$), the infection would eventually die out. For examples, note that for the delta variants of the COVID-19 $R_0 = 5.1$ (meaning that a person with delta variant of COVID-19 would infect 51 persons among 100 susceptible persons upon contact), and R_0 drops down to 2.2 during 7–13 June 2021.

To sense the severity of COVID-19, note that $R_0 = 1.51$ for Ebola virus and $R_0 = 2.06$ for Zika virus, Van den Driessche (2017) for details. The initial rate R_0 excludes new cases produced by the secondary cases. It is affected by several factors including the rate of contacts in the host population, the probability of infection being transmitted during contact, the duration of infectiousness. In this process to assess how fast the endemics might spread, health researchers resort to estimating the so called "Herd immunity", which refers a proportion of the population (or the herd) that have been vaccinated (or are immune), resulting in protection for susceptible (unvaccinated) individuals. The larger the number of people who are immune in a population, the lower is the chance for a susceptible person to get infected. It is more difficult for diseases to spread. In other words, the herd immunity threshold is the proportion of a population that need to be immune in order for an infectious disease to become stable. When the proportion of the population excessive of $p = 1 - \frac{1}{R_0}$, they need to be immunized to slow down or extinguish the spread of the infection, where the ratio $\frac{1}{R_0}$ is recognized as *endemic equilibrium level.* Synthesizing the initial rate R_0 as a product $R_0 = \overline{c}\tau T$, we could interpret that $\overline{c}$ is the *rate of contact* by an infected person per unit time, T is the *transmissibility*, and τ measures the *duration* of the life of the endemics (based on differential-difference equation approach).

This definition fits well the COVID-19 pandemics. Wherever and/or whenever the COVID-19 virus started (as it is not clear yet, according to the Intelligence Committee's Report), the rate, R_t, far exceeds one in many nations around the world, and the COVID-19 was classified as *Pandemics.* The Intelligence Committee (IC)'s report concluded that "the virus that causes COVID-19, probably emerged and infected humans through an initial small-scale exposure that occurred no later than November 2019 with the first known cluster of COVID-19 cases arising in Wuhan, China in December 2019" and "the IC remains divided on the most likely origin of COVID-19. All agencies assess those two hypotheses are plausible: natural exposure to an infected animal and a laboratory-associated incident".

This article captures and illustrates how were the COVID-19 transmissibility in regular as well as anomalous incidences in nations within the continents: *Africa, America, Asia, Europe,* and *Oceanic.* A death index, $\Re$, is developed and estimated to portray the proportion of COVID-19 cases ended in deaths among nations in all seven continents. Such an index helps to compare the pandemic patterns of the COVID-19 and secure practical strategies for success against the COVID-19 pandemic.

6.2 Patterns of COVID-19 Cases Versus Deaths Including Exceptional Incidences

Assume that the number, X, of COVID-19 in a time at a location is a random number generated by a chance-oriented mechanism with the *COVID-19 infection risk*, $0 < p < 1$ for anyone to contract the COVID-19 virus in a nation. Note that $(1 - p)$ portrays the level of immunity. The selection of an appropriate probability model depends on its data variance.

We notice that

$Var(X|r, p) > E(X|r, p)$, where $r \geq 1$ is an unknown *accumulation parameter.*

Let $E(X|r, p) = \mu_x = \frac{r_{case} p}{(1-p)} = r_{case}\{odds_{case}\}$ be the expected number of COVID-19 cases (see Fig. 6.1 for its dynamic nonlinear configuration), where the $Odds_{case} = \frac{p}{(1-p)}$ denotes the odds of contracting COVID-19. In a possibility in which the volatility, $Var(X|r_{case}, p)$ is more than the expected, $E(X|r, p)$ number of cases, then the pattern of the COVID-19 cases is likely to be an inverse binomial type.

In the beginning of the pandemic, a first identified proto case must have transmitted to others indiscriminately, and hence the $\hat{r}_{case} = \frac{\overline{x}^2}{s_x^2 - \overline{x}} = 1$ in regular incidences of COVID-19 cases, which is data evidence echoes in the configuration of $z = \frac{\overline{x}(1+\overline{x})}{s_x^2} - 1$ around zero (Fig. 6.2). Also, realize that in anomalous nations, the volatility, s_x^2, is exceedingly larger, and consequently, the estimate accumulation parameter, $\hat{r}$, is near one.

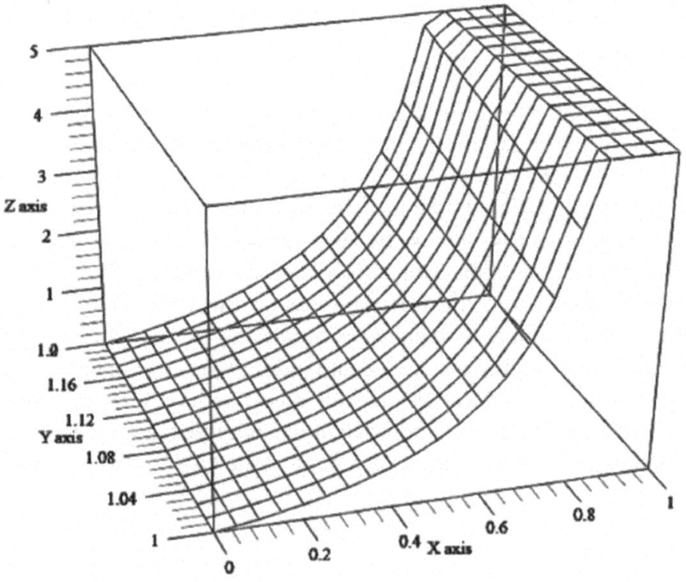

Fig. 6.1 Nonlinear configuration of mean in terms of odds

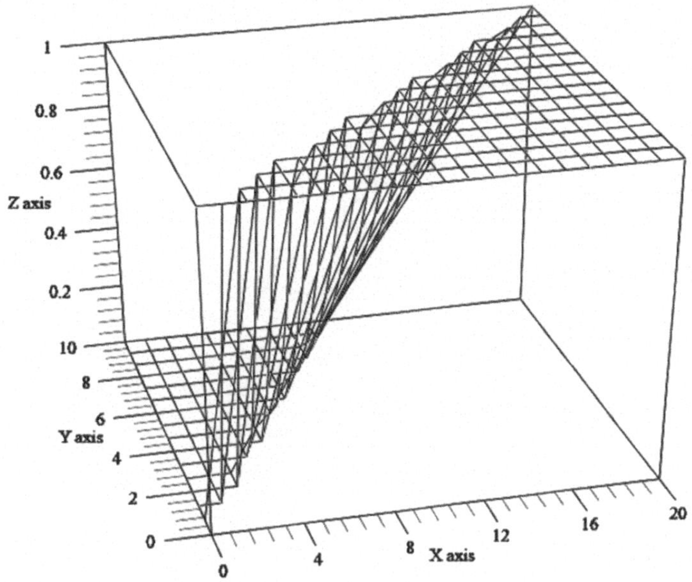

Fig. 6.2 Configuration of $z = \frac{\overline{x}(1+\overline{dx})}{s_x^2} - 1$

Figure 6.3a, b give the patterns of the average and spread number of COVID-19 cases in a location, where *x*, *y*, *z* axes represent $r,\ p,\ \mu_x$, respectively.

Likewise, let $\sigma_x^2 = \frac{r_{\text{case}}p}{(1-p)^2} = r_{\text{case}}\{\text{odds}_{\text{case}}\}\{1 + \text{odds}_{\text{case}}\}$ be the volatility of the cases. One such probability pattern is *inverse binomial model* (6.1), where $0 < 1 - p < 1$ denotes the *immunity level* of not contracting COVID-19. The natural estimate of the odds is $\widehat{\text{Odds}}_{\text{case}} = \left(\frac{s_x^2}{\overline{x}} - 1\right)$ and $\hat{r}_{\text{case}} = \left(\frac{\overline{x}^2}{s_x^2 - \overline{x}}\right)$. The probability of noting *x* number of COVID-19 cases at a location during a time interval is therefore a reparametrized inverse binomial probability.

$$\Pr\left(X = x \middle| \mu_x, \sigma_x^2\right) = \frac{\left(x + \frac{\mu_x^2}{\sigma_x^2 - \mu_x^2} - 1\right)!}{x!\left(\frac{\mu_x^2}{\sigma_x^2 - \mu_x^2} - 1\right)!}\left(\frac{\mu_x}{\sigma_x^2}\right)^{\left(\frac{\mu_x^2}{\sigma_x^2 - \mu_x^2}\right)}\left(1 - \frac{\mu_x}{\sigma_x^2}\right)^x;$$

$$x = 0,\ 1,\ 2, \ldots..,;\ \mu_x > 0;\ \sigma_x^2 > 0 \qquad (6.1)$$

The natural (that is, the maximum likelihood) estimates of the average and variance parameters in the model (6.1) are respectively $\hat{\mu}_x = \overline{x}$ and $\hat{\sigma}_x^2 = s_x^2$. We call $v_x = \frac{\mu_x}{\sigma_x^2}$ as the ratio of the *average* to *vulnerability* for a location to contract the COVID-19.

In an extreme situation with the volatility parameter, σ_x^2, is extremely large compared to μ_x, the expression (6.1) changes to a geometric model.

$\Pr(X = x|\mu_x, \sigma_x^2 \to \infty) = \left(\frac{\mu_x}{\sigma_x^2}\right)\left(1 - \frac{\mu_x}{\sigma_x^2}\right)^x;\ x = 0,\ 1,\ 2, \ldots..,$ with an estimated Odds of getting COVID-19 is $\widehat{\text{Odds}}_{\text{case}} = \left(\frac{s_x^2}{\overline{x}} - 1\right)$ if $\hat{r}_{\text{case}} \neq 1$ and

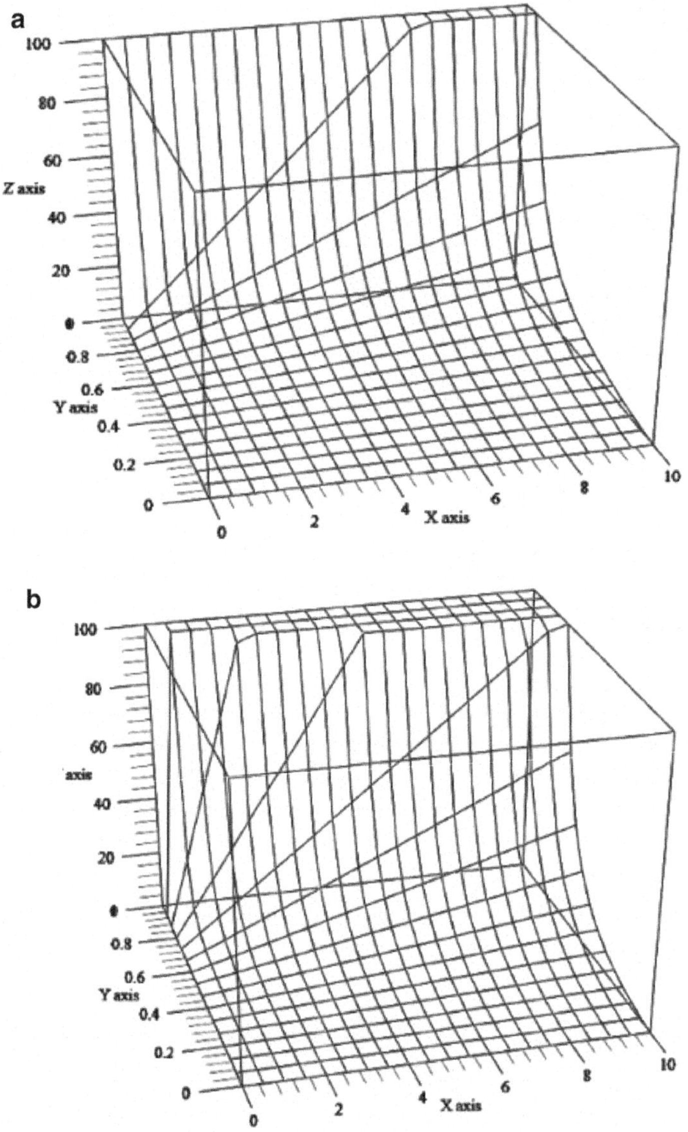

Fig. 6.3 **a** Expected number of COVID-19 cases **b** Volatility of COVID-19 cases

$\widehat{\text{Odds}}_{\text{case}} = \overline{x}$ if $\hat{r}_{\text{case}} = 1$. See Fig. 6.4 the performance of $\widehat{\text{Odds}}_{\text{case}}$ in the z-axis, the volatility, s_x^2, in x-axis and $\overline{x}$ in the y-axis (see Fig. 6.4a, b for comparison). Now, let us discuss the stochastic pattern of COVID-19 deaths.

Realize that not everyone who is infected with the COVID-19 virus dies as the person is likely to receive treatment in a hospital, home, or nursing center. For such a COVID-19 patient, let the chance to survive the pandemic is $0 < 1-\phi < 1$, meaning

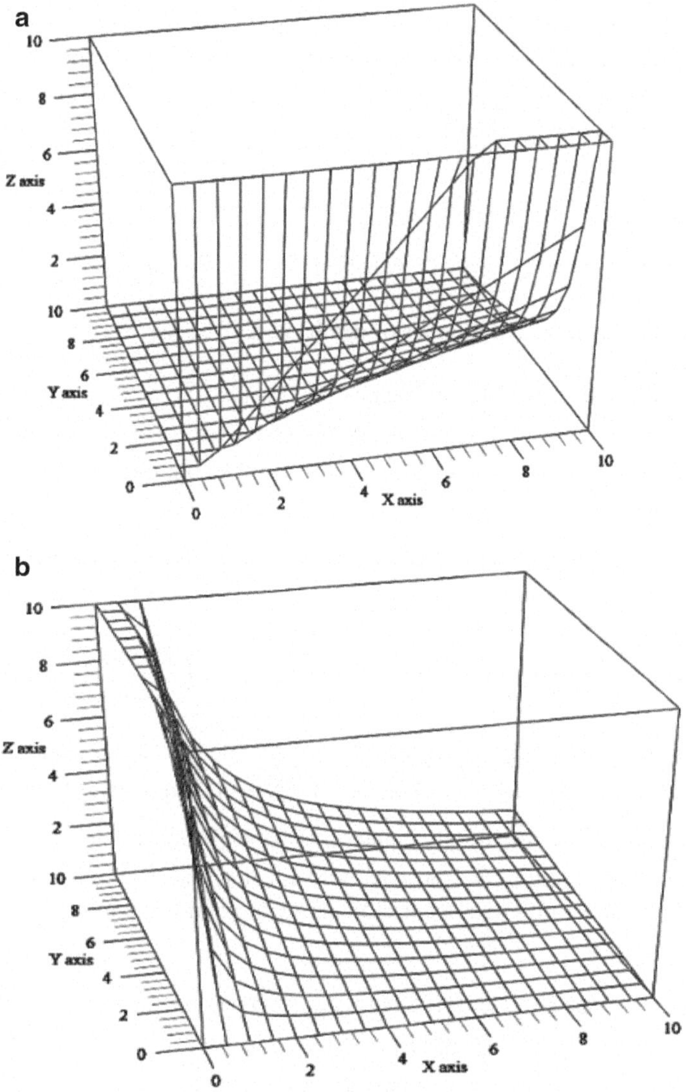

Fig. 6.4 **a** $\text{Odds}_{\text{case}}$, $r > 1$ contracting COVID-19 **b** $\text{Odds}_{\text{case}}$, $r = 1$ contracting COVID-19

that the chance of dying from COVID-19 is ϕ. The number, Y, of COVID-19 deaths is regressively predictable due to its correlated relationship with the number, x of COVID-19 cases in a location. Recall that *accumulated cases* who are vulnerable to death is r_{case}. Note that the volatility σ^2_{cases} is a function, $\mu_{\text{cases}}\left(1 + \frac{\mu_{\text{cases}}}{r}\right)$ of the average number, μ_{cases}, of the COVID-19 cases. Likewise, the odds of death are $\widehat{\text{Odds}}_{\text{death}} = \left(\frac{s^2_y}{\overline{y}} - 1\right)$ if $\hat{r}_{\text{case}} \neq 1$ and $\widehat{\text{Odds}}_{\text{death}} = \overline{y}$ if $\hat{r}_{\text{case}} = 1$. Also, note that their estimates are $\hat{\mu}_{\text{case}} = \overline{x}$, $\hat{\mu}_{\text{death}} = \overline{y}$, $\hat{\sigma}^2_{\text{case}} = s^2_{\text{case}}$, $\hat{\sigma}^2_{\text{death}} = s^2_{\text{death}}$, $\hat{r}_{\text{case}} = \frac{\overline{x}^2}{(s_x^2 - \overline{x})}$, $\widehat{\text{Odds}}_{\text{case}} = \left(\frac{s_x^2}{\overline{x}} - 1\right)$, and $\widehat{\text{Odds}}_{\text{death}} = \left(\frac{s_y^2}{\overline{y}} - 1\right)$. In anomalous situation, the accumulation number changes to $r_{\text{case}} = 1$. Consequently, a *stochastic pattern* (6.2) emerges for the number of deaths. That is,

$$\Pr(Y = y | d_{\text{cases}}, \phi) = \frac{(r_{\text{case}} + y - 1)!}{y!(r_{\text{case}} - 1)!} (1 - \phi)^{r_{\text{case}}} \phi^y; \quad y = 0, 1, 2, ...; \; 0 < \phi < 1 \tag{6.2}$$

The conditional expected number of deaths is $E(Y = y | r_{\text{case}}, \phi) = r_{\text{case}} \frac{\phi}{(1-\phi)} = r_{\text{case}}(\text{Odds}_{\text{death}})$.

The conditional volatility in deaths is $\text{Var}(Y = y | r_{\text{case}}, \phi) = (1 + \text{Odds}_{\text{death}}) E(Y = y | r_{\text{case}}, \phi)$. The estimate of such *volatility* is $\left(\frac{s^2_{\text{death}}}{\overline{y}}\right)$. This points out that the volatility is proportionally increasing with the conditional expected number of deaths. The estimate of the conditional expected number is $\hat{E}(Y \big| \hat{r}_{\text{case}}, \hat{\phi}) = \frac{\overline{x}^2}{(s_x^2 - \overline{x})} \left(\frac{s^2_{\text{death}}}{\overline{y}} - 1\right)$. The unconditional expected number of the number of COVID-19 deaths is,

$$E(Y) = E_x E(Y = y | d_{\text{case}}, \phi) = r_{\text{case}}(\text{Odds}_{\text{death}}) = \left(\frac{\text{Odds}_{\text{death}}}{\text{Odds}_{\text{case}}}\right) E(X) \tag{6.3}$$

From (6.3), we notice that the expected number of deaths is a factor multiple of the expected number of cases and such a factor is $\left(\frac{\text{Odds}_{\text{death}}}{\text{Odds}_{\text{case}}}\right)$. In other words, the ratio $\frac{E(Y)}{E(X)}$ of their expectations is structurally equal to $\left(\frac{\text{Odds}_{\text{death}}}{\text{Odds}_{\text{case}}}\right)$. Proceeding further to obtain the unconditional volatility, we provoke a result that $\text{Var}(Y) = E_X\{\text{Var}(Y|x)\} + \text{Var}_X\{E(Y|x)\}$, according to a result in Shanmugam and Chattamvelli (2015). That is,

$$\begin{aligned} \text{Var}(Y) &= E_x\{(1 + \text{Odds}_{\text{death}}) E(Y = y | r_{\text{case}}, \phi)\} \\ &\quad + \text{Var}_x\{E(Y = y | r_{\text{case}}, \phi)\} \\ &= r_{\text{case}}(\text{Odds}_{death})(1 + \text{Odds}_{\text{death}}) \end{aligned}$$

because $\text{Var}(X) = (1 + \text{Odds}_{\text{case}}) E(X)$.

The above expression simplifies to

$$\text{Var}(Y) = \left\{\frac{\text{Odds}_{\text{death}}(1 + \text{Odds}_{\text{death}})}{\text{Odds}_{\text{case}}(1 + \text{Odds}_{\text{case}})}\right\} \text{Var}(X) \tag{6.4}$$

meaning that volatility, Var(Y) in COVID-19 deaths proportionally connects to the volatility in the COVID-19 cases. Their ratio, $\frac{\text{Var}(Y)}{\text{Var}(X)}$, of the volatility in death to the volatility in cases is

$$\left\{\frac{\text{Odds}_{\text{death}}(1+\text{Odds}_{\text{death}})}{\text{Odds}_{\text{case}}(1+\text{Odds}_{\text{case}})}\right\}.$$

In this juncture, we wonder what proportion of the COVID-19 cases end up in death? Such a proportion is $\frac{Y}{X}$. What is the expected value of such a proportion? What is the volatility of such a proportion? To answer these questions, we import the two expressions on page 8 of Blumenfeld (2001) and they are.

$E\left(\frac{Y}{X}\right)=\frac{E(Y)}{E(X)}\left[1+\frac{\text{Var}(X)}{E(X)}\right]$ and $\text{Var}\left(\frac{Y}{X}\right)=\left[\frac{E(Y)}{E(X)}\right]^2\left[\frac{\text{Var}(Y)}{E(Y)}+\frac{\text{Var}(X)}{E(X)}\right]$.

Because of (6.3) and (6.4), we note (see Fig. 6.5 for the configuration in which $\text{Odds}_{\text{death}}$ in y-axis and $\text{Odds}_{\text{case}}$ in x-axis) that

$$E\left(\frac{Y}{X}\right)=\left(\frac{\text{Odds}_{\text{death}}}{\text{Odds}_{\text{case}}}\right)(2+\text{Odds}_{\text{case}}) \tag{6.5}$$

and

$$\begin{aligned}\text{Var}\left(\frac{Y}{X}\right)&=\left(\frac{\text{Odds}_{\text{death}}}{\text{Odds}_{\text{case}}}\right)^2[2+\text{Odds}_{\text{case}}+\text{Odds}_{\text{death}}]\\&=\left(\frac{\text{Odds}_{\text{death}}}{\text{Odds}_{\text{case}}}\right)\left[E\left(\frac{Y}{X}\right)+\text{Odds}_{\text{death}}\right]\end{aligned} \tag{6.6}$$

In other words, from (6.5), we realize that the expected percent of COVID-19 cases end up in deaths is $(2+\text{Odds}_{\text{case}})$ times the percent ratio, $\left(\frac{\text{Odds}_{\text{death}}}{\text{Odds}_{\text{case}}}\right)$ of their odds. From the regression type relation (6.6), we learn that the volatility in the percent of

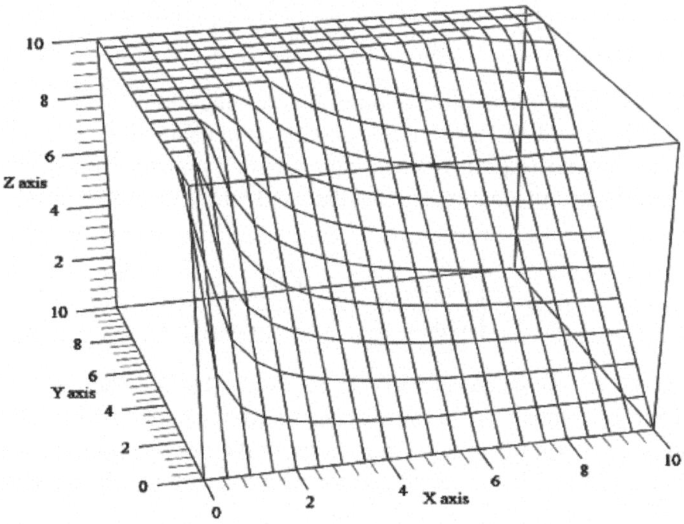

Fig. 6.5 Expected value, $E\left(\frac{Y}{X}\right)$

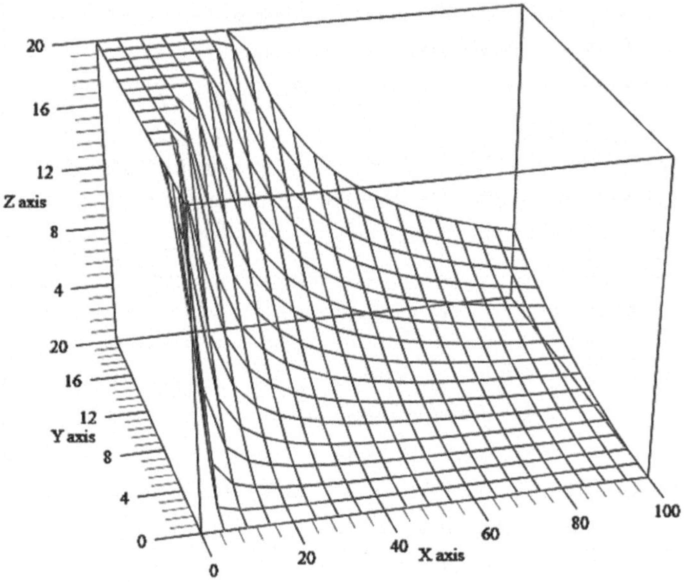

Fig. 6.6 Volatility, $\mathrm{Var}(\frac{Y}{X})$

COVID-19 cases ending up in deaths increases along with an increase in the expected amount (see Fig. 6.6 for their configuration).

Integrating (6.6) and (6.5), we note

$$\mathrm{Var}\left(\frac{Y}{X}\right) = \left\{\frac{E(\frac{Y}{X}) + \mathrm{Odds}_{\mathrm{death}}}{(2 + \mathrm{Odds}_{\mathrm{case}})}\right\} E\left(\frac{Y}{X}\right)$$

meaning that the volatility is proportionally increasing along with the increase in the expected percent of COVID-19 cases ending up in death. Such a proportionality is attributed as an *index* to portray the uniqueness of a transmissibility versus death situation that prevailed in a nation against the pandemic COVID-19 (see Fig. 6.7 for patterns).

$$\Re = \left\{\frac{\mathrm{Odds}_{\mathrm{death}}}{\mathrm{Odds}_{\mathrm{case}}} + \frac{\mathrm{Odds}_{\mathrm{death}}}{(2 + \mathrm{Odds}_{\mathrm{case}})}\right\}$$

In the anomalous incidence of COVID-19 cases (that is, when the volatility is extremely large in comparison to the average incidence), the estimate of (6.5) and (6.6) should be dealt with $\hat{r} = 1$ in computing. In the next section, the *death index of the pandemic COVID-19* is calculated for nations within each of the continents: *Africa, America, Asia, Europe,* and *Oceanic.*

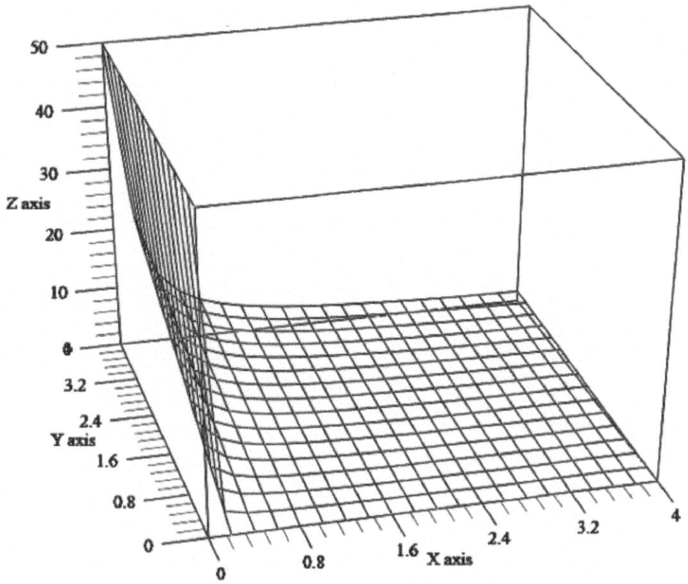

Fig. 6.7 Pattern of the index, $\Re$

6.3 How Did Nations Fight Uniquely Against COVID-19 Cases or Deaths?

In this section, let us examine the patterns of the COVID-19 cases and deaths within each: *Africa, America, Asia, Europe,* and *Oceanic* continent. The results pertinent to our themes alone are displayed in the Tables 6.1, 6.2, 6.3, 6.4, and 6.5. Because the data variance is larger (sometimes much larger) than the data average, which is the characteristic property of the negative binomial model, the probability pattern (6.1) is confirmed to be the underlying model for the collected data on the number of COVID-19 cases as well as the number of deaths. The accumulation parameter, r, the Odds $\text{Odds}_{\text{case}} = \left(\frac{\sigma_x^2}{\mu_x} - 1\right)$ of contracting COVID-19, Odds $\text{Odds}_{\text{death}} = \left(\frac{\sigma_y^2}{\mu_y} - 1\right)$ of dying due to COVID-19, the proportion, $\Re = \frac{\text{Var}(\frac{Y}{X})}{E(\frac{Y}{X})}$ of the volatility to the average percent, $(\frac{Y}{X})$ of the, COVID-19 deaths in comparison to the cases, and the bivariate pattern of the data on Y and X is computed and displayed in the tables below. We screen and select the nations whose estimated proportion, $\hat{\Re}$, is above 0.10 at the end of each Table.

Table 6.1 Summary of accumulation parameter, Odds of COVID-19 cases and deaths, correlation, and index (Africa)

Nation	$\hat{r} = \frac{\overline{x}^2}{s_x^2 - \overline{x}}$	$\widehat{\text{Odds}}_p = \left(\frac{s_x^2}{\overline{x}} - 1\right)$	$\widehat{\text{Odds}}_\phi = \left(\frac{s_y^2}{\overline{y}} - 1\right)$	$\hat{\Re}$
Algeria	0.726	299	9	0.05
Angola	0.556	91	5	0.07
Benin	0.028	35	3	0.01
Botswana	0.085	277	59	0.04
Burkina Faso	0.202	23	6	0.11
Burundi	0.091	29	0	0
Cameroon	0.051	150	50	0.03
Central African	0.011	16	2	0
Chad	0.1	8	1	0.04
Camaros	0.088	6	2	0.08
Congo	0.096	26	6	0.05
Democratic Congo	0.385	86	38	0.34
Djibouti	0.181	18	1	0.03
Dominica	0.041	8	1	0.02
Egypt	1.502	520	16	0.1
Equatorial Guinea	0.027	19	15	0.04
Eritrea	0.181	10	12	0.46
Eswatini	0.193	76	7	0.04
Ethiopia	0.838	523	12	0.04
French Guinea	0.499	65	1	0.02
Gabon	0.212	52	1	0.01
Gambia	0.119	14	4	0.08
Ghana	0.368	184	-1	0
Kenya	0.798	360	9	0.04
Libya	0.666	532	11	0.03
Morocco	1	1334	35	0.03
Mozambique	0.303	216	10	0.03
Niger	0.089	10	3	0.06
Nigeria	0.574	306	13	0.05
Rwanda	0.202	141	6	0.02
Senegal	0.353	103	4	0.03
Somalia	0.183	32	13	0.15
South Africa	1	4494	179	0.06
Sudan	0.409	63	39	0.51
Tunisia	1	1009	64	0.07
Uganda	0.048	179	214	0.11
Zambia	0.265	296	26	0.05
Zimbabwe	0.128	249	40	0.04

Note that the proportion, $\Re = \frac{\text{Var}\left(\frac{Y}{X}\right)}{E\left(\frac{Y}{X}\right)}$, is more in Democratic Congo, Eritrea, Somalia, Sudan, and Uganda in African continent and such a finding is not trivial without our data modeling and analysis

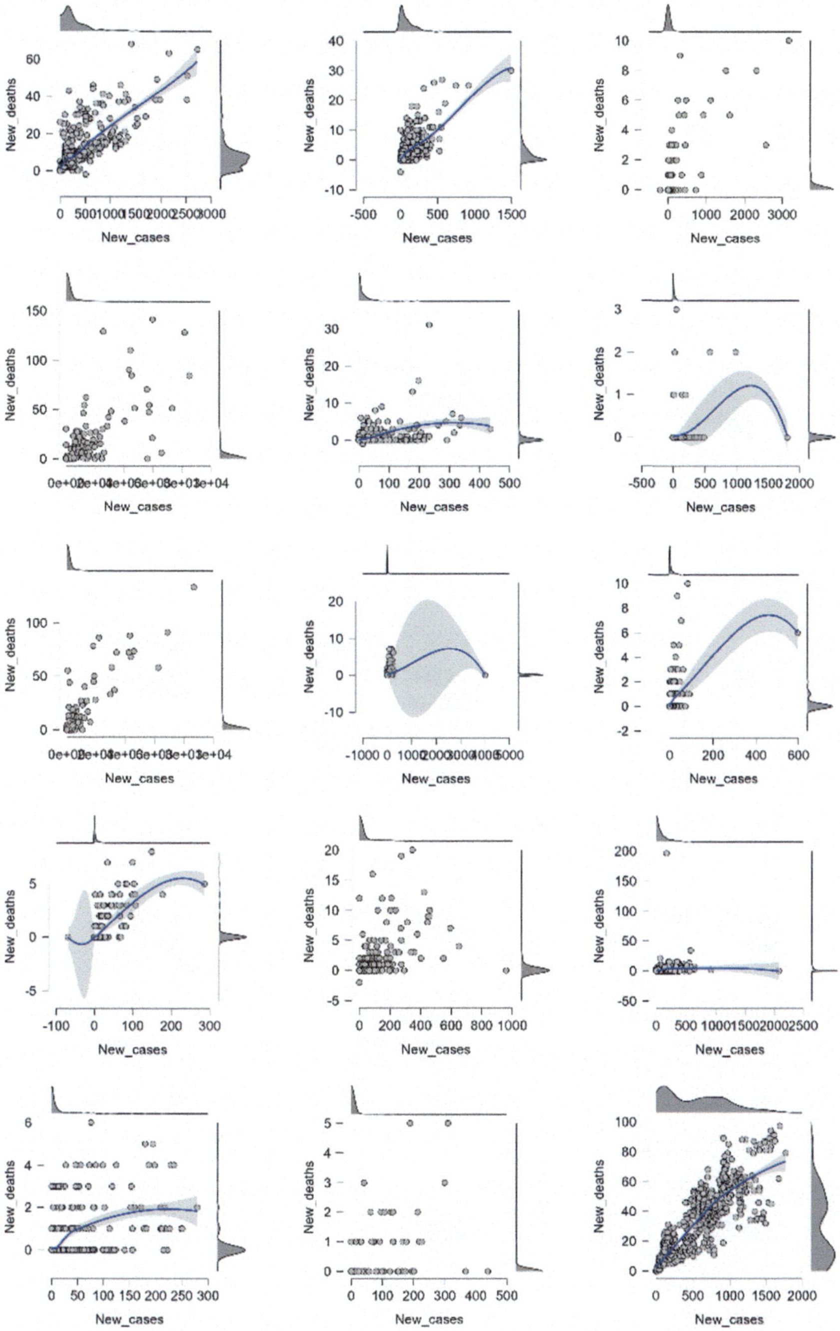

60
40
20
0
0 5001000150020002500 3000
New_deaths
New_cases
40
30
20
10
0
-10
-500 0 500 1000 1500
New_deaths
New_cases
10
8
6
4
2
0
0 1000 2000 3000
New_deaths
New_cases
150
100
50
0
New_deaths
New_cases
30
20
10
0
0 100 200 300 400 500
New_deaths
New_cases
3
2
1
0
-500 0 500 1000 1500 2000
New_deaths
New_cases
100
50
0
New_deaths
New_cases
20
10
0
-10
New_deaths
New_cases
10
8
6
4
2
0
-2
0 200 400 600
New_deaths
New_cases
5
0
-5
-100 0 100 200 300
New_deaths
New_cases
20
15
10
5
0
-5
0 200 400 600 800 1000
New_deaths
New_cases
200
150
100
50
0
-50
New_deaths
New_cases
6
4
2
0
0 50 100 150 200 250 300
New_deaths
New_cases
5
4
3
2
1
0
0 100 200 300 400 500
New_deaths
New_cases
100
80
60
40
20
0
0 500 1000 1500 2000
New_deaths
New_cases

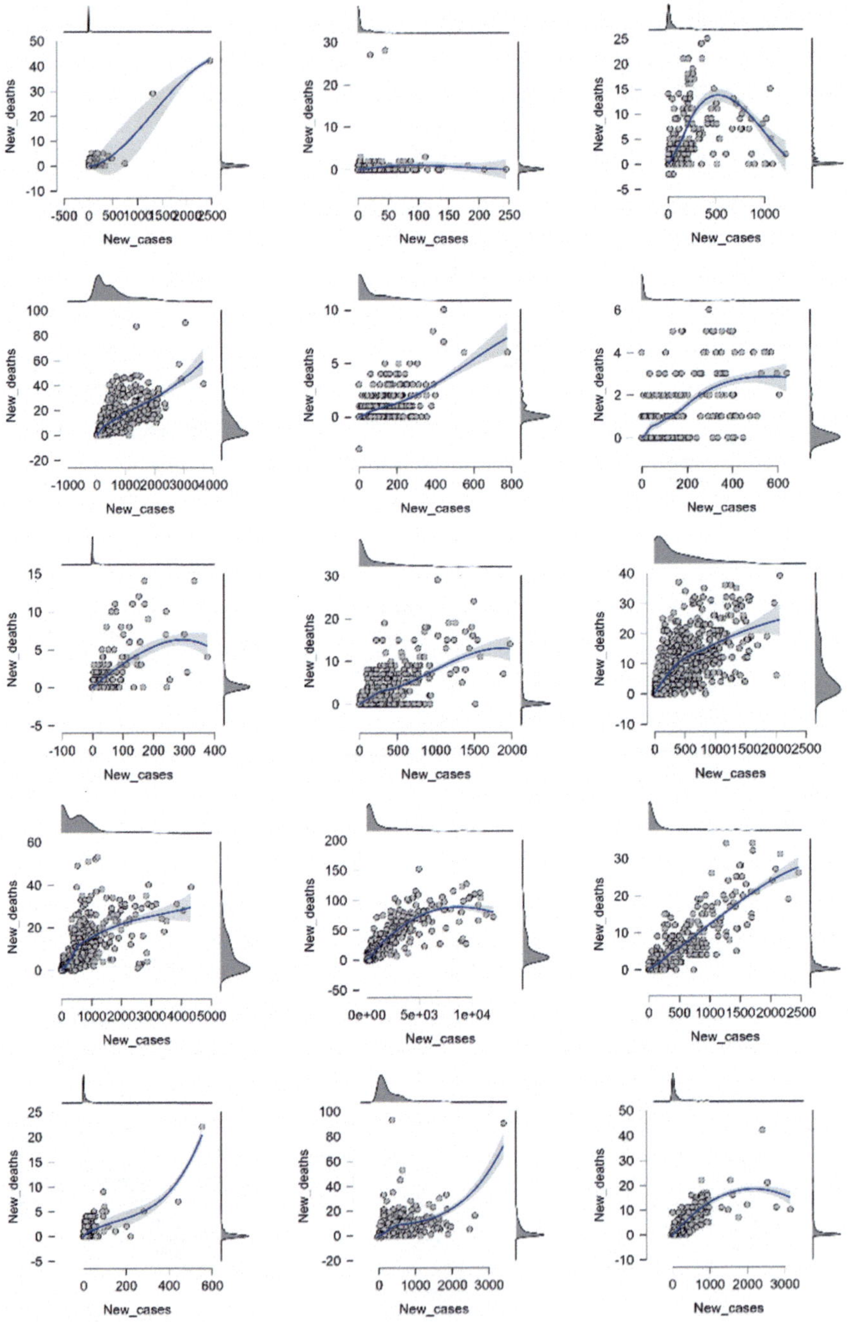
New_deaths
New_cases

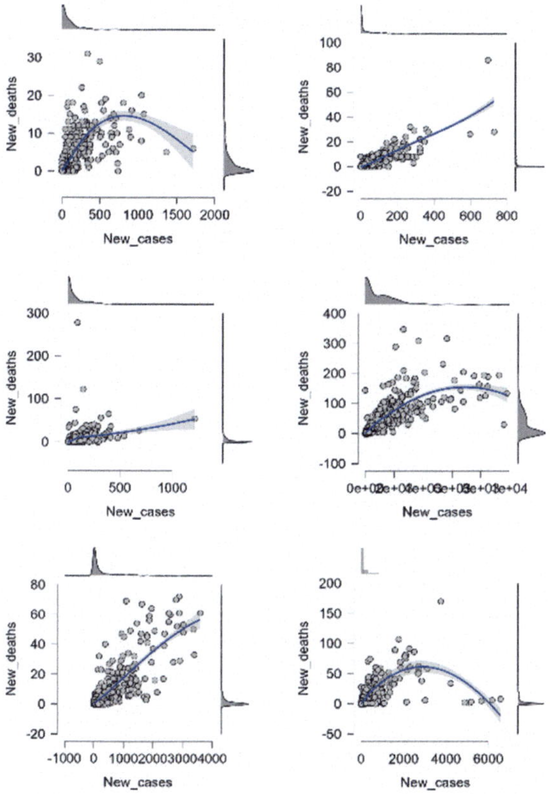
New_deaths
New_cases

Table 6.2 Mean and variance of the number of new COVID-19 cases and deaths (America)

Nation	$\hat{r} = \frac{\bar{x}^2}{s_x^2 - \bar{x}}$	$\hat{\text{Odds}}_p = \left(\frac{s_x^2}{\bar{x}} - 1\right)$	$\hat{\text{Odds}}_\phi = \left(\frac{s_y^2}{\bar{y}} - 1\right)$	$\hat{\Re}$
Argentina	7520	7520	270	0.07
Bahamas	0.52	61.75	7	0.22
Barbados	0.17	223.62	0	0
Belize	0.25	171.56	2	0.02
Bermuda	0.09	87.625	3	0.07
Bolivia	1.09	713.47	152	0.43
Bonaire	0.12	33.75	0	0
Brazil	1	31,090	752	0.05
Canada	1	2581	51	0.04
Cayman Islands	0.03	363.64	0	0
Chile	1	2503	84	0.07
Columbia	1	9147	164	0.04
Costa Rica	1.1	723.2	7	0.02
Cuba	1	1351	49	0.07
Curacao	0.09	271.33	2	0.01
Dominic Republic	1.62	356.4	7	0.04
Ecuador	0.72	1093.9	2751	5.03
El Salvador	0.19	893.94	2	0
Grenada	0.05	172.5	6	0.07
Guadeloupe	0.03	2788.6	40	0.03
Guatemala	1	873	25	0.06
Haiti	0.41	88.444	4	0.09
Honduras	0.98	540.08	18	0.07
Jamaica	0.48	265.13	7	0.05
Malawi	0.21	413.99	16	0.08
Mexico	1	5496	252	0.09
Panama	0.87	774.95	11	0.03
Paraguay	0.64	1008.3	49	0.1
Peru	1	3163	245	0.15
Puerto Rica	1	357	8	0.04
Trinidad	0.41	275.97	8	0.06
Uruguay	1	564	33	0.12
USA	1	69,794	731	0.02
Venezuela	1.7	361.8	4	0.02

Note that the proportion, $\Re = \frac{\text{Var}\left(\frac{Y}{X}\right)}{E\left(\frac{Y}{X}\right)}$, is more in Bahamas, Bolivia, Ecuador, Peru, and Uruguay in American continent and such a finding is not trivial without our data modeling and analysis

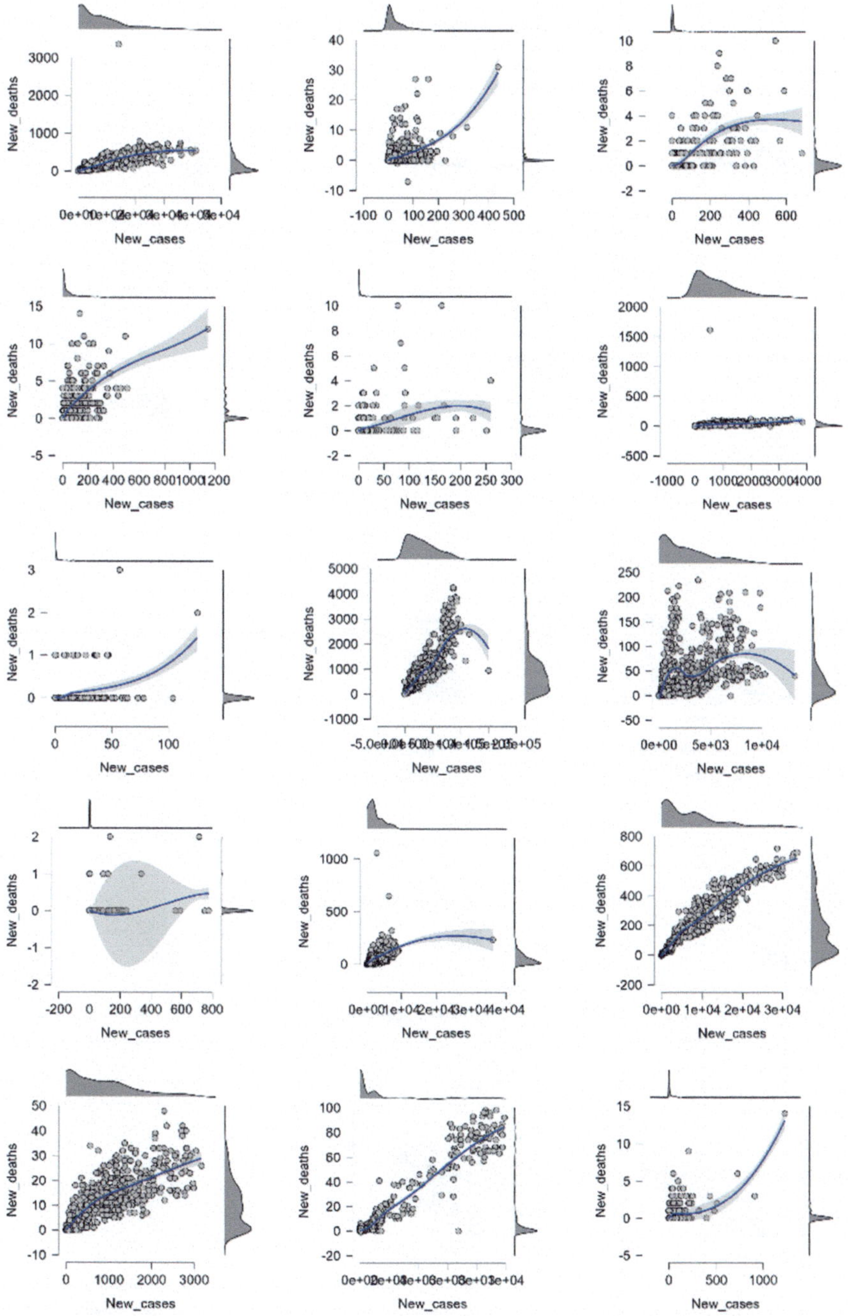
New_deaths
New_cases

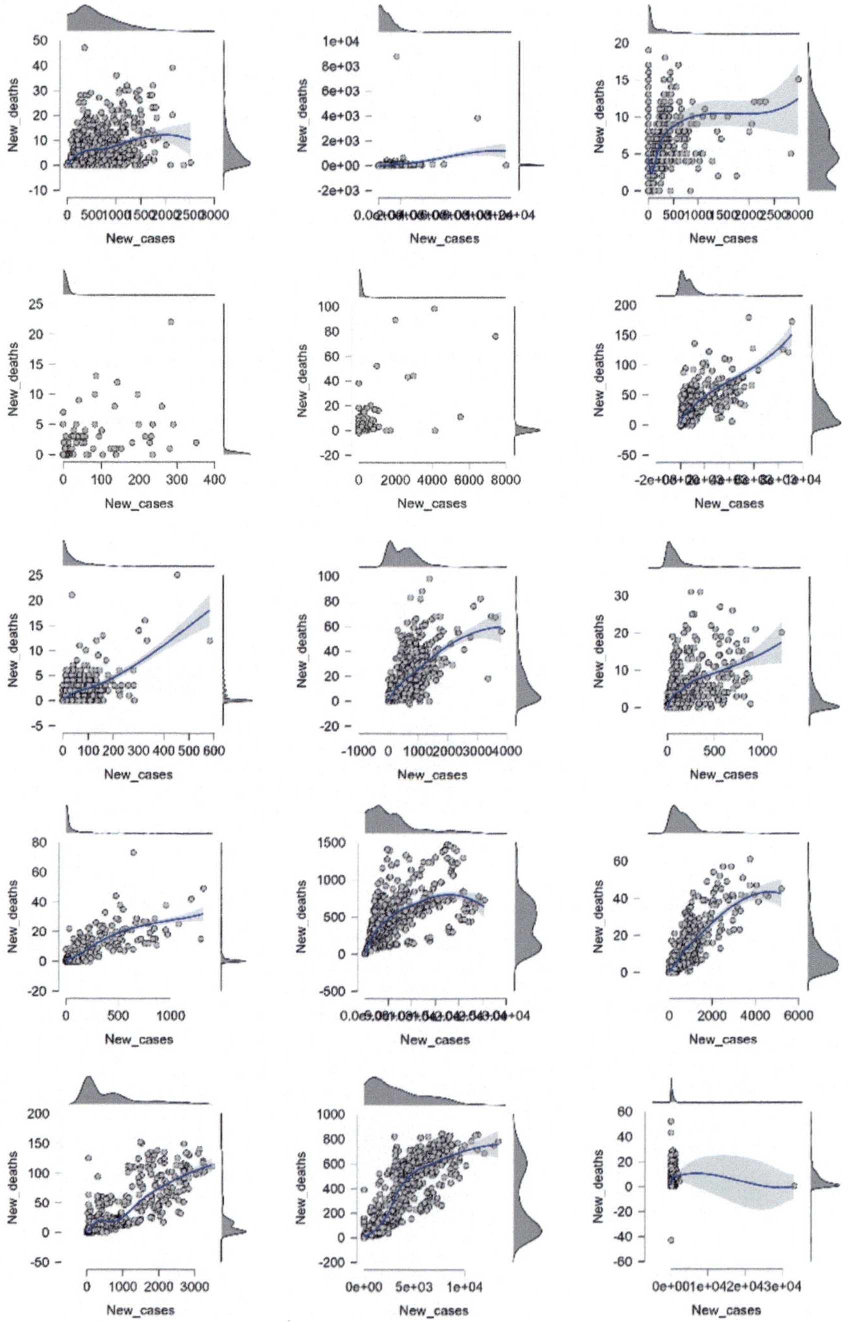
New_deaths
New_cases

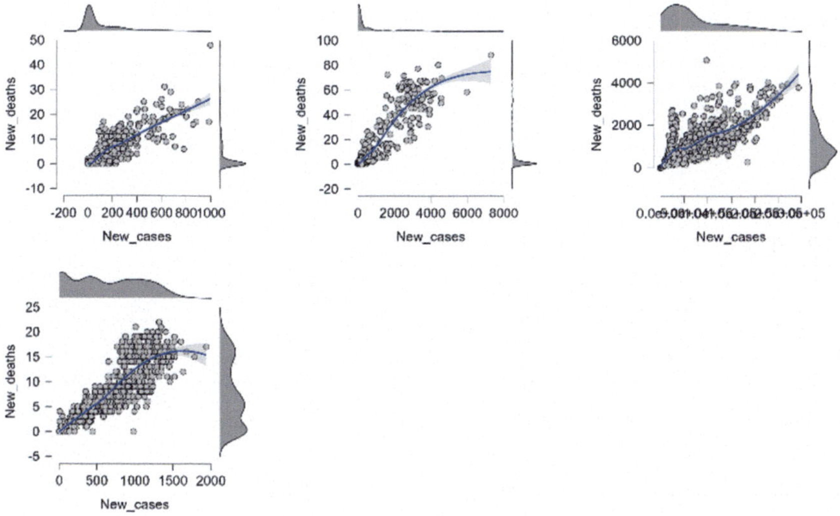

Table 6.3 Mean and variance of the number of new COVID-19 cases and deaths (Asia)

Nation	$\hat{r} = \frac{\bar{x}^2}{s_x^2-\bar{x}}$	$\widehat{\text{Odds}}_p = \left(\frac{s_x^2}{\bar{x}} - 1\right)$	$\widehat{\text{Odds}}_\phi = \left(\frac{s_y^2}{\bar{y}} - 1\right)$	$\hat{\Re}$
Afghanistan	0.31	713.1	34.8	0.097
Azerbaijan	1	850	14	0.033
Bahrain	0.61	637.3	6.5	0.02
Bangladesh	1	2216	71.49	0.064
Bhutan	0.15	19.67	0	0
Brunei	0.13	156.1	0.625	0.008
Cambodia	0.36	467.8	13.25	0.057
China	0.06	2924	346.8	0.237
India	1	48,682	1341	0.055
Indonesia	1	5974	601.1	0.201
Iran	1	8539	118.5	0.028
Iraq	1	2929	22.94	0.016
Israel	1	1896	15.27	0.016

(continued)

Table 6.3 (continued)

Nation	$\hat{r} = \frac{\bar{x}^2}{s_x^2 - \bar{x}}$	$\hat{\text{Odds}}_p = \left(\frac{s_x^2}{\bar{x}} - 1\right)$	$\hat{\text{Odds}}_\phi = \left(\frac{s_y^2}{\bar{y}} - 1\right)$	$\hat{\Re}$
Japan	1	2425	36	0.03
Jordan	1	1430	32	0.045
Kazakhstan	1	1492	650.8	0.872
Korea Rep	1	752	17	0.045
Kuwait	1.28	451.9	3.333	0.015
Kyrgyzstan	0.6	433.1	52.67	0.243
Lao	0.19	665.5	4.1	0.012
Lebanon	1	968	38.58	0.08
Malaysia	1	3785	167.5	0.088
Maldives	0.28	458.6	2.033	0.009
Mongolia	0.39	1400	12.89	0.018
Myanmar	1	739	148.7	0.402
Nepal	1	1157	63.25	0.109
Oman	0.52	821.1	14.33	0.035
Pakistan	1	1808	34.88	0.039
Philippines	1	3978	87.64	0.044
Qatar	0.78	438.8	2.275	0.01
Reunion	0.06	1454	9.66	0.013
Saudi Arabia	0.68	1130	11.25	0.02
Singapore	0.22	1784	8.418	0.009
Sri Lanka	1	808	89.4	0.221
Syria	0.76	91.25	2.897	0.063
Thailand	1	3055	122.8	0.08
Vietnam	1	2024	245.7	0.243
Yemen	0.38	36.43	7	0.374

Note that the proportion,$\Re = \frac{Var(\frac{Y}{X})}{E(\frac{Y}{X})}$, is more in China, Indonesia, and Kazakhstan. Myanmar, Nepal, Sri Lanka, Vietnam, and Yemen in Asian continent and such a finding is not trivial without our data modeling and analysis

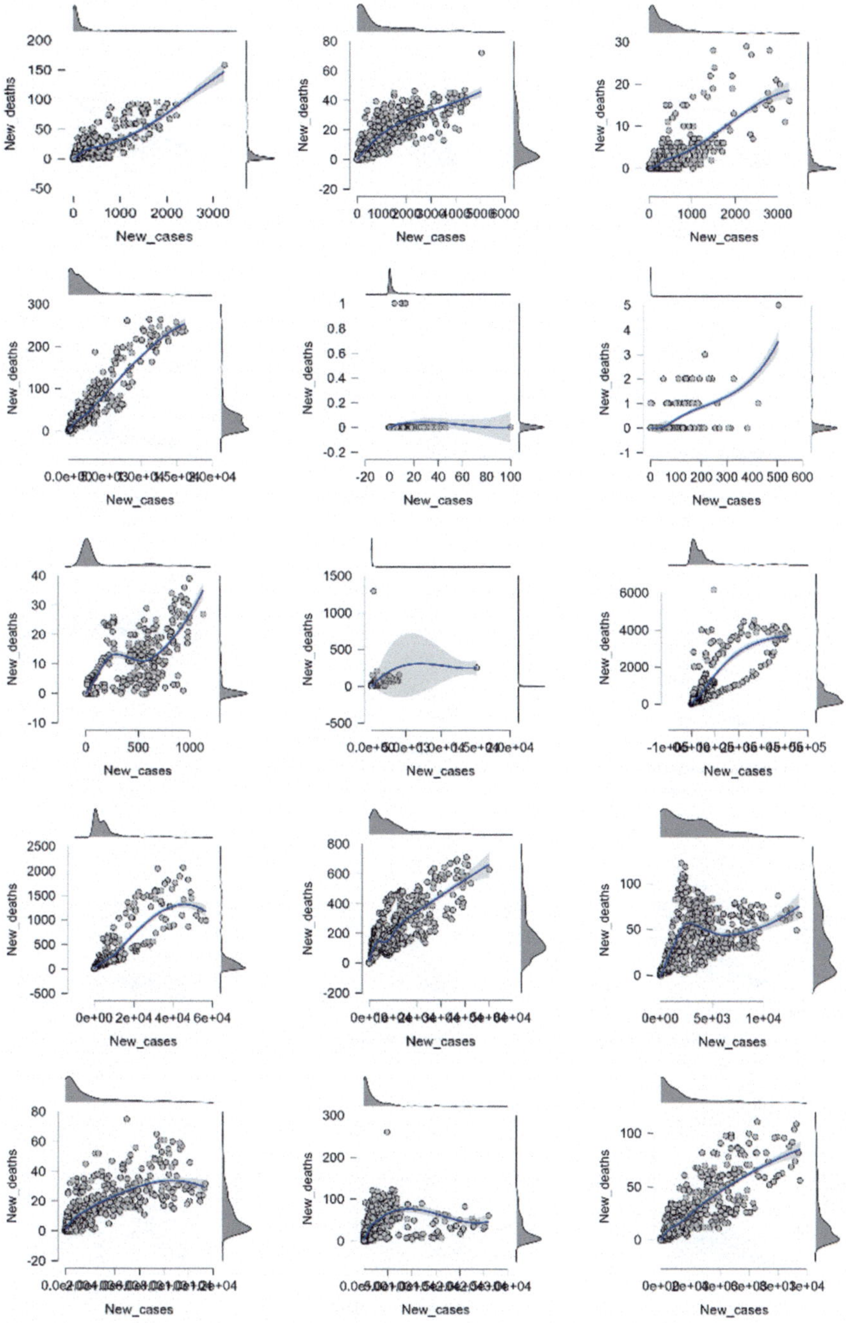
New_deaths
New_cases
New_deaths
New_cases
New_deaths
New_cases
New_deaths
New_cases
New_deaths
New_cases
New_deaths
New_cases
New_deaths
New_cases
New_deaths
New_cases
New_deaths
New_cases
New_deaths
New_cases
New_deaths
New_cases
New_deaths
New_cases
New_deaths
New_cases
New_deaths
New_cases
New_deaths
New_cases

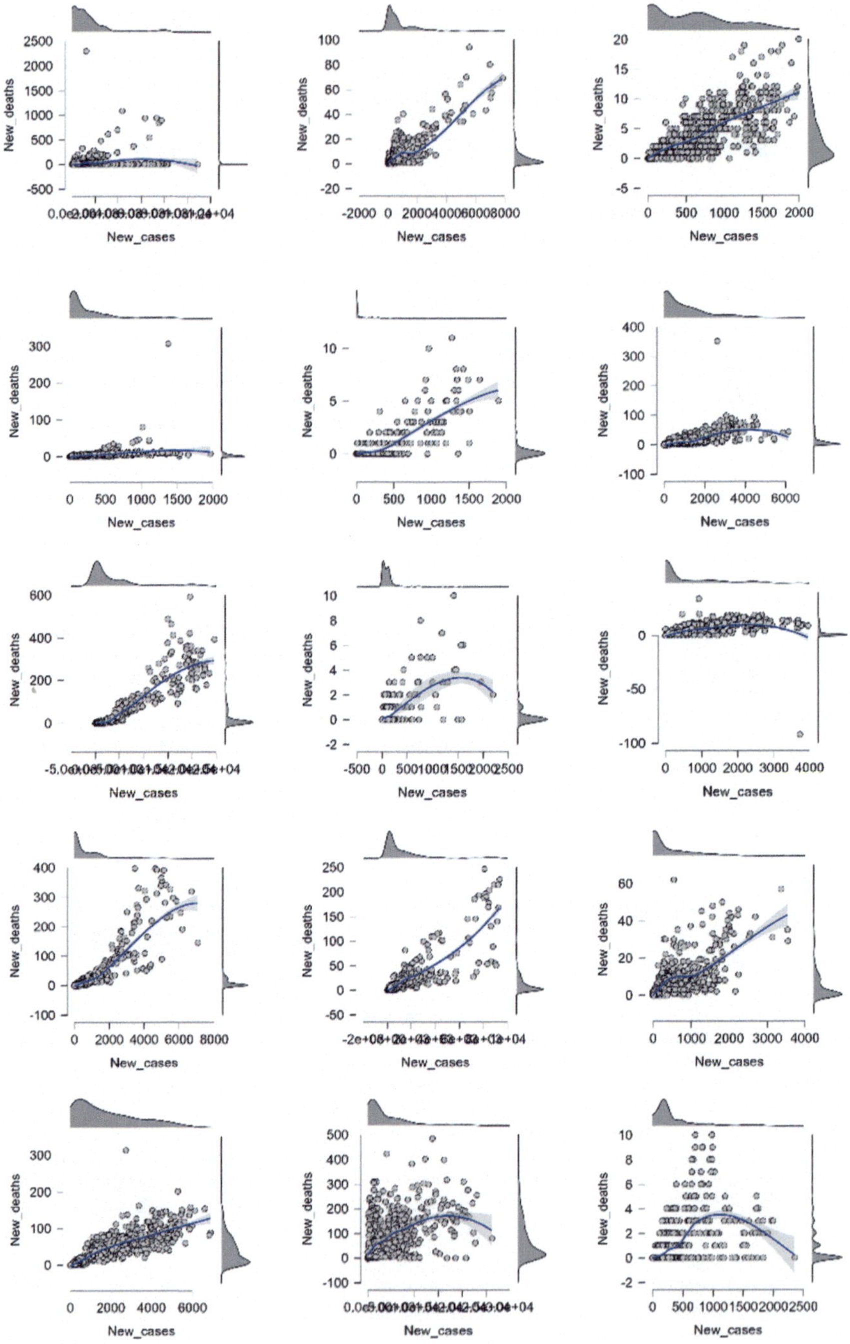
New_deaths
New_cases
2500
2000
1500
1000
500
0
-500
100
80
60
40
20
0
-20
-2000 0 2000 4000 6000 8000
-5
0 500 1000 1500 2000
300
200
100
10
5
400
-100
0 2000 4000 6000
600
-500 0 500 1000 1500 2000 2500
-50
0 1000 2000 3000 4000
0 2000 4000 6000 8000
250
150
50
60
40
20

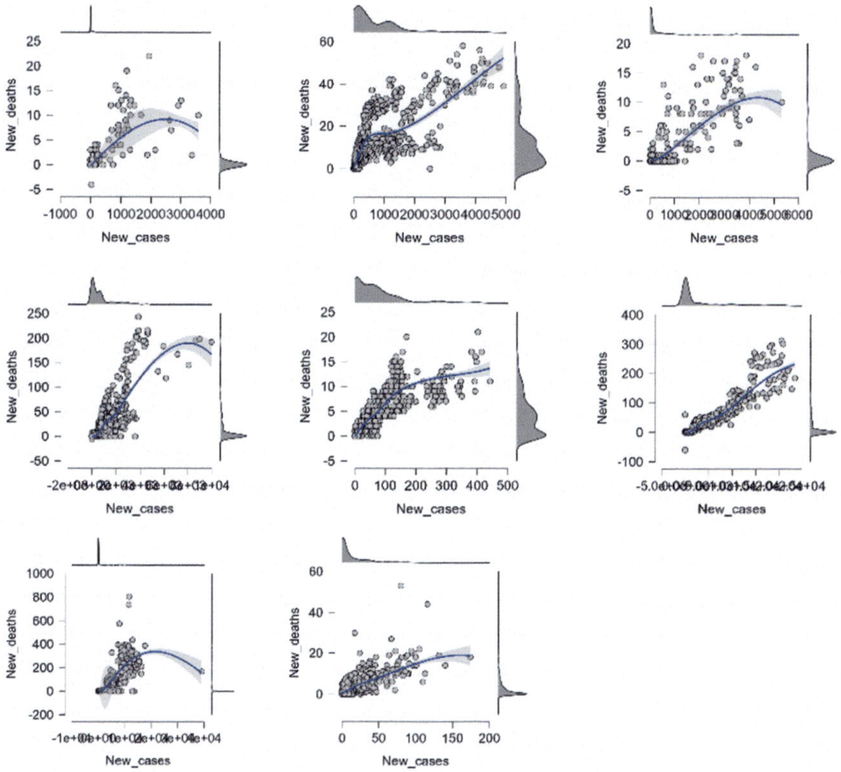
New_deaths
New_cases
New_deaths
New_cases
New_deaths
New_cases
New_deaths
New_cases
New_deaths
New_cases
New_deaths
New_cases
New_deaths
New_cases
New_deaths
New_cases

Table 6.4 Mean and variance of the number of new COVID-19 cases and deaths (Europe)

Nation	$\hat{r} = \frac{\bar{x}^2}{s_x^2 - \bar{x}}$	$\hat{Odds}_p = \left(\frac{s_x^2}{\bar{x}} - 1\right)$	$\hat{Odds}_\phi = \left(\frac{s_y^2}{\bar{y}} - 1\right)$	$\hat{\Re}$
Albania	0.81	353.37	5	0.03
Andorra	0.25	110.04	0.778	0.01
Armenia	0.79	605.31	11.91	0.04
Austria	1	1727	36.11	0.04
Belarus	1.86	508.41	2	0.01
Belgium	1	2760	85.13	0.06
Bosnia	0.76	519.23	23.61	0.09
Bulgaria	1	1007	74.54	0.15
Croatia	1	922	27	0.06
Cyprus	0.65	308.27	1.176	0.01
Czechia	1	3319	93.94	0.06
Denmark	1	800	10.43	0.03
Estonia	0.56	578.62	4.333	0.01
Finland	0.81	361.55	2.5	0.01
France	47.8	14.916	326.7	41.2
Georgia	1	1259	23.94	0.04
Germany	1	9275	343.6	0.07
Gibraltar	0.23	43.3	2.286	0.1
Greece	1	B24	33.96	0.05
Hungary	1	1694	116.3	0.14
Ireland	1	886	30.25	0.07
Italy	1	7375	266.8	0.07
Jersey	0.25	87.864	0.727	0.02
Kosovo	0.39	578.95	7	0.02
Latvia	0.47	779.17	14.83	0.04
Lithuania	0.61	1145.7	16.44	0.03
Moldova	1.01	516.2	11.46	0.04
Montenegro	1.04	214.43	2.958	0.03
Netherlands	1	4077	48.25	0.02
New Macedonia	0.75	409.63	12.1	0.06
Norway	0.37	1230.1	11.01	0.02
Poland	1	5443	283.8	0.1
Portugal	1	1683	96.62	0.11
Romania	1	2516	116.1	0.09
Russia	1	14,169	299.7	0.04
Serbia	1	1790	21	0.02

(continued)

Table 6.4 (continued)

Nation	$\hat{r} = \frac{\overline{x}^2}{s_{x}^2 - \overline{x}}$	$\hat{\text{Odds}}_p = \left(\frac{s_x^2}{\overline{x}} - 1\right)$	$\hat{\text{Odds}}_\phi = \left(\frac{s_y^2}{\overline{y}} - 1\right)$	$\hat{\Re}$
Slovakia	1	1106	52.81	0.1
Slovenia	0.55	1120.9	20.2	0.04
Spain	1	7526	230.2	0.06
Sweden	1	1742	38.29	0.04
Switzerland	1	1596	40.75	0.05
Turkey	1	12,735	76.76	0.01
UK	1	15,333	468.5	0.06
Ukraine	1	5019	218.1	0.09

Note that the proportion, $\Re = \frac{\text{Var}\left(\frac{Y}{X}\right)}{E\left(\frac{Y}{X}\right)}$, is more in Bulgaria, France, Hungary, Poland, Portugal, and Slovakia in European continent and such a finding is not trivial without our data modeling and analysis

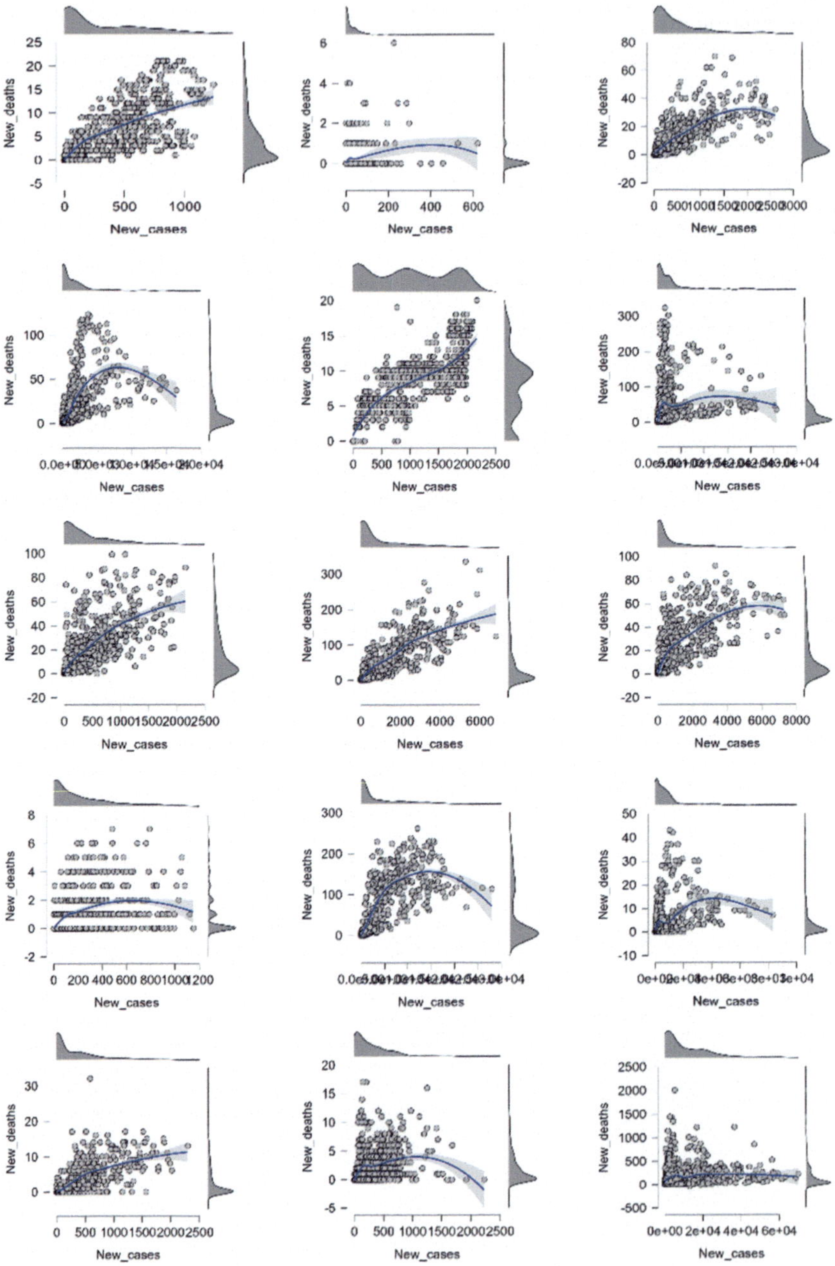
New_deaths
New_cases
New_deaths
New_cases
New_deaths
New_cases
New_deaths
New_cases
New_deaths
New_cases
New_deaths
New_cases
New_deaths
New_cases
New_deaths
New_cases
New_deaths
New_cases
New_deaths
New_cases
New_deaths
New_cases
New_deaths
New_cases
New_deaths
New_cases
New_deaths
New_cases
New_deaths
New_cases

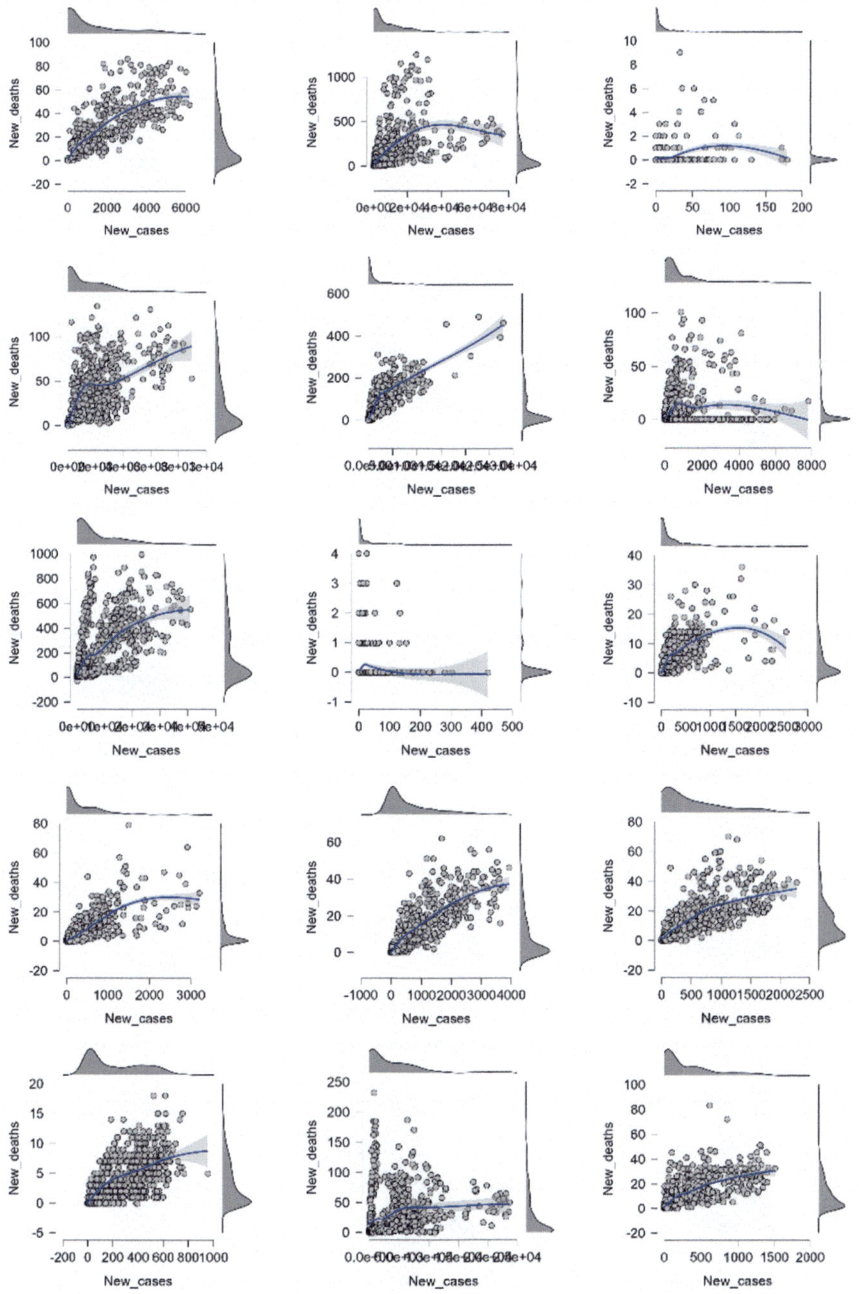
New_deaths
New_cases
New_deaths
New_cases
New_deaths
New_cases
New_deaths
New_cases
New_deaths
New_cases
New_deaths
New_cases
New_deaths
New_cases
New_deaths
New_cases
New_deaths
New_cases
New_deaths
New_cases
New_deaths
New_cases
New_deaths
New_cases
New_deaths
New_cases
New_deaths
New_cases
New_deaths
New_cases

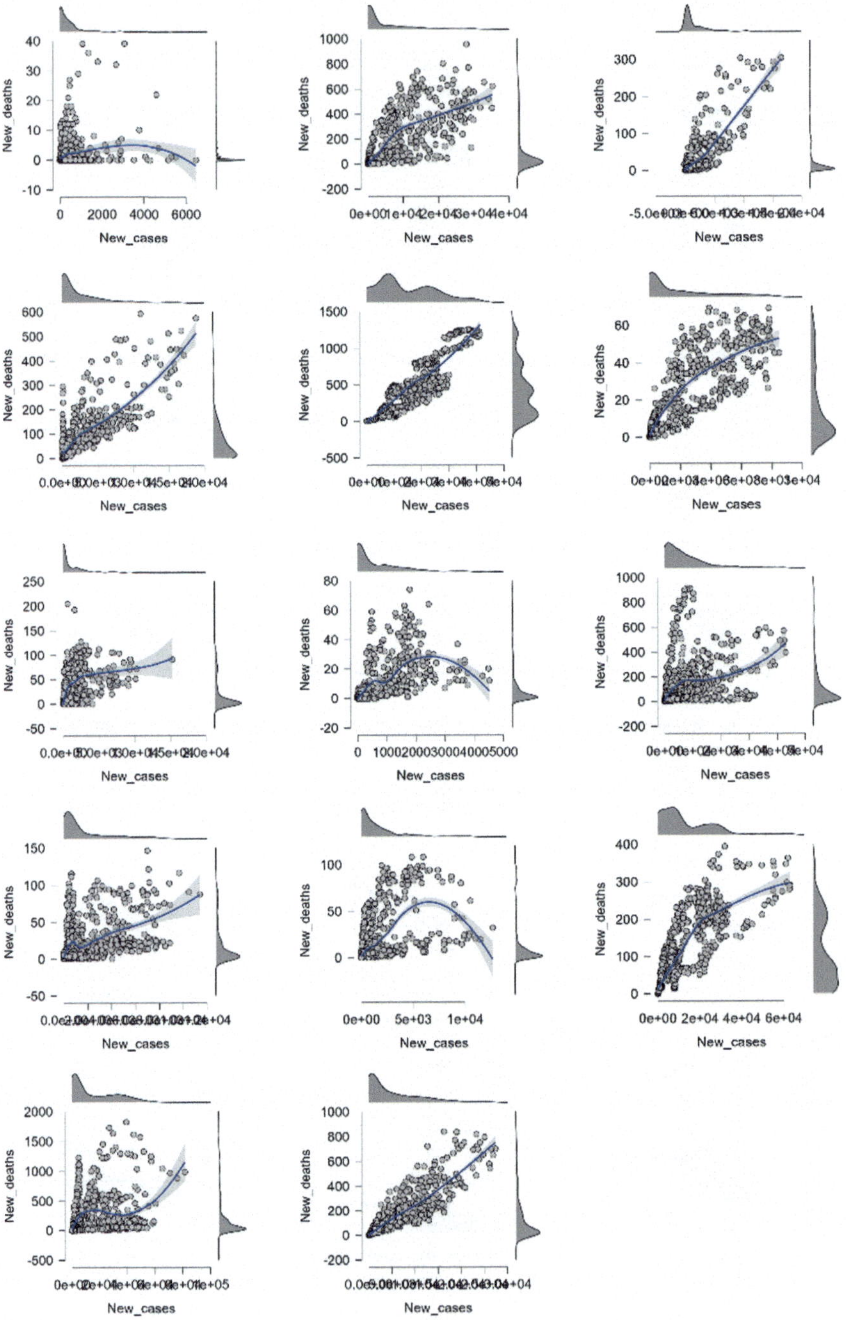
New_deaths
New_cases
40
30
20
10
0
-10
0
2000
4000
6000
1000
800
600
400
200
0
-200
300
200
100
0
600
500
400
300
200
100
0
1500
1000
500
0
-500
60
40
20
0
250
200
150
100
50
0
-50
80
60
40
20
0
-20
1000
800
600
400
200
0
-200
150
100
50
0
-50
100
50
0
0e+00
5e+03
1e+04
400
300
200
100
0
0e+00
2e+04
4e+04
6e+04
2000
1500
1000
500
0
-500
1000
800
600
400
200
0
-200

Table 6.5 Mean and variance of the number of new COVID-19 cases and deaths (Oceanic)

Nation	$\hat{r} = \frac{\bar{x}^2}{s_x^2 - \bar{x}}$	$\hat{\text{Odds}}_p = \left(\frac{s_x^2}{\bar{x}} - 1\right)$	$\hat{\text{Odds}}_\phi = \left(\frac{s_y^2}{\bar{y}} - 1\right)$	$\hat{\Re}$
Antigua	0.06	80.6	6	0.15
Australia	0.28	1163.9	9.67	0.02
British Virgin	0.01	332.75	3.8	0.02
Cabo Verde	0.69	77.778	0.47	0.01
Fiji	0.12	634.05	9.04	0.03
French Polynesia	0.07	963.13	14.7	0.03
New Caledonia	0.04	457.24	8.15	0.04
New Zealand	0.18	96	0.83	0.02

Note that the proportion, $\Re = \frac{Var(\frac{Y}{X})}{E(\frac{Y}{X})}$, is more in Antigua, in Oceanic continent and such a finding is not trivial without our data modeling and analysis

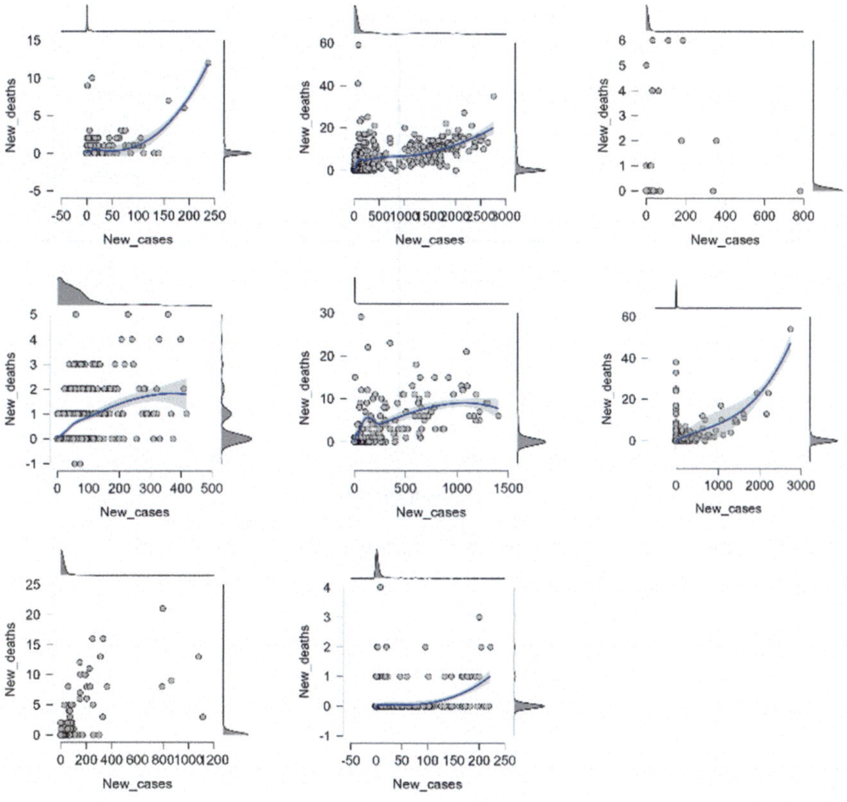

6.4 Conclusion with Comments

Finally, we distinguish with comments on how each nation dealt the transmissibility and death aspects of the pandemic COVID-19 using the *principal component analysis* (PCA). The proximities of the nations are displayed in Figs. 6.7, 6.8, 6.9, 6.10, and 6.11. In A's Eritrea, Uganda, Equatorial Guiana, Sudan, Somalia, Democratic Congo, and Camaros fell out of the trend by the other countries. In American continent, the nation's Argentina and Ecuador fell out of the trend by the other countries. In Asian continent, the nation's Kazakhstan, Myanmar, Indonesia, and China fell out of the trend by the other countries. In European continent, only the nation France fell out of the trend by the other countries. Interestingly, in the Oceanic continent, no nation fell out of the trend by the other countries. Further research work could explore the factors that caused the above-mentioned nations to deviate from the trend of other nations and such a knowledge would be a valuable addition for the effective handling of any future worldwide pandemic, if any (Fig. 6.12).

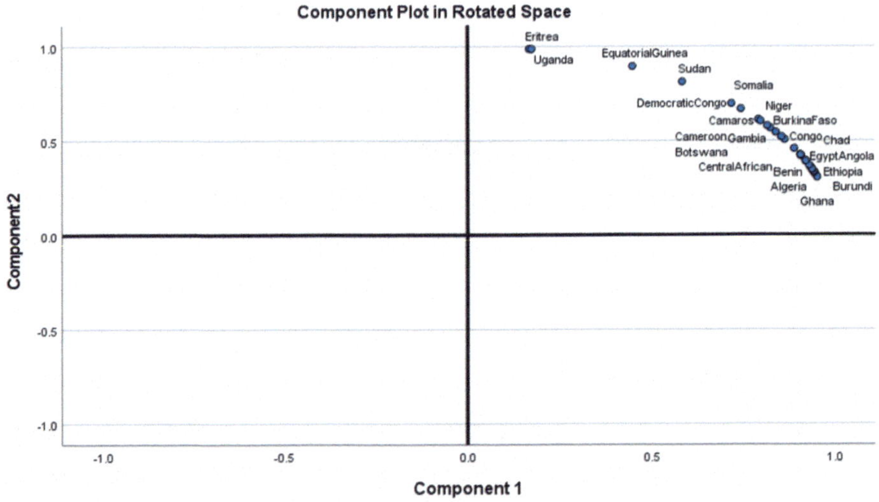

Fig. 6.8 Proximity of the African nations' performance dealing with COVID-19

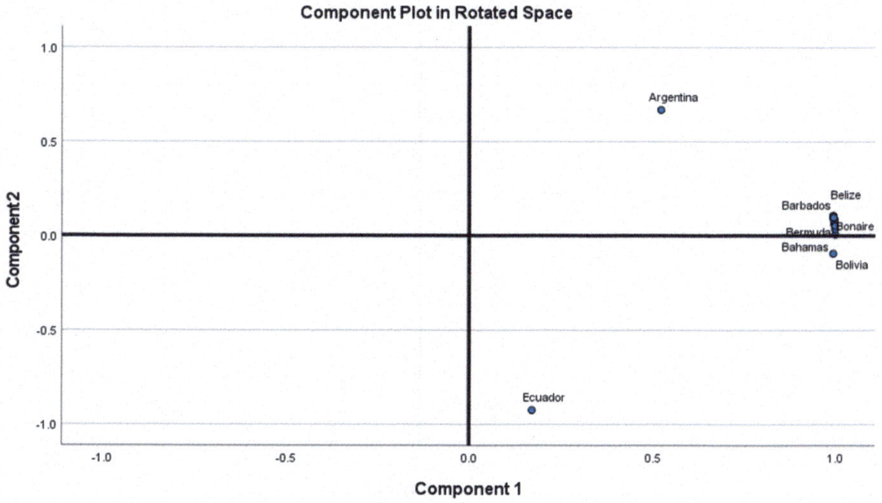

Fig. 6.9 Proximity of the American nations' performance dealing with COVID-19

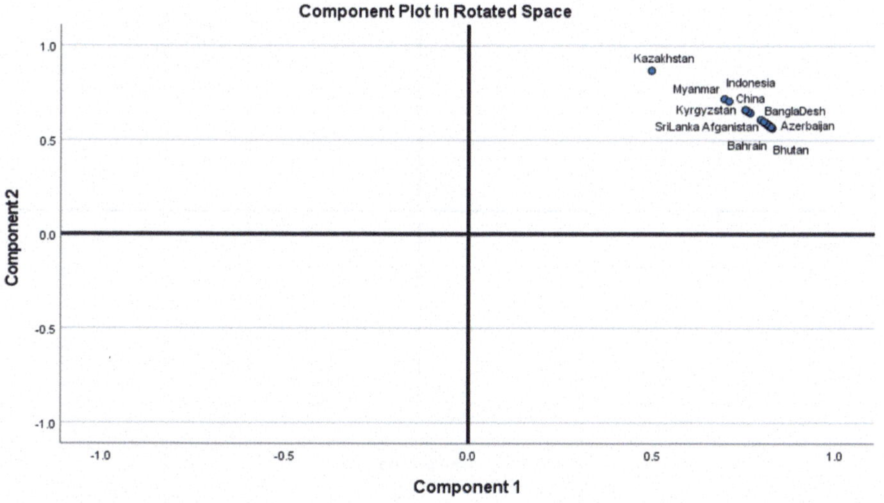

Fig. 6.10 Proximity of the Asian nations' performance dealing with COVID-19

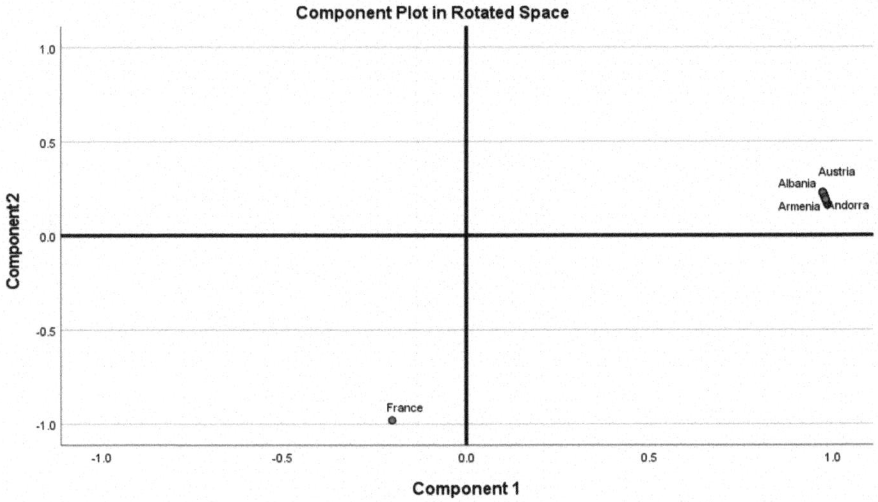

Fig. 6.11 Proximity of the European nations' performance dealing with COVID-19

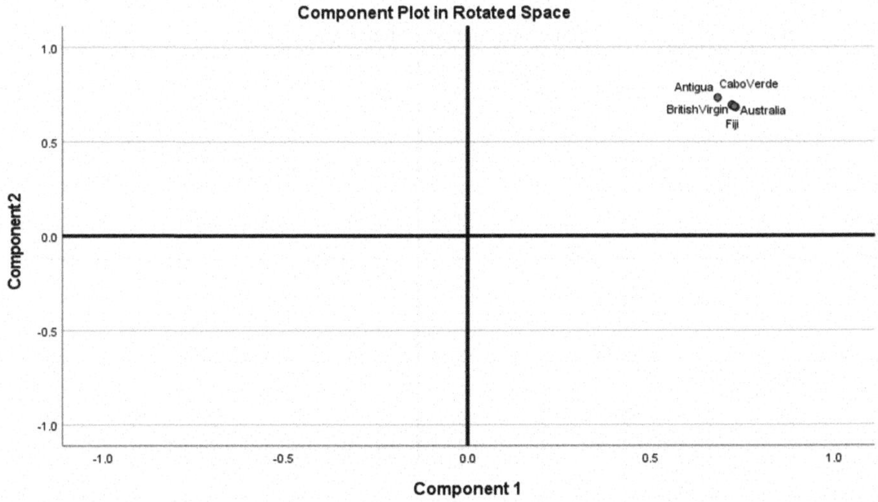

Fig. 6.12 Proximity of the Oceanic nations' performance dealing with COVID-19

References

Agrawal S, Garg R, Kumar N, Prabhakaran M (2020) A practical model for collaborative databases: securely mixing, searching and computing. In: European Symposium on Research in Computer Security. Springer, Cham, pp 42–63

Barratt H, Kirwan M, Shantikumar S (2018) Epidemic theory (effective & basic reproduction numbers, epidemic thresholds) & techniques for analysis of infectious disease data (construction & use of epidemic curves, generation numbers, exceptional reporting & identification of significant clusters). Health Knowl

Blumenfeld D (2001) Operations research calculations handbook. CRC Press

KFF (2023) Global COVID-19 tracker. https://www.kff.org/coronavirus-covid-19/issue-brief/global-covid-19-tracker/

Milligan GN, Barrett AD (2015) Vaccinology: an essential guide. Wiley Blackwell, Chichester, West Sussex, p. 310. ISBN 978-1-118-63652-7. OCLC 881386962

Shanmugam R (2014) Over/under dispersion sometimes necessitates modifying. Inter J Ecol Econ Statis 34(3):37–42

Shanmugam R (2021) Restricted prevalence rates of COVID-19's infectivity, hospitalization, recovery, death in the USA and their implications. J Health Informat Res, 1–18

Shanmugam R, Chattamvelli R (2015) Statistics for scientists and engineers. John Wiley Inter-Science Publication, Hoboken, NJ, USA. ISBN: 978-1-118-22896-8

Shanmugam R, Singh KP (2021) Structural zero data of COVID-19 discovers exodus probabilities. J Multidiscip Healthc 14:1443–1449

Shanmugam R, Ledlow G, Singh KP (2021) Predicting COVID-19 cases with unknown homogeneous or heterogeneous resistance to infectivity. PLoS ONE 16(7):e0254313

Shanmugam R, Ledlow G, Singh KP (2022) Stochasticity among victims of COVID-19 pandemic. J Multidiscip Healthc 15:1–10

Umakanthan S, Sahu P, Ranade AV, Bukelo MM, Rao JS, Abrahao-Machado LF, Dhananjaya KV (2020) Origin, transmission, diagnosis and management of coronavirus disease 2019 (COVID-19). Postgrad Med J 96(1142):753–758

Van den Driessche P (2017) Reproduction numbers of infectious disease models. Infect Disease Model 2(3):288–303

Worldometer (2023) World population. Retrieved from http://worldometers.info. Accessed on May 10, 2023

Ramalingam Shanmugam Ramalingam Shanmugam received Ph.D. degree in statistics from Temple University. In 2021, he was elected to be Fellow of the American Statistical Association (ASA). Since 2016, he is Honorary Professor of international studies in the School of Health Administration at Texas State University. He is passionate to motivate and teach students including how to critically think. His undergraduate students nominated Ram Shanmugam to the Honor Society of Phi Kappa Phi. He has published over 177 research articles in frontline journals. In 1984, he was elected to be Fellow of the International Statistical Institute. He is Book Review Editor for the Journal of Statistical Computation and Simulation. He has published five books with titles Data Guided Healthcare Decision Making (Cambridge University Press). Continuous Distributions Volumes I and II in Engineering and the Applied Sciences (Morgan & Claypool Publishers), Discrete Distributions in Engineering and the Applied Sciences (Morgan; Claypool Publishers), Generating Functions in Engineering and the Applied Sciences (Synthesis Lectures on Engineering, Morgan & Claypool Publishers), and Statistics for Scientists and Engineers (John Wiley Inter-Science Publication).

Karan P. Singh received Ph.D. degree from the University of Memphis. He is Professor and Chair in the Department of Epidemiology and Biostatistics at The University of Texas at Tyler School of Medicine. He has over 25 years of leadership experience in schools of public health, medicine, and other health-related schools/departments, ranging from Department Chair to Director of centers and graduate programs. He has a strong background as Student-focused Administrator at both research-oriented and teaching-oriented institutions. He has served as Senior Scientist in the Comprehensive Diabetes Center, the Comprehensive Center for Healthy Aging, the Center for Outcomes and Effectiveness Research, and Education and the Center for AIDS Research at the University of Alabama at Birmingham. He has also served as Chair or Member on national, international, state, and local committees, including DSMBs, NCI GCSC Steering Committee, and RePORT India Consortium Phase II Clinical Epi Working Group. He has given conference plenary or keynote presentations nationally and internationally. He is a Fellow of the American Statistical Association, the Modeling and Simulation Society of Australia and New Zealand, and the Indian Society for Medical Statistics. He is also an Elected Member of the International Statistical Institute. He has received many awards in mentoring, research, and service.

Chapter 7
Assessing Health Inequalities of Diabetes Care Through the Application of the Bio-ecology Theory

Alan Shaw

Abstract The National Institute for Health and Care Excellence (NICE) guidelines for the management of diabetes state that structured diabetes education should be offered to every person and their carer(s) at or around the time of diagnosis, with annual reinforcement and review. In 2016, the UK's Health and Social Care Information Centre's National Diabetes Audit for England identified only 6% of newly diagnosed Type 2 diabetics attended a course. Diabetes UK has called for radical improvements to the provision. This study attempts to determine why the uptake has been so poor and then offer possible solutions. The study utilised Bronfenbrenner's bio-ecology theory and was made up of four phases: phase one, a pilot study of health educators to identify why patients were not attending the courses. Phase two a qualitative review, using thematic analysis, of patients on their views of structured education. Phase three a census investigating the provision of structured education. It compared the 152 Primary Care Trusts (PCTs) with the new 194 Clinical Commissioning Groups (CCGs) in England. Phase four is a qualitative review using thematic analysis of healthcare professionals (HCPs) on their reasons for providing the care they did. NHS England has a decentralisation approach to managing diabetes structured education in England. There is a lack of awareness of these programmes amongst patients. This is driven by the proliferation of courses provided by NHS England and the budget restrictions to promote them. The quality of diabetes structured education and the ability of patients to attend varied by PCT/CCG, creating a non-inclusive service. In this example, it was established that centralising elements of the diabetes structured education programmes like branding, marketing, course development and programme management could alleviate many of the problems that NHS England currently faces and increase patient engagement. Such a move would also reduce costs and help bridge the current budget deficit. This chapter demonstrates how researchers can utilise Bronfenbrenner's bio-ecology theory to investigate healthcare management processes. More specifically, it is an example

A. Shaw (✉)
Chair of the Retail Institute Special Interest Research Group and Health Research Lead for the Business School, Leeds Business School, Leeds Beckett University, Leeds, UK
e-mail: alan.shaw@leedsbeckett.ac.uk

E. Çetin and H. Özen (eds.), *Healthcare Policy, Innovation and Digitalization*,
Accounting, Finance, Sustainability, Governance & Fraud: Theory and Application,
https://doi.org/10.1007/978-981-99-5964-8_7

of investigating patients, their careers, healthcare professional and policy all in one study. It also addresses a common debate amongst healthcare managers whether systems should be centralised or decentralised.

Keywords Bio-ecology theory · Health inequalities · Centralisation · Diabetes structured education

7.1 Introduction

England's Department of Health and Social Care (DHSC) projected spending for 2024/25 is £176.4 billion (The Kings Fund 2022). After removing the impact of COVID-19, it can be seen that a comparison of the historical and predicted budgets demonstrates a continuing upward trajectory. This is compounded by the fact that many of England's National Health Service (NHS) providers are in deficit (Codery 2022). It places additional pressure on an already over-stretched service. The DHSC (2012) identified that seventy percent of the spending is attributed to the 15 million people living with a long-term chronic condition: i.e. one that cannot, at present, be cured but is controlled by medication and/or other treatment/therapies (DHSC 2012). Demographic changes in England's society mean that the number of people living with a long-term chronic condition (LCC) is also likely to grow (Rechel et al. 2009). It is, therefore, not surprising to see that the UK Government and the DHSC are predicting future health budgets will become unsustainable if issues like the management of long-term chronic conditions are not addressed accordingly (DHSC 2012).

A key initiative launched in 2009 by the DHSC was the Quality, Innovation, Productivity and Prevention (QIPP) programme (Alakeson 2011). It was designed to increase productivity, eliminate waste and drive-up clinical quality. The QIPP target was to save £20bn by 2014 (DHSC 2010a). It should be noted that the actual results are still unclear (DHSC 2015): despite not fully understanding the impact, the UK Government has doubled its efficiency target to 2.2% in an attempt to free up £4.75 billion to fund other health priority areas (HM Treasury 2022). The QIPP programme has twelve work streams deriving from three categories: 'commissioning', 'providers', and 'supporting' (Alakeson 2011). The management of LCCs is one of the said work streams and is found within the 'commissioning' category. This LCC work stream was built on three underlying principles (DHSC 2011): risk profiling to understand the population's needs, the introduction of neighbourhood care teams to provide 'joined-up' and 'personalised services' and finally the development of self-care/shared decision-making (DHSC 2012). It is the self-care/shared decision-making work stream that this study will focus on. The DHSC (2012) believe that self-care/shared decision-making will allow the NHS to systematically transfer knowledge and power to patients: empowering them to maximise self-management and choice. The key mechanism for delivering self-care/shared decision-making is structured health education. Based on the facts identified above, the following objectives were set for this research:

RO1. *To understand if the concept of structured health education within NHS England was flawed.*
RO2. *If it is identified that RO1 was flawed, could the issue(s) be addressed?*

A single case study focusing on the examination of the effectiveness of diabetes structured education programmes within NHS England was used. It should also be noted that on 1 April 2013, England's Health and Social Care Act 2012 replaced the country's Primary Care Trusts (PCTs) with Clinical commissioning groups (CCGs) (Wenzel and Robertson 2019). These CCGs were designed to streamline the commissioning process by placing General Practitioners (GPs)[1] at the forefront of the commissioning process. This led to research question 3:

RO3. *Has the introduction of CCGs improved the provision of self-management within NHS England?*

This study addressed the three research questions using Bronfenbrenner's (2005) bio-ecology theory. As will be demonstrated, such a theory can be used to understand better the problems found within the health domain. It also overcomes the common denunciation of scholars who apply the ecology theory (Tudge et al. 2016) by utilising Bronfenbrenner's final iteration, the PPCT model.

7.2 Diabetes Structure Health Education

As mentioned, self-care and shared decision-making is a process that empowers patients to self-manage their conditions; it requires cultural shifts from healthcare professionals (HCPs) and patients: such modifications are achieved through structured health education (DHSC 2012). Another term associated with the process is self-management. Self-management is defined as the practice that gives those people living with long-term conditions the tools, skills and support needed to improve their own well-being (Self-management UK 2016). Structured education is the key enabler for self-management, and there is overwhelming evidence demonstrating that such programmes can benefit patients and the healthcare process (De Sliva 2011; Purdy 2010 are just some examples). A large number of conditions fall into the LCC category, but only a small quantity of these can be credited as major cost drivers within NHS England's budget. They include diabetes, chronic obstructive pulmonary disease (COPD) and coronary heart disease (The NHS Confederation 2012).

In England, diabetes is managed using a National Service Framework (NSF): this is a set of national standards for the treatment and care of people living with

[1] A General Practitioner is a family doctor who is the main point of contact for general healthcare for NHS patients.

diabetes (DHSC 2010a). There is a category within the said framework that states every patient and their carers should be offered structured education (DHSC 2001 and NHS England 2014). These structured education programmes are designed to help those individuals living with diabetes (and their families) to improve their knowledge, skills and confidence in the condition (NHS England 2014). The objective is to enable individuals living with diabetes to take increasing control of the management of their diabetes and reduce the burden on the already over-stretched health system (NHS England 2014). As identified earlier, there is overwhelming evidence demonstrating how structured health education can benefit patients (De Sliva 2011; Purdy 2010; Singh 2005). Unfortunately, Diabetes UK (2016, p. 19) identified that the number of patients engaging with this service was too small: 'only 2 per cent of people newly diagnosed with Type 1 diabetes and 6 per cent of people newly diagnosed with Type 2 diabetes were recorded as attending structured education'. This finding supports the author's concerns relating to the delivery of structured education within NHS England and is the justification for focusing purely on diabetes. It should be noted that this study started at the end of 2013 (more specifically, focusing on the financial year-end of 2012/13 for the PCTs in NHS England), then again in 2020 (financial year-end of 2019/20). It allowed the author to compare the PCT results with CCG results.

7.3 Bronfenbrenner's Bio-ecological Theory

Before establishing how the bio-ecology theory can be applied to this chapter, it will be prudent first to explain its history. Bronfenbrenner (2005) began challenging developmental theory in 1974; he encouraged scholars and practitioners to view the broader ecological implications when considering the learning abilities of a child. In 1977, Bronfenbrenner introduced the concept known as the *ecology of human development*, which utilised a framework consisting of four nested systems (see Fig. 7.1). Tudge et al. (2009, 2016) argue that many scholars mistakenly use this model as Bronfenbrenner's defining ecological theory, but as history shows, Bronfenbrenner continued to develop his theory until his death in 2005. This demonstrates that Bronfenbrenner, 'was a very self-reflective theorist and fairly frequently noted the changing nature of his theory' (Tudge et al. 2009, p. 199), making the final iteration the ideal option to use.

In the late eighties, more prominence was given to the individual, with Bronfenbrenner (1989) asking scholars to consider how genetic differences could influence the developing child; he also introduced time as a factor. In the nineties, the model transitioned from the nested systems into the *General Ecological Model* (Bronfenbrenner 1993; Bronfenbrenner and Morris 1998), it then progressed into the 'Process, Person, Context and Time' (PPCT) concept, which proposed that the development of an individual is linked to the context (as identified in Fig. 7.1); the *process* used in an individual's development (i.e. the *life-course* changes over time and specific traits identified in a person). The *process* (or proximal process as it is formally known) is a

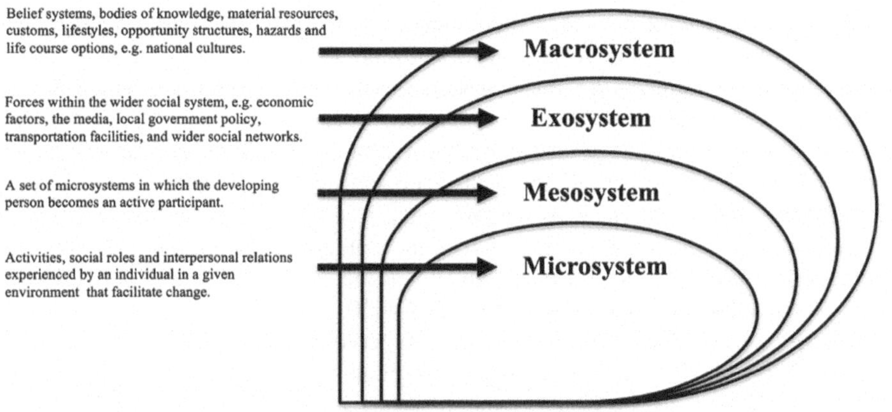

Fig. 7.1 The nested levels of the ecology theory (adapted from Bronfenbrenner (1977))

continuing form of interaction in the person's immediate environment which should also manifest itself within the *microsystem* (Bronfenbrenner 2005). The *process* concept is plural and will vary as a function of the person and the environment they are presented in. Concerning this research, the *process* is the *diabetes structured education* programme provided to patients living with diabetes. The person equates to 'the individual repertoire of biological, cognitive, emotional and behavioural characteristics', Lerner (2005, p. xv). The *life-course* changes refer to sequential transitions that happen to an individual after major events (Clausen 1986). In relation to this research, the *person* is the patient living with diabetes (either type 1 or type 2). The *context* builds on Bronfenbrenner's (2005) earlier framework, which is the one depicted in Fig. 7.1. The *microsystem* relates to all activities, social roles and interpersonal relations experienced by an individual in a given environment that facilitates change. For a person living with diabetes, these could be the interactions between the individual and their diabetes healthcare professional and the interactions between the individual and the structured education programme's health educators. These are the three microsystems considered for this research; in reality, there could be many more, but for parsimonious reasons, it is prudent for researchers to limit their outlook on what could be considered the most important.

The mesosystem is a connection of a set of microsystems in which the developing person becomes an active participant. In the case of this research, it is the link between the three microsystems discussed earlier. The *exosystem* are forces within the wider social system. They include economic factors, the media, local government policy, transportation facilities and wider social networks. This research has limited its outlook to focus only on the policies instigated by NHS England and the various PCTs and CCGs. Finally, there is the *macrosystem*, which includes belief systems, bodies of knowledge, material resources, customs, lifestyles, opportunity structures, hazards and life-course options, e.g. national cultures. This study restricted its research to focus on the impact ethnicity and structured education.

The *time* element of the PPCT model has two aspects to it: 'what happens over the course of both ontogenetic[2] and historical time' (Rosa and Tudge 2013, p. 254). Lerner (2005) articulates the chronosystems as follows; *microtime*, which is the 'continuity verses discontinuity within ongoing episodes of proximal process' (p. xvii); for this study, it was the start and stop times of the Structured education course; *mesotime* which is 'the periodicity of these episodes across broader time intervals, such as days and week' (p. xvii) for this study, it was the length of the Structured education course; and *macrotime*, which is 'the changing expectations and events in the larger society' (p. xvii), for this study, it was the comparison of the deliveries of the self-management programme in 2012 and 2020.

Readers will no doubt observe that the bio-ecology framework is very comprehensive. There are potentially many options a scholar can investigate; careful consideration should be made on which elements to focus on, and these should be guided by the research objectives.

7.4 Method

This research began in 2012 with a pilot study and then a series of investigations which included patient surveys (the *process, person, microsystem and macrosystem* elements of the PPCT model), patient interviews (again, the *process, person, microsystem and macrosystem* elements of the PPCT model), healthcare professional (HCP) interviews (the *process and mesosystem* element of the PPCT model), policy reviews (the *exosystem*element of the PPCT model) and a census which looked at the delivery of diabetes education within the 152 PCTs that existed in England (the *process and time* elements of the PPCT model). It concluded in 2020 with another census of the 194 CCGs (the *process and time* elements of the PPCT model), which also looked at the delivery of diabetes education. The pilot study was a focus group of 12 health educators and social marketers from the Expert Patients Programme (a Community Interest Company that was part of the NHS and was created to deliver patient education).[3] The purpose of this focus group was to help me establish a framework of questions to be used in the semi-structured interviewing of the patients. The discussions in the pilot study were built around understanding the perceptions these health educators and social marketers had on why patients were engaging or not with the diabetes structured health education courses.

The patient's sampling frame was from Diabetes UK's membership database: they produced an advert in their member's magazine calling for participants. These participants were invited to complete a simple questionnaire that requested demographic information and the option to take part in a one-to-one telephone interview regarding diabetes structured education. The survey provided the author with the ability to filter

[2] Bronfenbrenner referred to ontogenetic as the development that occurs as a function of experience rather than as a function of the genetic make-up of an individual (see Lambert and Johnson 2011).

[3] Now Self-Management UK (see www.selfmanagementuk.org/).

out those individuals deemed suitable for the interview: a purposive sample of 30 candidates was made up of an equal number of Type 1 and 2 diabetics, gender, race, employment status and areas of residence in England (six from the Northwest, North East, South West, South East and London). A thematic analysis approach was used to review the answers of the semi-structured interviews with the patients.

The census was initiated through a Freedom of Information request (see Justice (2012)). Its objective was to establish the delivery quantity and types of providers for structured diabetes education amongst the 152 PCTs and 194 CCGs over a period of one year. Each PCT and CCG were asked to list what they delivered, who delivered it and how many participants they delivered to. The policy analysis was restricted to documents provided by the Department of Health and Social Care, NHS England and the National Institute for Health and Care Excellence (NICE). These were used to identify the guidance being given to the PCTs and CCGs on the management of diabetes.

The findings from the patient's semi-structured interviews and the census were then used as a framework for developing the semi-structured schedule for the HCPs. Essentially, the objective was to establish why 'the NHS was doing what it was doing' and whether HCPs had the same perceptions about structured education as those living with the condition. The questioning of the healthcare professionals and administrators used a theoretical saturation approach: it alternated between targeting individuals from PCTs that were perceived (in the patient's mind) to have provided a well-structured education service and those that had not.

Ethics approval for this study was obtained through the author's research establishment, Diabetes UK and the NHS through their Integrated Research Approval System (IRAS) (although the NHS said that their approval was not required as the study would be using Diabetes UK's members and it was non-clinical).

7.5 Results

The census of the PCTs and CCGs was based on a Freedom of Information Request; this meant that each organisation had a legal obligation to respond (Justice 2012). The provision of Type 1 structured education can be seen in Table 7.1.

Table 7.1 Type 1 structured education delivery profile

Type 1 delivery profile	PCTs		CCGs	
	Number	%	Number	%
Delivered internal	41	27	58	30
Commissioned externally	25	16	38	20
Did not know or did not deliver	86	57	98	50
Total	152		194	

The study found that 43% of the PCTs and 50% of the CCGs either delivered diabetes structured education programmes themselves or commissioned the programmes externally. More concerning was the fact that 57% of PCTs and 50% of CCGs either did not deliver or did not know what structured education programmes were delivered.

There was one nationally identified Type 1 structured education course: DAFNE. This accounted for 75% of the courses delivered by the PCTs and CCGs. There were a further 18 other courses provided by the various organisations (see Appendix 7.1). The duration of the local variant courses varied from a session delivered in one day to five sessions held over five weeks.

A similar profile was identified for the delivery of Type 2 diabetes structured education (see Table 7.2). A total of 57% of the PCTs and 58% of the CCGs either delivered self-management programmes themselves or commissioned the programmes externally for delivery. About 43% of PCTs and 42% of CCGs either did not deliver or did not know what structured education programmes were delivered.

With Type 2 education, there were two national courses identified, DESMOND and XPERT, with an additional ten local variants on offer. Again, like Type 1 diabetes, the duration of each type of course varied. Most of the PCTs and CCGs did not give their patients a choice in programmes to attend. Appendix 7.2 provides a summary of the courses that were on offer during the period of research.

Looking at the engagement of those individuals living with diabetes: a total of 93 people agreed to be interviewed, 49 living with Type 1 and 44 living with Type 2. These were filtered down to 32 individuals (see Table 7.3): it should be noted that the original purposive target was 30, but it was difficult to engage with the Black, Minority and Ethnic (BME) community. For the purposes of this study, the definition of the BME community only referred to those individuals from South Asia; this is because the prevalence of diabetes is much higher in this community (Hanif et al. 2014): the 2011 census for England and Wales identified 726,000 people could not speak English well. The majority were from South Asia (ONS 2013). One hundred thirty-eight thousand of that population had no understanding of English. The study had initially only recruited three individuals but managed to find an extra two towards the end of the study.

Each individual took part in the semi-structured interviews and a summary of the core themes in relation to why these individuals chose not to engage with diabetes structured education can be seen in Table 7.4.

Table 7.2 Type 2 structured education delivery profile

Type 2 delivery profile	PCTs		CCGs	
	Number	%	Number	%
Delivered internal	71	47	83	43
Commissioned externally	16	10	30	15
Did not know or did not deliver	65	43	81	42
Total	152		194	

Table 7.3 Descriptive analysis of the patient participants

	Male	%	Female	%	Employed	%	Unemployed/ retired	%	White/ European	%	BME	%	Attended a Course	%	Did not attend a course	%	Totals	%
Type 1	7	44	9	56	7	44	9	56	15	94	1	6	8	50	8	50	16	50
Type 2	8	50	8	50	7	44	9	56	12	75	4	25	8	50	8	50	16	50
Total	15	47	17	53	14	44	18	56	27	84	5	16	16	50	16	50	32	

Note BME refers to individuals from the Black, Minority and Ethnic community. In this study, only those from South Asia and Southeast Asia were classified into this group

Table 7.4 A summary of the axial and core themes associated with why individuals chose not to engage with diabetes structured education

Initial theme	Axial theme	Core theme
I really cannot afford to take time off at the moment even though I know that I should	Taking time off work	Flexibility
Taking time off is not a problem for me, my boss is very flexible and as long as I make the time up it is ok		
I don't want my boss to know I have diabetes; my job is very demanding physically and I don't want him to think I cannot do it		
I can't take time off work, if the course was in the evening or weekends then I could do it	Availability of courses	Flexibility
I tried to get myself on the XPERT course but was told that they don't do it in my area		
The time of the course is not convenient for me, they are not very flexible		
Before your survey I was not aware that these courses existed	Awareness	Promotions
I did not know that I should have been given a place on the course		
I have heard about DESMOND but I have not heard of any of the others you mentioned		

It appears that a lack of flexibility (i.e. in most cases, they only provided courses between 9 am and 5 pm during a working week) was the primary reason why the participants did not want to engage in the programme. Unsurprisingly, those who were retired or unemployed did not have a problem with the course timing restrictions. None of the participants who were interviewed was aware that language-specific courses existed. Many of the employed participants found the evenings and weekend courses better for them, but as identified earlier, not ever PCT offered the same profile of options. From the analysis, it was identified that six Type 1 and five Type 2 diabetics interviewed, who had stated that they had not attended a course, confirmed that they were given an opportunity of participating in a course, but the attendance option was restricted to a weekday between the hours of 9 am and 5 pm: this either clashed with work or childcare commitments. Four of the sixteen who did not attend a course stated that they were unaware that these courses existed or that they and their carers should be given the opportunity of attending one. In contrast, nine of the sixteen who attended a course stated that they were given the selection of attending a day, evening or weekend option. The final individual who did not attend had heard of the programme and had been given a place but thought that it would not be appropriate. In her words:

> I have been told that that the recipes you are taught are very English, but this is not what I or my family would eat. I did not want to take up a place if it was going to be no good for me.
>
> Participant 28: Type 2 (BME)

Following the analysis of the census and the patient's review, a thematic analysis of Healthcare Professionals (HCPs) thoughts on why their PCTs were successful/

unsuccessful in engaging patients to attend diabetes structured education courses were initiated. Each PCT was grouped as either 'good' or 'bad' at providing diabetes health education courses: this was based on a simple criterion of 'good' being those that had provided at least one course and 'bad' as those that either did not know what they were providing or did not provide any diabetes health education courses. A 'good' and 'bad' provider was approached in consecutive order to maintain a balanced view because a theoretical saturation method was used to manage the sample size. A semi-structured interview schedule was developed to identify why each PCT opted to adopt their given strategy (or not, as the case may be). A total of 12 HCPs were interviewed (six 'good' and six 'bad'): each lasting between forty-five minutes and an hour. It should be noted that in the process, twenty-nine different PCTs were contacted, with seventeen saying that they were either too busy or simply did not want to engage with the research.

The underlying theme that led me to conclude that I had reached theoretical saturation was 'budgetary constraints': based on the guidance of O'Reilly and Parker (2013). The semi-structured questions used in the survey looked at why PCTs opted to adopt the strategies they did. It also established why courses were run at particular times, what variants were provided (i.e. language-specific courses), the provisions and support given to those individuals living with diabetes and what level of engagement was had with carers. Examples of the responses from elements of the schedule are listed below.

> We felt that DESMOND offered the best value for us: our diabetes nurses and dietitians have all been trained to deliver this package. They are very familiar with it and I am happy with the outcomes… We simply could not afford to retrain our staff on another course.
>
> Diabetes Nurse 3
>
> We have a number of issues with staffing levels at the moment, I am having to use agencies staff, to be honest it is just not on my radar at the moment…. Outsourcing is not really an option, my budget is being eaten away with people costs, we can give our patients guidance when we see them…
>
> Diabetes Nurse 4
>
> By developing our own course we have removed the need to pay out licence and educator fees. I really do not see the need for that we are following the NICE guidelines and really empowering our patients.
>
> Clinical Lead 1
>
> I work with our diabetes team to decide what is best, we work with the Expert Patients Programme who delivery the XPERT diabetes programme… they are given targets which they need to delivery… I have to balance the books, there are a range of long-term conditions that I have to manage and I need to make sure they all get an equal bite at the Cherry…
>
> Commissioner 1

The 'budgetary constraints' theme continued throughout the interviews; many of the HCPs would have liked to have provided the patients (and their carers) more but felt their hands were tied, primarily through the financial restrictions that were placed upon them.

7.6 Discussion

NICE (2004, 2008) guidelines for the management of diabetes (Type 1 and Type 2) state that structured diabetes education should be offered to every person and their carer(s) at or around the time of diagnosis, with annual reinforcement and review. The results from this research demonstrate that a significant number of PCTs and CCGs either 'did not know' or 'did not deliver' structure education to individuals living with diabetes in their area. This discovery is supported by Diabetes UK (2016). It could be argued that such findings question the legitimacy of the DHSC's policies and the commitment of the Primary care providers in the management of diabetes. One explanation, as identified during the interviews with HCPs, is that clinicians do not see this type of treatment as a priority: more credence is given to drug therapy solutions. This view was only a snapshot of opinions taken from a small cohort of HCP, although it should be noted that there is a body of research that already supports such a premise (Glasgow et al. 2003; Jerant et al. 2005; Jones et al. 2000). A possible solution to this apathy of compliance with non-pharmacotherapy could be to increase the education of the staff, but, as Jerant et al. (2005) identified, the paradigm shift that would be required to make such a change could take a generation to transpire.

Despite the indifference, it was noted that nineteen different Type 1 and twelve different Type 2 structured education courses were on offer to patients living in England: these courses varied in length and quality, some lasting half a day whilst others provided participants with six separate two-to-three-hour sessions over a six-week period. Depending on the patient's place of residence, attendance was either restricted to the hours of between 9 am to 5 pm during a weekday or participants were given a wider option: i.e. attending a weekday, evening or weekend session. Some areas also had multilingual and single-sex options available. It appears that those PCTs offering restricted course timings were the ones who struggled to engage with patients. This may help explain the problems of engagement identified by Diabetes UK (2016).

It should also be noted that each of the thirty-one different course variants had its own branding, marketing processes and educational materials. Such a large variety of courses may also explain why the uptake by those individuals living with diabetes was so poor: the ability to promote the courses becomes very difficult in a competitive environment, resulting in poor awareness amongst the targeted market. Taking into consideration the budget limitations, one could question why there was a need for such a proliferation of diabetes structured education courses within NHS England. Developing a single brand (albeit with a variety of variants like single-sex, language-specific or hard-of-hearing courses) could widen the awareness and reduce cost. Interestingly, one of the HCP participants who was interviewed had stated that the reason they had developed an in-house version of the programme was to avoid paying an annual license fee. Another argued that by developing their own programme, they could control their own quality systems, which would bring down costs. Both respondents believed that their approach would save money which could be invested back into the patients. This argument does have some merits, but it demonstrates a move away from a key tenet of the NHS, which is to provide evidence-based therapies (De Brún 2013). Many of these 'in-house' diabetes structured education programmes did not have any evidence to support their effectiveness. This must bring into question the motivations for running 'in-house versions' instead of the Nationally recognised programmes; health providers would not be allowed to introduce alternative pharmacotherapies just because costs were an issue without any evidence to support it, so should it not be the case that structured education adopts the same premise?

The results show that the key driver in the decision-making process for designing diabetes structured education was the budget allocation. Many PCTs and CCGs in NHS England had taken the decision to provide much shorter courses within restricted time windows. This finding demonstrates that there is a divide in the quality of service that patients receive, giving rise to the popular quote used in the British press that 'healthcare is a post-code (zip code) lottery'. This goes against the whole ethos of public health, where 'no one individual is intrinsically superior to, or worth more than another' (Graham 2010, p. 152). On reflection, it could be argued that the processes adopted have created an element of 'social exclusion' amongst the population of diabetics living in England. Before continuing, the term 'social exclusion' needs further explanation.

7.7 Social Exclusion

Paradoxically, the issues relating to social exclusion are confusing because different terms can be used to highlight what essentially is the same problem; they include inclusion/exclusion, equality/inequality and equity/inequity. This is compounded by the fact that there are no accepted definitions (Hayes et al. 2008). Most link social exclusion to individuals who are unemployed, have low incomes, live in poor housing, have poor health or come from minority ethnic groups (Farrer et al. 2015). However, if we take a historical perspective, it can be seen that social exclusion was associated with low social integration, lack of participation and/or powerlessness (Levitas 2005). These roots were born from the French Republican idea of universal rights (Shortall and Warner 2010). Social inclusion/exclusion is more common in Europe than in the USA (Shortall and Warner 2010); this may be another reason why there is a lack of clarity on these terms. The USA tends to use the expression 'social disparities', focusing on race and socioeconomic factors (Asada et al. 2013). What should be accepted is that all public health policies must have, as their foundations, the principle of equality: i.e. there should be no one individual who is better off than another (Graham 2010). NHS England has a set of priorities in relation to equality: these were developed from the DHSC's Business Plan for 2012–13 and now form part of their objectives for 2016 (DHSC 2012). They include better health outcomes for all and improved patient access (DHSC 2012). Arguably this means that there should be no differentials by geography for any National programmes: a premise shared by Pearce et al. (2003).

Equality or inequality should be classed as measurable quantities that are simple to determine (Kawachi and Kennedy 2002; Young 2001). Harper et al. (2010) builds on this further by stating that such a measurement should be considered as 'value free' in as much as no figure of 'social justice' should be attributed to it: one simply compares the number of a certain group that has received a service against another. Taking this into account and using the results of this study, we can see that only 27% of trusts had a measurement of key performance indicators for Type 1, and 46.7% of trusts had a measurement of key performance indicators for Type 2 (more specifically relating to the management of diabetes structured education). It demonstrates that even though equality has been set as a key priority for NHS England, its assessment across the total spectrum of healthcare provision is not being considered. It also brings another dimension to the discourse of social exclusion: one where social standing and background play no credence in the outcome. If there was a need to put a value on the 'social justice' aspect, then equity and inequity should be considered: here, a normative judgement or physiological commitment is used to value a difference (Harper et al. 2010). The measurement of health inequalities is completed by comparing the variances in health of individuals and/or the society of which they are apart (Harper et al. 2010). This study was only able to assess the differences of the provision of diabetes structured education, no review was done at the health differences associated with each participant: further studies should be initiated to consider this.

Alternatively, we could consider an Index of Disparity (Harper et al. 2010), although Thomson et al. (2006) argue that such analysis should be limited to Health Equity. Arguably the two are indistinguishable, so either approach will be equally valid, although it is fair to say that the concepts are multidimensional and complex, as agreed by Foster and Sen (1997). Building on the work of O'brien and Penna (2008), this study proposes that the terms social inclusion, equality and equity should be considered a dualistic concept, with their respective parts of the dynamic being social exclusion, inequality and inequity.

With this in mind, what can we draw from the study? It appears to challenge the traditional view that social exclusion relates to those who are unemployed, have low incomes, live in poor housing or come from minority ethnic groups. The results demonstrate that some of these disadvantaged groups may be offered services that are far superior to those who are classed as emanating from a higher social standing, the distinguishing factor being, what PCT and CCGs (geographical area) they belong to. Farrer et al. (2015) have identified that the UK and many other EU countries have plans to reduce health inequalities, but with the pressures growing on health budgets, is it not the case that the practice of social exclusion is hitting the broader elements of society? There is clear evidence from this study that the detailed processes identified in care pathways are being ignored by some PCTs and CCGs in favour of reducing costs rather than tackling inequalities.

7.8 Centralising the Process

Based on the discussions above, one could conclude that the key driver for the differences in services provided was a direct result of the decentralisation policy used in managing diabetes structured education. This decentralisation, which Slatman et al. (2007, p. 7) defined as 'the transfer of authority and power from higher to lower levels of government or from national to sub-national levels', has given PCTs and CCGs the opportunity to focus their efforts on other priorities, which arguably is a direct result of the funding constraints they are experiencing. It may also be why those issues of equality highlighted earlier are prevalent in the process. Such a concern is not new, many scholars have already highlighted that decentralisation causes inequities (De Vries 2000; Jommi and Fattore 2003; Koivusalo 1999 and Vrangbæk 2007a, 2007b). That is not to say that decentralisation is all bad: there is a long history that has demonstrated its benefits. Oates's theorem (1972)[4] is one; it contends that local authorities are more efficient than national authorities in providing services. There are even more recent examples with Foster and Plowden (1996) demonstrating that enhancements are achieved through goodwill, knowledge and enthusiasm that is generated locally with local autonomy. Jommi and Fattore (2003) identified better accountability, Jervis and Plowden (2003) argument that decentralisation delivers health strategies that are focused on local needs. Finally, Smullen (2015) argued that 'bottom-up peer dialogue' and feedback of decentralisation create a more superior 'proceduralisation' process.

So, is it now the case that we can dismiss all arguments for adopting a decentralisation strategy? The answer must be no because, conceivably, we would need to consider each on a case-by-case basis. Another problem is that decentralisation is not clearly defined: this argument is backed by Bankauskaite and Saltman (2007), they believe that decentralisation should be considered as a multilevel concept: at its extreme, there are two ends of a continuum (decentralisation and centralisation) which, in essence, is easy to understand, but the middle element of the continuum is the one aspect that causes confusion. Could it be that the extremes of the continuum can never truly be achieved? This premise is supported by Vrangbæk (2007a) who believes that central bodies will always hold a certain amount of control: a good example to consider is the work of Frumence et al. (2014), they identified that the late disbursement of funds in the decentralised health services in Tanzania compromises the whole process but having some form of central control alleviated it.

In addition to the funding issues highlighted above, the question of control can be linked to what could be classed as the 'struggle of power and recognition'. This link was identified by Scharpf (1997) and Vrangbæk (2007b), and based on their findings, a possible reason why so many different diabetes health education courses

[4] Oates (1972) provided an insightful analysis of the trade-off between centralisation and decentralisation by contrasting efficient internalisation of inter-jurisdictional spillovers through centralisation and efficient matching of local policies to local tastes through decentralisation. This analysis culminated in the celebrated 'Oates' Decentralization Theorem', delineating conditions under which centralized or decentralised provision of public goods is efficient. (Bloch and Zenginobuz 2012).

exist in England is because certain 'actors' (i.e. the HCP linked to the management of diabetes) wanted to create a programme credited to themselves: namely a vanity project. This proliferation does seem wasteful as many of the PCTs and CCGs were duplicating work by re-branding courses and producing materials which could be shared across NHS England. Mintzberg (1992) contended that central control would provide a standardisation of processes and products. This would be ideal for diabetes structured health education because not only would it provide cost savings through the maximisation of economies of scale, also it would mean specialist requirements for language-specific courses, braille formats and/or single-sex options can be produced and delivered at minimal costs. Specialist teams could be held centrally and travel around the country as and when needed: such courses may not be offered at the same frequency as the standard courses, but at least every location would have the opportunity of engaging with such services. There are also the issues of the efficacy of each course: NICE (2015) have stipulated that diabetes structured education must be supported with evidence and as stated earlier, but it is only the National programmes that have this, so utilising them would be the solution. This option for tackling the diabetes structure education problem is advocated by Morgan et al. (2006, p. 67): they believe that "centralised reviews can make important, 'positive' contributions to decision-making by raising the evidentiary basis for decisions". The concept of centralising with NHS England is not new, the Change for Life programme is one that is centrally managed, where Trusts can all access the same marketing and education collateral (Prescriber 2007). The results of the Change for Life programme exceeded the Department of Health's initial targets in the first year (DHSC 2010b). Centrally funded courses would remove the need for Trusts to develop their own courses to negate the need for paying a license fee. It also ensures equity is achieved.

NHS England's National Specialised Commissioning Team (NSCT) manages approximately 60 specialised services (Coles et al. 2012). In this case, a small number of teams are located in various parts of the country to deliver therapies to conditions which are considered as rare, where the costs of providing a country-wide service would be too prohibitive. Diabetes does not fit such a profile; it accounts for approximately 10% of NHS England's annual spending (NHS England 2015). There is also a prediction that the condition (adults with Type 2 diabetes) is set to rise to 4.6 million or 9.5% of the adult population by 2030 (Gatineau et al. 2014): so adapting the concept initiated by the NSCT can only be a good thing.

Does this mean that the concept of decentralisation should be abandoned across the whole of NHS England? Again, the answer must be no, as such a move would be counterproductive: the devolution of power to frontline staff is one of the founding principles of the NHS reforms (Peckham et al. 2005). The diabetes management team could still control their own destinies, they simply draw down shared resources, but they need to demonstrate set quality standards and engagement criteria that should be agreed upon by all parties.

This research had identified that a number of trusts relied on their Diabetes Specialist Nurses to administer the process of engaging and coordinating with patients. This, they found to be too onerous and demotivating. The creation of a central hub to administer patients on the various courses could be controlled centrally. There already exists a body that could take on this role: Self-Management UK, formally the Expert Patients Programme and part of the NHS, now a charity in its own right delivering a variety of health education programmes (Self-management UK 2016).

The proposal thus is not to fully centralise diabetes structured education in England but to support Trusts (now CCGs) in delivering programmes that maximises the reach to patients, carers and potential patients; does not affect patient safety; has equity across the country and across the various target groups (market segments); is able to adhere to the NICE guidelines; is cost-effective; is able to continually innovate; and is fully transparent and auditable. Such a proposal should not mean that the programmes would be reduced to a single option for the Type 1 and Type 2 conditions. Local variants can be added to the core framework; an example could be a more diverse discussion of global recipes in cosmopolitan areas. There will need to be full consultation with HCPs, commissioners, patients and current providers. The author would also advocate looking at other examples, namely the NHS Supply Chain's generic specifications project, where a proposal was made to introduce a plan to develop a set of generic national specifications classed as 'clinically acceptable' and 'fit for purpose' for seven wound care and dressing categories: see the work of Guest et al. (2015) and Fronzo (2016).

7.9 Conclusion

It is hoped that readers will now appreciate that using Bronfenbrenner's (2005) bio-ecology theory as a framework for research allows scholars to analyse a domain from a more holistic perspective: i.e. one that includes many of the stakeholders and environmental factors that ultimately influence behaviour. Arguably it provides a more complete and comprehensive review of the facts, making the results more robust and trustworthy. That said, it would be impractical to review all the likely touch points, so researchers must decide which are the most appropriate for their research. Returning to the three research objectives, it can be concluded that:

> **RO1 & RO2**. *The concept of diabetes structured education is not flawed because of the existing evidence that demonstrates its efficacy. That said, the management of diabetes structured education is flawed: too many of the organisations in NHS England are not applying the strict framework in which they were designed to be delivered.*
>
> **RO3**. *The introduction of CCGs has not improved the provision of diabetes structured education within NHS England.*

Possible reasons for these findings relate to the financial pressures faced by NHS England. The ability of the QIPP programme instigated to facilitate these savings is questionable unclear (DHSC 2015). Could it be that specific fundamental organisational changes are just not being considered? It was difficult to fully justify such a bold statement through this limited research, but there were some clear issues that the NHS needs to address whether savings are to be made. The argument for decentralising national health education programmes is poor because the results have identified that this process does not work. The two key reasons for this are as follows:

1. A disparity in the services being offered and
2. The awareness of these services.

These reasons were a consequence of budget restrictions levied on each PCTs and CCGs.

There were some positives identified: successfully PCTs and CCGs were those who offered a variety of options for their patients. Centralising elements of the programme like branding, marketing, course development and programme management could help elevate many of the problems that NHS England currently faces. It is unlikely that this move alone will make any substantial dents in the budget deficit, but it should overcome the inequality issues identified and increase patient empowerment, which, as other studies have shown, will have an impact on the deficit.

Acknowledgements The author would like to thank Diabetes UK with their support in this research. The author would also like to that Anna Heyman, the research assistant who diligently coordinated the results of this research.

Conflict of Interest Readers should note that the author is currently a trustee (volunteer) at X-PERT Health, one of the national providers of diabetes structured education.

Appendix

See Tables 7.5 and 7.6.

Table 7.5 The range of type 1 structured education courses delivered in England

Course	Description
DAFNE	Dose adjustment for normal eating. A five-day course running usually run Monday to Friday (9 am–5 pm)
KATIE	Kent adult type 1 Education: very similar to DAFNE
BERTIE	Bournemouth type 1 intensive education: four, 6-h session run over four weeks
InSight	A diabetes education course for individuals living with type 1: four, 3–4-h sessions run over four weeks
T1IDE	Trafford type 1 intensive diabetes education: four, 6-h sessions run over four weeks
Skills for Life	A type 1 educations, eight 6-h sessions run over eight weeks
REACCT	Re-education and carbohydrate counting: two, 2-to-3-h sessions delivered held six weeks apart
EDWARD	Education for diabetes without a restrictive diet: four, 5.5-h sessions delivered held four weeks apart
BHITE	Brighton and hove intensive type 1 education programme, four 7-h sessions run over four weeks
GATTO	Guy's and Tommy's type 1, 4 full-day sessions held over a period of f 4 weeks for 6 to 8 people held three times each year
BHICEP	Bart's and Homerton insulin & carbohydrate education programme, four full-day sessions held over a period of four weeks for 6 to 10 people
Carb Counting	This is a single 3-h session focusing specifically on food management
STILE	Shropshire titration of insulin and lifestyle education, 4 full-day sessions held over a period of 4 weeks for 6 to 10 people
BDEC	Bournemouth diabetes and endocrine centre have an e-learning platform to support individuals
TIFA	Torbay insulin food adjustment course, four half-day sessions held over a period of four weeks
Freedom4Life	Type 1 diabetes education, five half-day sessions that run over 5 weeks
BITES	Brief Intervention in type 1 diabetes, education for self-efficacy, a three-day course
BEND 1	Basic education for newly diagnosed diabetes type 1, four, three-hour sessions that run over four weeks
CHOICE	Carbohydrate and insulin calculation education—for those with type 1 diabetes

Table 7.6 The range of type 2 structured education courses delivered in England

Course	Description
DESMOND	Diabetes education and self-management for on-going and newly diagnosed. Either delivered as a 6-h, one-day session or two, 3-h half-day sessions. There are four variants including language-specific courses for Punjabi, Urdu, Bengali and Gujarati
XPERT	Designed for people with type 2 diabetes; however, it is also open to individuals with type 1. It has six, 2.5-h sessions held over a period of six weeks
DOTTIE	Doncaster type 2 information education, two, 3-h half-day sessions
EDDI	Education for diabetes for individuals newly diagnosed with type 2 diabetes, two half-day sessions delivered a week apart
DEREK	Diabetes education and revision in East Kent, one, 4-h session
Right Start	Structured education for people with type 2 diabetes, four, two-hour sessions delivered over four weeks
Spotlight Diabetes	Structured education for people with type 2 diabetes, two, two-hour sessions delivered over two weeks
Living with Diabetes	Structured education for newly diagnosed people with type 2 diabetes, one, one-day session with a half-day follow up 3 months later
Good2Go	Structured education for people with type 2 diabetes, a one-day course
HARRIET	Harrogate Initiative providing education on type 2 diabetes, three half-day sessions delivered over 6 weeks
DEAL	Diabetes education awareness for life
BEND 2	Basic education for newly diagnosed diabetes type 2, four, three-hour sessions that run over four weeks

References

Alakeson V (2011) Quality, innovation, productivity and prevention (QIPP) and personal health budgets, (Accessed 21 May, 2022), [available at http://www.in-control.org.uk/resources/health/qipp-and-personal-health-budgetspdf.aspx]

Asada Y, Yoshida Y, Whipp AM (2013) Summarizing social disparities in health. Milbank Q 91(1):5–36

Bankauskaite V, Saltman RB (2007) Central issues in the decentralization debate. In: Saltman RB, Bankauskaite V, Vrangbæk K (eds) Decentralization in health care. Open University Press, Berkshire

Bloch F, Zenginobuz U (2012) Oates' decentralization theorem with household mobility. hal-00657823. https://hal.science/hal-00657823/document

Bronfenbrenner U (1977) Toward an experimental ecology of human development. Am Psychol 32(7):513

Bronfenbrenner U (1993) Ecological models of human development. In: Gauvain M, Cole M (eds) Readings on the development of children. Freeman, New York, NY, pp 37–43

Bronfenbrenner U (2005) Making human beings human: bioecological perspectives on human development. Sage, London

Bronfenbrenner U, Morris PA (1998) The ecology of developmental processes. In: Damon W, Lerner RM (eds) Handbook of child psychology: vol. 1. Theoretical models of human development, 5th edn. Wiley, New York, NY, pp 793–829

Bronfenbrenner U (1989) The developing ecology of human development: paradigm lost or paradigm regained. Biennial Meeting of the Society for Research in Child Development, Kansas City, MO
Clausen JA (1986) The life course: a sociological perspective. Prentice Hall, Upper Saddle River, NJ
Codery S (2022) Efficiency asks risk damaging patient care. Accessed 17 May, 2022. Available at https://nhsproviders.org/news-blogs/blogs/efficiency-asks-risk-damaging-patient-care
Coles S, Haire K, Kenny T, Jessop EG (2012) Monitoring access to nationally commissioned services in England. Orphanet J Rare Dis 7(1):85
De Brún C (2013) Finding the evidence: a key step in the information production process. NHS England, London
De Silva D (2011) Evidence: helping people help themselves. The Health Foundation, London
De Vries MS (2000) The rise and fall of decentralization: a comparative analysis of arguments and practices in European countries. Eur J Polit Res 38(2):193–224
DHSC (2001) National service framework for diabetes. Department of Health, London
DHSC (2010b) Change for life, one year on. Department of Health, London
DHSC (2012) Long term conditions compendium of information, 3rd edn. Department of Health, London
DHSC (2010a) The NHS quality, innovation, productivity and prevention challenge: a guide for clinicians. Accessed 11 Aug, 2022. Available at http://webarchive.nationalarchives.gov.uk/20130107105354/ http://www.dh.gov.uk/prod_consum_dh/groups/dh_digitalassets/%40dh/%40en/%40ps/documents/digitalasset/dh_113807.pdf
DHSC (2011) QIPP workstreams: long term conditions. Accessed 11 Aug, 2022. Available at http://webarchive.nationalarchives.gov.uk/20130107105354/ http://www.dh.gov.uk/en/Healthcare/Qualityandproductivity/QIPPworkstreams/DH_115448
DHSC (2015a) Policy paper 2010 to 2015 government policy: NHS efficiency. Accessed 24 Aug, 2022. Available at https://www.gov.uk/government/publications/2010-to-2015-government-policy-nhs-efficiency/2010-to-2015government-policy-nhs-efficiency
Diabetes UK (2016) State of the nation: time to take control of diabetes. Accessed 11 June, 2022. Available at https://diabetes-resources-production.s3-eu-west-1.amazonaws.com/diabetes-storage/2017-08/DiabetesUK%20State%20of%20the%20Nation.pdf
England NHS (2015) Annual report and accounts 2014–15. Her Majesty's Stationery Office, London
NHS England (2014) Action for diabetes. Accessed 17 June, 2022. Available at https://www.england.nhs.uk/ourwork/qual-clin-lead/diabetes-prevention/action-for-diabetes/
Farrer L, Marinetti C, Cavaco YK, Costongs C (2015) Advocacy for health equity: a synthesis review. Milbank Q 93(2):392–437
Foster CD, Plowden FJ (1996) State under stress hollow: can the hollow state be good government? Open University Press, Maidenhead
Foster JE, Sen A (1997) On economic inequality, expanded. Clarendon Press, Oxford
Fronzo C (2016) Questions about the NHS supply chain's generic specifications project arise at TVS 2016. J Wound Care 25(5):225–226
Frumence G, Nyamhanga T, Mwangu M, Hurtig A (2014) The dependency on central government funding of decentralised health systems: experiences of the challenges and coping strategies in the Kongwa district, Tanzania. BMC Health Serv Res 14(1):39
Gatineau M, Hancock C, Holman N, Outhwaite H, Oldridge L, Christie A, Ells L (2014) Adult obesity and type 2 diabetes. Public Health England, London
Glasgow RE, Davis CL, Funnell MM, Beck A (2003) Implementing practical interventions to support chronic illness self-management. Joint Comm J Qual Patient Saf 29(11):563–574
Graham H (2010) Where is the future in public health? Milbank Q 88(2):149–168
Guest JF, Vowden P, White R, Bowen G, Iversen C, Anderson R (2015) Letters—response to call to action: the proposed national formulary for wound care. J Wound Care 24(12):553–559
Hanif W, Khunti K, Bellary S, Bharaj H, Karamat MA, Patel K, Patel V (2014) Type 2 diabetes in the UK South Asian population. An update from the South Asian health foundation. Accessed

24 Aug, 2022. Available at http://www.sahf.org.uk/sites/default/files/publications/Type%202%20Diabetes%20in%20the%20UK%20South%20Asian%20population.pdf
Harper S, King NB, Meersman SC, Reichman ME, Breen N, Lynch J (2010) Implicit value judgments in the measurement of health inequalities. Milbank Q 88(1):4–29
Hayes A, Gray M, Edwards B (2008) Social inclusion: origins, concepts and key themes. Commonwealth of Australia, Canberra
Jerant AF, von Friederichs-Fitzwater MM, Moore M (2005) Patients' perceived barriers to active self-management of chronic conditions. Patient Educ Couns 57(3):300–307
Jervis P, Plowden W (2003) The impact of political devolution on the UK's health services. The Nuffield Trust, London
Jommi C, Fattore G (2003) Regionalization and drugs cost-sharing in the Italian NHS. Euro Observer 5(3):1–4
Jones A, Pill R, Adams S (2000) Qualitative study of views of health professionals and patients on guided self-management plans for asthma. BMJ 321(7275):1507–1510
Justice (2012) Freedom of information act 2000. (c.36). Her Majesty's Stationery Office, London
Kawachi I, Kennedy BP (2002) The health of nations: why inequality is harmful to your health. New Press, New York
Koivusalo M (1999) Decentralisation and equity of healthcare provision in Finland. BMJ 318(7192):1198
Lambert MC, Johnson LE (2011) Ontogenetic development. In Goldstein S, Naglieri JA (Eds), Encyclopedia of child behavior and development (p. 1037). Springer. https://doi.org/10.1007/978-0-387-79061-9_2018
Lerner RM (2005) Forward—Urie Bronfenbrenner: career contributions of the consummate developmental scientist. In: Bronfenbrenner U (ed) Making human beings human: bioecological perspectives on human development. Sage, London, pp ix–xxvi
Levitas R (2005) The inclusive society?: Social exclusion and New Labour. Palgrave Macmillan, Basingstoke
Mintzberg H (1992) Structure in fives: Designing effective organizations. Prentice Hall, Upper Saddle River, NJ
Mintzberg H (1993) Structure in fives: designing effective organizations. Prentice-Hall, London
Morgan S, McMahon M, Mitton C (2006) Centralising drug review to improve coverage decisions. Appl Health Econ Health Policy 5(2):67–73
NICE (2004) Type 1 diabetes: diagnosis and management of type 1 diabetes in children, young people and adults. National Institute for Health and Care Excellence, London
NICE (2008) Type 2 diabetes NICE clinical guideline 66. National Institute for Health and Care Excellence, London
NICE (2015) Diabetes in adults: quality standard. Accessed 10 June, 2022. Available at https://www.nice.org.uk/guidance/qs6/chapter/quality-statement-1-structured-education
O'Brien M, Penna S (2008) Social exclusion in Europe: some conceptual issues. Int J Soc Welf 17(1):84–92
Oates WE (1972) Fiscal Federalism. New York: Harcourt Brace Jovanovich
Oates WE (2008) On the theory and practice of fiscal decentralization. In: Auerbach AJ, Shaviro DN (eds) Institutional foundations of public finance: economic and legal perspectives. Harvard University Press, Cambridge, Massachusetts, pp 166–192
ONS (2013) Office for national statistics: language in England and Wales 2011. Accessed 11 June, 2022. Available at https://www.ons.gov.uk/peoplepopulationandcommunity/culturalidentity/language/articles/languageinenglandandwales/2013-03-04
O'Reilly M, Parker N (2013) Unsatisfactory saturation: a critical exploration of the notion of saturated sample sizes in qualitative research. Qual Res 13(2):190–197
Pearce D, Groom B, Hepburn C, Koundouri P (2003) Valuing the future. World Econ 4(2):121–141
Peckham S, Exworthy M, Powell M, Greener I (2005) Decentralisation as an organisational model for health care in England. NHS Service Delivery and Organisation R&D Programme, London
Prescriber (2007) Change for life. J Prescribing Med Manag 18(12):55–55

Purdy S (2010) Avoiding hospital admissions. What does the research evidence say? King's Fund, London

Rechel B, Doyle Y, Grundy E, McKee M (2009) How can health systems respond to population ageing? Policy brief 10. World Health Organization, Copenhagen

Rosa EM, Tudge J (2013) Urie Bronfenbrenner's theory of human development: its evolution from ecology to bioecology. J Fam Theory Rev 5(4):243–258

Saltman R, Busse R, Figueras J (2007) Decentralization in health care: strategies and outcomes. McGraw-Hill Education (UK), Maidenhead

Scharpf FW (1997) Games real actors play: actor-centered institutionalism in policy research. Westview Press, Boulder, Colorado

Self-management UK (2016) What is self-management? Accessed 19 June, 2022. Available at http://selfmanagementuk.org/what-is-self-management

Shortall S, Warner ME (2010) Social inclusion or market competitiveness? A comparison of rural development policies in the European Union and the United States. Soc Policy Adm 44(5):575–597

Singh D (2005) Transforming chronic care: evidence about improving care for people with long-term conditions. Accessed 17 June, 2022. Available at http://www.download.bham.ac.uk/hsmc/pdf/transforming_chronic_care.pdf

Smullen A (2015) Not centralisation but decentralised integration through Australia's national mental health policy. Aust J Public Adm 75(3):280–290

The Kings Fund (2022) The NHS budget and how it has changed. Accessed 17 May, 2022. Available at https://www.kingsfund.org.uk/projects/nhs-in-a-nutshell/nhs-budget

The NHS Confederation (2012) Investing in emotional and psychological wellbeing for patients with long-term conditions: a guide to service design and productivity improvement for commissioners, clinicians and managers in primary care, secondary care and mental health. The NHS Confederation, London

Thomson GE, Mitchell F, Williams M (2006) Examining the health disparities research plan of the National Institutes of Health: unfinished business. National Academies Press, Washington

HM Treasury (2022) New efficiency drive to cut £5.5 billion of government waste. Accessed 17 May, 2022. Available at https://www.gov.uk/government/news/new-efficiency-drive-to-cut-55-billion-of-government-waste

Tudge JR, Mokrova I, Hatfield BE, Karnik RB (2009) Uses and misuses of Bronfenbrenner's bioecological theory of human development. J Fam Theory Rev 1(4):198–210

Tudge JR, Payir A, Merçon-Vargas E, Cao H, Liang Y, Li J, O'Brien L (2016) Still misused after all these years? A re-evaluation of the uses of Bronfenbrenner's bioecological theory of human development. J Fam Theory Rev 8(4):427–445

Vrangbæk K (2007a) Key factors in assessing decentralization and recentralization in health systems. In: Saltman RB, Bankauskaite V, Vrangbæk K (eds) Decentralization in health care. Open University Press, Berkshire, pp 63–76

Vrangbæk K (2007b) Towards a typology for decentralization in health care. In: Saltman RB, Bankauskaite V, Vrangbæk K (eds) Decentralization in health care. Open University Press, Berkshire, pp 44–62

Wenzel L, Robertson R (2019) What is commissioning and how is it changing? Accessed 17 May, 2022. Available at https://www.kingsfund.org.uk/publications/what-commissioning-and-how-it-changing

Young IM (2001) Equality of whom? Social groups and judgments of injustice. J Polit Philos 9(1):1–18

Alan Shaw is Vice-Chair of the Retail Institute Special Interest Research Group. He is also Senior Lecturer in Digital Marketing, Founder of Strategic Planet, Member of the Social Media Research Foundation, and Trustee at X-PERT Health. His main research interests are in social marketing, social media marketing, social network analysis, and social listening. Much of his research focuses on the health sector because of his experience of working within epilepsy, diabetes, Crohn's disease, inborn metabolic disorders, and wound management sectors. He has been involved in patient advocacy groups and helped the Expert Patients Program transition from being part of the NHS to becoming a community interest company (CIC) and then a charity. He has over 25 years of experience working in the private sector, much of this was in health. Some notable deliverables were (1) the role out of the concept home delivery for prescription products related to individuals living with Phenylketonuria (PKU) in the UK and several European countries, (2) the development of a digital "Keto Calculator" to support the management of pediatric epilepsy, and (3) the design of a patient management system to support the concept home delivery for prescription products in Australia. He is currently working on several social listening projects using both qualitative and quantitative processes. He is also able to design and deliver social listening in-house company workshops using platforms like NodeXl and Gephi or provide support for any structural equation modeling ventures. His education and qualifications are as follows:

2016: Ph.D. in Social Marketing, University of Hull.

2015: Fellow of the Higher Education Academy.

2010 Postgraduate Diploma in Marketing, Chartered Institute of Marketing.

2009: Danone Marketing University (internal course focusing on marketing).

1999–2001: MSc, Logistics and Supply Chain Management, Cranfield University.

1983–1987: Portsmouth Poly. B.Sc. (Hons) Engineering with Business Studies.

Chapter 8
The Computational Perspective on Internalized and Simplex-Structured Motivation

Ali Ünlü

Abstract Self-determination theory (SDT), introduced by Deci and Ryan, is a popular theory of motivation. Applications of SDT are numerous and include areas like health care, health professions education, and digital health. Over the recent years, the author published a series of quantitative papers on SDT. To the author's knowledge, these contributions have remained relatively unrecognized in the SDT community. However, the methodology developed therein can be useful to the field. With the present work, the author reviews, exemplifies with data, and mathematically describes that methodology, in a coherent manner. The focus of this chapter is on computational as well as mathematical aspects. For the investigation of motivation internalization and simplex structure, the author recapitulates the convex decomposition, or constrained regression, model and assembles the computation steps of the convex quadratic program. The author also contributes to the mathematical foundations of the methodology. In particular, mathematical definitions are proposed for the in SDT important concepts of theoretical closeness of regulations, cumulative internalities of regulations along the motivation continuum, and simplex structure of motivation. The idea is to consider a linear order on the set of regulations, take the induced geodesic distance, form a linear motivational structure, and posit that these distances, as measure of closeness, are compatible with the shares of the convex decomposition model. By examples, the author shows how the method can be used for the exploratory data analysis of simplex structure. In particular, for a given intermediate regulation, the author employs the method to estimate the theoretically closer regulation of the two neighbor regulations, contiguous to it. In addition, the technique was applied in a systematic empirical study. The study compared science teaching in a classical school class versus an expeditionary outdoor program. Succinctly, the main results of this study are recapped. In the internal and external shares of identified regulation, the science teaching formats did not differ. The teaching formats differed in the internalization of introjected regulation, which was more strongly external motivation in the outdoor program. The simplex structure of SDT could basically be

---

A. Ünlü (✉)
School of Social Sciences and Technology, Technical University of Munich, Munich, Germany
e-mail: aligalibuenlue@gmail.com

E. Çetin and H. Özen (eds.), *Healthcare Policy, Innovation and Digitalization*, Accounting, Finance, Sustainability, Governance & Fraud: Theory and Application,
https://doi.org/10.1007/978-981-99-5964-8_8

supported in the study data. The statistical computing and graphics environment R is powerful. Throughout this chapter, computations were made in R, with the package SDT. The functions internalization and simplex of the package SDT were used for computations of the internalization shares and simplex structure shares, respectively. Finally, this chapter concludes with a few general ideas about the motivation theory and with personal suggestions for modifications of it.

Keywords Self-determination theory · Motivation · Internalization · Simplex structure · Convex decomposition · Constrained regression · Quadratic program · Linear order · Geodesic distance · Linear motivational structure · R · Indeterminacy in closeness · Amotivation · Basic psychological needs

8.1 Self-Determination Theory

> It is when you give of yourself that you truly give. For what are your possessions but things you keep and guard for fear you may need them tomorrow?
>
> Gibran Khalil Gibran, The Prophet

Self-determination theory (SDT) is a successful and popular theory of motivation, of what drives people to act (Deci and Ryan 1985, 2000, 2008; Gagné et al. 2022; Ryan 2019; Ryan and Deci 2000, 2002, 2017, 2020). Based on SDT, researchers have made significant contributions to a plethora of application areas including health, specifically health care, health professions education, and digital health (Gorin et al. 2014; Halvari et al. 2017; Keenan et al. 2021; Kusurkar et al. 2012; Kusurkar and Croiset 2015; Morbée et al. 2021; Nelson et al. 2022; Nicholas et al. 2021; Orsini et al. 2016; Smit et al. 2017; Williams et al. 2007; Wuyts et al. 2021). As an example, SDT can be employed to theoretically identify and select useful constructs and measures to inform, change, or maintain health behavior in effective ways, thus providing for a principled approach to health behavior interventions (Bellg 2003; Block et al. 2016; Gillison et al. 2019; Ng et al. 2012; Ntoumanis et al. 2021; Ryan et al. 2008; Silva et al. 2014; Teixeira et al. 2020).

A common observation across these works is the importance of autonomous or uncontrolled, in particular intrinsic, motivation. Empirically corroborated in SDT publications, the more internalized or autonomous motivation is, ideally intrinsic motivation, the more effective and abiding the outcomes and behaviors are in general. For example, you live healthier, grow, and feel well mentally. SDT posits that such internalized or intrinsic motivation arises from the satisfaction to sufficient degree, of three basic psychological needs, innate and universal (see also Conesa et al. 2022). These psychological needs are competence, autonomy, and relatedness. Competence is the need for being good at something, feeling effective, autonomy refers to the perceived choices and control over one's decisions, actions, or behaviors, and relatedness stands for being connected to others through positive relationships, feeling understood and cared for.

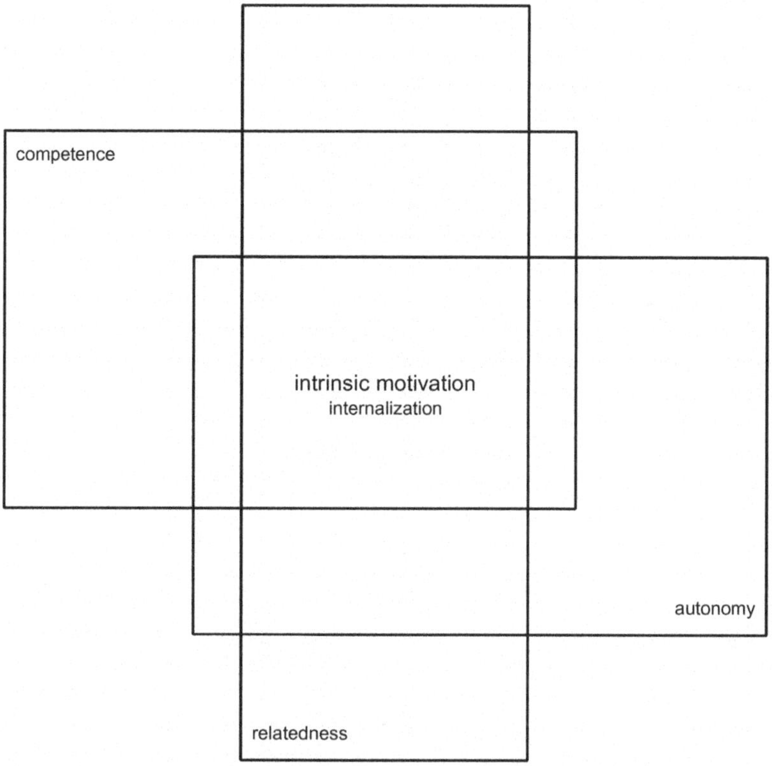

Fig. 8.1 Basic psychological needs and motivation

A pictorial representation of this fundamental postulate is shown in Fig. 8.1, where it can be seen that satisfaction of the basic psychological needs for competence, autonomy, and relatedness facilitates internalized or intrinsic motivation..

From the above discussion, it becomes obvious that intrinsic motivation plays a pivotal role. What is intrinsic motivation? Deci (1975) initiated SDT with research on intrinsic motivation. This type of motivation relates to internal factors for human behavior. Internal factors can be, for example, enjoyment, genuine interest, inherent satisfaction, or curiosity, natural intrinsic motives and needs within humans. Intrinsic motivation encompasses behaviors that are characterized by such internal factors. Slangily speaking, behaviors described by such phrases as "truly choose and want to" or "love to" are associated with intrinsic motivation. For example, "I do math for fun, for its own sake." Intrinsic motivation is fully internal, the most self-determined or autonomous, the terms being synonymously used, and uncontrolled form of motivation.

Intrinsic motivation, which is important in practice, along with the other motivation types discussed below are summarized in Table 8.1.

Table 8.1 Motivation continuum and the basic concepts of self-determination theory

Behavior	Nonself-determined	Self-determined			
Motivation type	Amotivation	Extrinsic motivation			Intrinsic motivation
Regulation type	Non-regulation	External regulation	Introjected regulation	Identified regulation	Intrinsic regulation
Locus of causality	Impersonal	External	Somewhat external	Somewhat internal	Internal
Form	Uncontrolled Non-autonomous	Controlled Non-autonomous	Controlled Non-autonomous	Quasi-uncontrolled Quasi-autonomous	Uncontrolled Autonomous

This representation is also called the self-determination continuum or self-determination theory's taxonomy of motivation

There can also be external factors impacting behavior. Examples are money, grades, sanctions, or reputation, imposed extrinsic forces outside a person acting on the person. Extrinsic motivation in the original sense represents behavior dictated by external factors. Behaviors phrased "I must" or "I should" are extrinsically motivated. For example, "I go to work to earn money." In SDT, researchers distinguish between this and other qualitative forms, so-called regulations, of extrinsic motivation (Table 8.1). In contrast, intrinsic motivation is not further differentiated but equated with intrinsic regulation. The stated definition describes the external regulation of extrinsic motivation. This type of regulation is fully external, and since external constraints are enforced, non-autonomous and controlled.

SDT also posits at least two other regulations of extrinsic motivation, the introjected regulation and identified regulation. Before I discuss these regulations briefly (for more information, see Deci and Ryan 2000), a remark is in order. A reason for postulating regulation types of extrinsic motivation is to account for qualitative gradations of extrinsic motivation. This is of empirical relevance. Imagine an extrinsically motivated student who, initially, is only externally moved. During work on a task, this student may value the activity more and more, thereby being gradually internalized. Increasingly, over time, this student may be moved by not only external factors but also internal factors. In other words, the extent of instrumentality, that is, dependence on external factors, may vary, which defines the distinct regulation types of extrinsic motivation. More generally, the regulation types across all motivations entail varying degrees of internalization. This is why the motivation continuum, shown in Table 8.1, is also referred to as the internalization continuum. Along that continuum, in particular, the more internalized extrinsic motivation is, the more autonomous and uncontrolled the behaviors are. I show later that the internalization degrees of extrinsic motivation, pertinent in practical settings, can be computationally examined by optimization.

From external to more internalized motivation, introjected regulation is a successor regulation to external regulation and the first state of internalized extrinsic motivation. It is internal and external motivation, but more external than internal. In SDT, researchers call this the "somewhat external" regulation. The successor regulation to introjected regulation is identified regulation, which is the second state of internalized extrinsic motivation. Identified regulation also is internal and external motivation, but more internal than external. The corresponding notion in SDT is the "somewhat internal" regulation. Introjected regulation can be behavior triggered by external societal expectations. For example, a student concerned about what the teacher may think of her can self-impose the internal controls of pride for success or shame for failure. Because of such restricting factors, introjected regulation is non-autonomous and controlled. In comparison, identified regulation entails higher degrees to which a person internally, and consciously, values the activity, but is still instrumental behavior with external value. In addition, compared to intrinsic regulation, identified regulation is not behavior enacted based on unforced internal motives, such as fun or enthusiasm. As an example of identified motivational behavior, a student may not enjoy doing math, but she may learn math, because personally she may recognize the importance and identify with the value of learning math for her studies in physics. Identified regulation is relatively volitional and uncontrolled, a quasi-autonomous and quasi-uncontrolled form of motivation.

Amotivation, or non-regulation, is understood as state of no motivation, the lack of any form of motivation. SDT describes amotivation with lack of value, interest, or perceived competence. The activity is of no relevance to the person. The person does not have any intention to act. For example, researchers have studied exercise amotivation among elderly people (e.g., Vlachopoulos et al. 2010) or environmental amotivation for why people fail to adopt protective measures toward environment (e.g., Pelletier et al. 1999). Empirically, amotivation is detrimental to outcomes such as mental well-being or bodily health. Amotivation is impersonal, in the sense that it lacks intentionality. No internal and external motives and controls drive and constrain behavior, thus amotivation, or non-regulation, may be viewed to be uncontrolled and the strongest non-autonomous form of the continuum.

8.2 Why the Computational Perspective on Motivation is Important

I want to motivate the computational perspective on motivation exemplarily with SDT. In light of the above discussion, see Table 8.1, the following useful questions naturally arise.

1. How can the degrees of internalization of extrinsic motivation, of the intermediate regulation types of identified regulation and introjected regulation, be computed from data?
2. How can the inherently linear structure of cumulative internalities along the motivation continuum, or simplex structure, be studied computationally?
3. How can a fast and frugal but informative measure be computed from data quantifying the extent of self-determined behavior, of self-determination?
4. Apparently computational in kind, how can the questions raised in 1, 2, and 3 be implemented effectively in data analysis software?

Why are questions 1 and 2 important? The internalizations of identified regulation and introjected regulation were vaguely expressed as somewhat internal and somewhat external, respectively, and remained undetermined. A solution to this problem was proposed by Ünlü and Dettweiler (2015), the paper reviewed in this chapter. They highlighted a useful special application of the convex quadratic program to SDT, as a mathematical representation for the problem of motivation internalization. Thus, they computed from inventory scores of well-elaborated questionnaires the internal and external motivation shares, as their degrees of internalization, of the intermediate regulation types of extrinsic motivation. The approach is based on modeling each of the intermediate regulations as a convex combination of the, by theory, fully internal and fully external poles of the internalization continuum, of intrinsic regulation and external regulation, respectively. That technique was also generalized to compute the simplex structure shares from test scores. What is the simplex structure? As can be seen from Table 8.1, the theory postulates that the regulation types form a chain or order of cumulative internalities along the motivation continuum. The regulations are considered to be theoretically closer to one another, if they are closer to one another in the posited chain or order. For example, external regulation and intrinsic regulation are theoretically least close, since they are most apart in the chain. A mathematical definition for this notion of closeness will be given later. Then, the simplex structure of SDT states that regulation types theoretically closer to one another are more strongly interrelated, where interrelated in SDT means statistically correlated. The simplex structure is an important property, an implicit assumption, deemed to be empirically necessary if the theory is acceptable. In SDT, this property is employed for scale construction and evaluation. In contrast to correlations used in SDT literature, in the alternative approach by Ünlü and Dettweiler (2015), confirming the simplex structure assumption in motivation data (essentially) translates to computing shares, based on the quadratic program, that are larger for regulation types closer to one another.

This chapter is structured as follows. In Sects. 8.1 and 8.2, I have reviewed the basic concepts of SDT and motivated, and now outline, the contents of this work. In Sects. 8.3 and 8.4, I address and answer the questions 1 and 2, respectively, in mathematical definitions or models, and in computational data analyses. Thereby, I review the internalization and simplex structure method based on the constrained regression problem, shown to be equivalent to and computable by the convex quadratic program, put forth by Ünlü and Dettweiler (2015). In particular, I introduce the mathematical concept of a linear motivational structure, which includes a mathematical definition of theoretical closeness of regulation types. Special linear motivational structures are simplex structures, which, thus, I also define mathematically. This is new and was not published in Ünlü and Dettweiler (2015). Only conceptually, I address the questions 3 and 4, as concluding remarks (Sect. 8.6). However, throughout, I use software code and functions of the package SDT (https://CRAN.R-project.org/package=SDT) for self-determination theory analysis in R (https://www.R-project.org; The R Core Team 2022), to illustrate technique and computations. R is the answer to question 4, which effectively implements not only the solutions to questions 1, 2, and 3, but also a very large number of statistical, computational, and graphical techniques. In Sect. 8.5, I briefly review the main findings of a systematic empirical study, on science learning motivation, as an interesting real application of the convex decomposition approach. Section 8.6 contains a summary of the chapter and concluding remarks.

8.3 Computation of Internalization

I describe the model and corresponding quadratic program for the computational analysis of the internalization degree of motivation.

Consider the motivation continuum, Table 8.1. Let $Y \in \{IdR, IjR\}$ be any of the intermediate regulation types (*IdR*: identified regulation; *IjR*: introjected regulation).

Denote the polar regulation types by $X_1 = InR$ (intrinsic regulation) and $X_2 = ExR$ (external regulation). These regulations vary in the extent to which they are internal motivation or external motivation. As already mentioned, Y is internal as well as external, X_1 is only internal, and X_2 only external. I posit that Y is a convex combination of X_1 and X_2, that is, $Y = \pi_1 X_1 + \pi_2 X_2$, with $\pi_1, \pi_2 \geq 0$ and $\pi_1 + \pi_2 = 1$. This is the basic model assumed for the motivation variables. I interpret the weights π_1 and π_2, reals in the unit interval [0, 1], to be the internal motivation and external motivation shares of Y, respectively. Under this model, Y is posited to be composed of $(\pi_1 \cdot 100)\%$ internal motivation and $(\pi_2 \cdot 100)\%$ external motivation. Obviously, the shares are complementary, $\pi_1 = 1 - \pi_2$ and $\pi_2 = 1 - \pi_1$.

The variables Y, X_1, and X_2 are the scores on the respective subscales of the test, that is, data. The parameters π_1 and π_2 are unknown constants and have to be estimated. How? The model used is a constrained regression model, with two inequality and one equality constraints. Estimation, by least-squares, is realized by minimizing, over all feasible values for the parameters π_1 and π_2, the sum of the squared differences between the observed values Y and predicted values $\pi_1 X_1 + \pi_2 X_2$. That is, if

$$D = \left\{\pi = (\pi_1, \pi_2)^T \in \mathbb{R}^2 : \pi_1, \pi_2 \geq 0, \pi_1 + \pi_2 = 1\right\},$$

where $.^T$ denotes the transpose, the estimate $\hat{\pi} = \left(\widehat{\pi_1}, \widehat{\pi_2}\right)^T$ for the parameter vector $\pi = (\pi_1, \pi_2)^T$ is defined by

$$\hat{\pi} = \arg\min_{\pi \in D} \sum (Y - (\pi_1 X_1 + \pi_2 X_2))^2.$$

This is a basic instance of the convex optimization problem (e.g., Boyd and Vandenberghe 2009). The objective function, the sum of squared differences as a function of the parameters, is convex, the feasible set D is convex. In particular, the existence and uniqueness of the above solution can be ensured, after verifying the Karush–Kuhn–Tucker conditions for the characterization of (global) optimality for this problem. Typically, there is no analytical solution, in simple closed form.

How to compute the solution numerically? The problem of interest is a quadratic program, that is, the objective function is of (convex) quadratic form and the inequality and equality constraint functions are affine. More precisely, $\hat{\pi}$ and D (as defined above) satisfy

$$\hat{\pi} = \arg\min_{\pi \in D} \frac{1}{2}\pi^T \mathrm{P}\pi - p^T \pi$$

and

$$D = \left\{\pi \in \mathbb{R}^2 : A^T \pi \geq b_1, a^T \pi = b_2\right\},$$

Table 8.2 Mean subscale scores in motivation variables for four fictitious subjects

Y	X_1	X_2
2.35	1.25	2.75
3.00	2.85	3.25
3.75	3.65	3.75
2.25	4.00	1.50

where for the motivation data, column vectors, Y, X_1, and X_2, I define $P = (X_1, X_2)^T(X_1, X_2)$, $p = (X_1, X_2)^T Y$, $A = I$ (identity matrix), $b_1 = (0, 0)^T$, $a = (1, 1)^T$, and $b_2 = 1$. Given this quadratic program formulation of the optimization problem in SDT variables, I can compute the solution by a standard numerical algorithm (e.g., Goldfarb and Idnani 1983). This is implemented in the R package SDT, with the function internalization of the package.

I illustrate the method by a toy numerical example. The data are fictitious and only intended to exemplify the computation steps. Let Y be one of identified or introjected regulation, X_1 and X_2 intrinsic regulation and external regulation, respectively. In Table 8.2, consider the aggregate subscale scores of four students, assumed to be averaged over all test items of a subscale of the test.

The Y values were made up from arbitrarily slightly changing any of the four values computed for the convex combination of X_1 and X_2 with $\pi_1 = 0.315$ and $\pi_2 = 0.685$. Thus, this choice of weights gives a good approximation of Y, with corresponding (rounded) values 2.28, 3.12, 3.72, and 2.29. In a sense, these values for π_1 and π_2 are the underlying parameters. I therefore expect the quadratic program minimizer to be close to these weights.

I have to find that complementary shares π_1 and π_2 in [0, 1] that minimize

$$(2.35 - 1.25\pi_1 - 2.75\pi_2)^2 + (3.00 - 2.85\pi_1 - 3.25\pi_2)^2$$
$$+ (3.75 - 3.65\pi_1 - 3.75\pi_2)^2 + (2.25 - 4.00\pi_1 - 1.50\pi_2)^2.$$

The P and p components of the quadratic program are given by

$$P = \begin{bmatrix} 1.25 & 2.85 & 3.65 & 4.00 \\ 2.75 & 3.25 & 3.75 & 1.50 \end{bmatrix} \begin{bmatrix} 1.25 & 2.75 \\ 2.85 & 3.25 \\ 3.65 & 3.75 \\ 4.00 & 1.50 \end{bmatrix}$$

$$= \begin{bmatrix} 39.0075 & 32.3875 \\ 32.3875 & 34.4375 \end{bmatrix}$$

and

$$p = \begin{bmatrix} 1.25 & 2.85 & 3.65 & 4.00 \\ 2.75 & 3.25 & 3.75 & 1.50 \end{bmatrix} \begin{bmatrix} 2.35 \\ 3.00 \\ 3.75 \\ 2.25 \end{bmatrix}$$

$$= \begin{bmatrix} 34.175 \\ 33.650 \end{bmatrix}.$$

Thus, I have to minimize

$$\frac{1}{2} \begin{bmatrix} \pi_1 \\ \pi_2 \end{bmatrix}^T \begin{bmatrix} 39.0075 & 32.3875 \\ 32.3875 & 34.4375 \end{bmatrix} \begin{bmatrix} \pi_1 \\ \pi_2 \end{bmatrix} - \begin{bmatrix} 34.175 \\ 33.650 \end{bmatrix}^T \begin{bmatrix} \pi_1 \\ \pi_2 \end{bmatrix}$$

subject to

$$\begin{bmatrix} 1 & 0 \\ 0 & 1 \end{bmatrix}^T \begin{bmatrix} \pi_1 \\ \pi_2 \end{bmatrix} \geq \begin{bmatrix} 0 \\ 0 \end{bmatrix},$$

and

$$\begin{bmatrix} 1 \\ 1 \end{bmatrix}^T \begin{bmatrix} \pi_1 \\ \pi_2 \end{bmatrix} = 1.$$

Running the function internalization of the package SDT on the motivation variables Y, X_1, and X_2 of the numerical example yields:

```
R> # subscale score vectors
R> Y <- c(2.35, 3.00, 3.75, 2.25) # intermediate regulation
R> X1 <- c(1.25, 2.85, 3.65, 4.00) # intrinsic regulation
R> X2 <- c(2.75, 3.25, 3.75, 1.50) # external regulation
R> # quadratic program minimizer
R> internalization(Y, X1, X2)
internal share external share
   0.2970012      0.7029988
```

Thus, the estimated regression weights are $\hat{\pi} = \left(\widehat{\pi_1} = 0.30, \widehat{\pi_2} = 0.70\right)^T$.

A plot of the data (bold line), fitted regression model (red triangle, $\widehat{\pi_1}$), and of two models for other choices of the regression weights (circle, $\pi_1 = 0.10$; square,

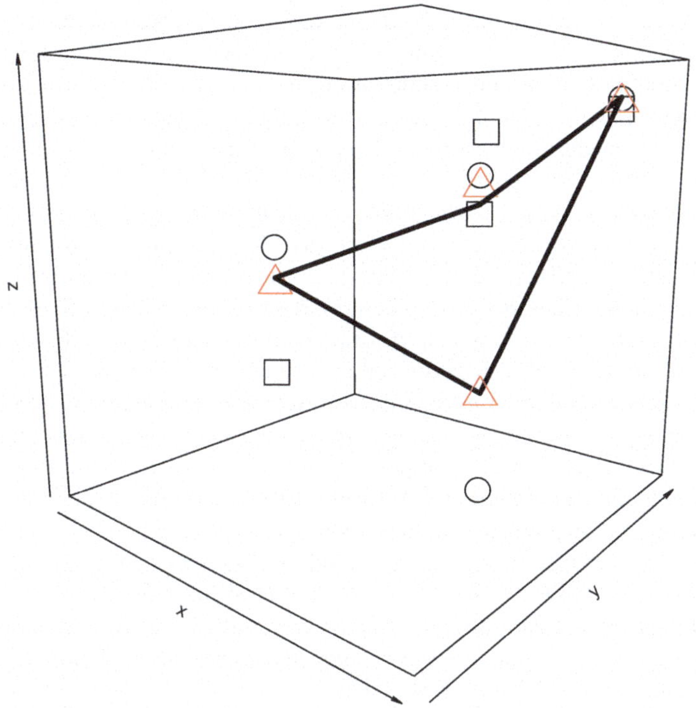

Fig. 8.2 Numerical example

$\pi_1 = 0.80$) is shown in Fig. 8.2. The axes x, y, and z stand for intrinsic, external, and intermediate regulations, respectively.

Figure 8.2 shows a three-dimensional plot of the observed motivation scores (bold line) and predicted values under three distinct models, the fitted model by technique (red triangle) and two models entailing convex combinations based on suboptimal coefficients (circle and square). The accurate fit of the solution can be clearly seen, since the red triangle points $\left(X_1, X_2, \hat{Y}\right)$, with predictions $\hat{Y}$, scatter in close vicinity of the net spanned by the data points (X_1, X_2, Y).

8.4 Simplex Structure Computation

The model in Sect. 8.3 for two predictors can be extended to more than two variables. The previous discussion translates straightforwardly (Ünlü 2019). The model is Y $= \pi_1 X_1 + \ldots + \pi_k X_k$, with integer $k \geq 2$, $\pi_1, \ldots, \pi_k \geq 0$, and $\pi_1 + \ldots + \pi_k = 1$. For $k = 2$, the prior model is a special case. This is a constrained regression model, which entails the same quadratic program, with respectively defined higher dimensional components. For example, $P = (X_1, \ldots, X_k)^T (X_1, \ldots, X_k)$ and $p =$

$(X_1, \ldots, X_k)^T Y$. I call any of the predictor variables X_i a base element and the set of all base elements $\{X_1, \ldots, X_k\}$ the reference system. Thus, the model posits that the target variable Y is decomposable into (jointly) convex components of a given reference system, with each of the regression weights of that convex decomposition representing the share of the target variable with respect to the corresponding base element.

The choice of a reference system of base motivation variables, according to which to convexly decompose a target motivation variable, is useful for the analysis of the simplex structure of SDT, see also Sect. 8.2. The simplex structure of motivation is an implicit assumption, deemed to be important for the plausibility of the theory. Thus, it is desirable to develop quantification methods for it. I describe one possible alternative, to study simplex structure computationally based on optimal shares of the convex decomposition model, compared to correlations among motivation variables, the common approach in SDT. As a descriptive data analysis technique, the method assumes that with more of the computed shares in accordance with theory, where this (essentially) means larger shares for theoretically closer regulation types, the evidence base for the simplex structure is stronger.

I define linear motivational structure, including closeness, mathematically. Denote by $ExR, IjR_1, \ldots, IjR_k, IdR_1, \ldots, IdR_l$, and InR the external, introjected, identified, and intrinsic regulation types, respectively, with different internalizations of introjected regulation (k gradations or classes) and identified regulation (l gradations or classes), the set of motivations M. I define the cumulative internalities along the SDT motivation continuum to be the linear order or chain $\sqsubseteq$ among the regulations,

$$ExR \sqsubseteq IjR_k \sqsubseteq \ldots \sqsubseteq IjR_1 \sqsubseteq IdR_1 \sqsubseteq \ldots \sqsubseteq IdR_l \sqsubseteq InR,$$

including transitive relation pairs such as $ExR \sqsubseteq IdR_1$, $ExR \sqsubseteq InR$, and $IjR_1 \sqsubseteq IdR_l$. For (theoretical) closeness, I take the covering relation or Hasse diagram of the linear order $\sqsubseteq$, viewed as an undirected graph. Then, essentially, I count the number of steps needed to go from one regulation to another along the chain, from left to right or right to left. This defines a distance d, which is the length of the graph geodesic, the geodesic distance, between any two regulations, as measure of their closeness in internalities. I call $(M, \sqsubseteq, d)$ a linear motivational structure. For details about orders, graphs, and distance, see, for example, Davey and Priestley (2002), Diestel (2017), and Rudin (1976), respectively.

In that model $(M, \sqsubseteq, d)$, for example, the distance between external regulation and intrinsic regulation is the largest possible distance $d(ExR, InR) = k + l + 1$, that is, the poles of the motivation continuum are least close to one another. In d, if $k > 1$, I would have at least one gradation or class of introjected regulation (always IjR_k) that is closer to external regulation than to any of the gradations of identified regulation. Similarly, if $l > 1$, I would have at least one class of identified regulation (always IdR_l) that is closer to intrinsic regulation than to any of the classes of introjected regulation. Also, if this representation is adequate, in monotonically decreasing values of the geodesic distance d, external regulation and intrinsic regulation are increasingly closer to $InR, IdR_l, \ldots, IdR_1, IjR_1, \ldots, IjR_k$ and

$ExR, IjR_k, \ldots, IjR_1, IdR_1, \ldots, IdR_l$, respectively. If $k = l = 1$, this would model the situation where the intermediate regulations are not further differentiated into gradual internalizations, each is represented by only one class. If these two are generic classes containing all possible forms of introjected regulation and identified regulation, respectively, those regulations may or may not be closer to the poles of the continuum than to each other, in any combination, depending on how much internal or external they are. Thus, it seems plausible to exclude that IjR is closer to ExR or IdR and that IdR is closer to InR or IjR. This is the case iff (if and only if) IjR and ExR, IjR and IdR, and IdR and InR are equally close. In the closeness measure d of the linear motivational structure, these equalities are reflected, $d(IjR, ExR) = d(IjR, IdR) = d(IdR, InR) = 1$. Thus, this model can be adequate for the representation of the situation in which I do not have specific knowledge about the intermediate regulations but treat them as single generic classes comprised of all possible forms.

I can have a mathematical definition of simplex structure. I call the linear motivational structure $(M, \sqsubseteq, d)$ a simplex structure iff for any convex decomposition

$$Y = \pi X_\pi + \mu X_\mu + R \tag{8.1}$$

of a target variable $Y \in M$ into predictors $X_\pi, X_\mu \in M$ and possibly other predictors of M subsumed in R, I have

$$\pi < \mu \text{ if } d(Y, X_\pi) > d\big(Y, X_\mu\big) \tag{8.2}$$

and

$$\pi = \mu \text{ if } d(Y, X_\pi) = d\big(Y, X_\mu\big). \tag{8.3}$$

For M, any (unordered) pair (multiset) of shares π and μ in [0, 1] with the property (8.1) is called a simplex structure pair, a pair of within-decomposition shares. For $(M, \sqsubseteq, d)$, a simplex structure pair of shares π and μ is called in accordance with theory iff π and μ satisfy the property (8.2) and property (8.3). Here, "theory" is represented by $(M, \sqsubseteq, d)$. (Then, I also say that the shares π and μ of the simplex structure pair are in accordance with theory). Obviously, $(M, \sqsubseteq, d)$ is a simplex structure iff all simplex structure pairs are in accordance with theory.

I give an example. Take as target variable external regulation $Y = ExR$ and convexly decompose it in the reference system of intrinsic regulation $X_1 = InR$, identified regulation $X_2 = IdR$, and introjected regulation $X_3 = IjR$ as the base elements. For the corresponding shares π_1, π_2, and π_3 to be (pairwise) in accordance with theory, as necessary conditions for the simplex structure, I must have $\pi_1 < \pi_2 < \pi_3$; under theory $(M, \sqsubseteq, d)$, it holds $d(ExR, InR) = 3$, $d(ExR, IdR) = 2$, and $d(ExR, IjR) = 1$. By interpretation, one may expect that π_1, the load of (fully external) external regulation on (fully internal) intrinsic regulation, should be essentially zero, or at least, relatively small. This is the case in concrete data. I can estimate these shares π_1, π_2, and π_3 in the accompanying empirical data set of the R package SDT, using the function simplex of the package.

```
R> # data set on learning motivation
R> attach(learning_motivation)
R> learning_motivation[1:3, 3:6]
    intrinsic   identified   introjected    external
1        4.0          4.5         3.75         3.75
2        1.8          2.5         2.00         3.00
3        3.0          3.0         3.00         2.25
R> # simplex structure analysis
R> simplex(external, intrinsic, identified, introjected)
base_regulation_1 share base_regulation_2 share base_
regulation_3 share
      0.0000000                 0.3388902                0.6611098
```

In shares, in accordance with theory, extrinsic regulation is most interrelated with introjected regulation $\pi_3 = 0.6611098$, followed by interrelation with identified regulation $\pi_2 = 0.3388902$, and with intrinsic regulation $\pi_1 = 0.0000000$. The computed shares are larger for theoretically closer regulation types. According to the method, descriptively, this provides some evidence for the simplex structure.

The method can be used for the exploratory inspection of the simplex structure. I outline how. Given four variables (the common use case), with one target variable and two or three base elements, I have in total sixteen convex decompositions among the variables, four decompositions each with three predictors and twelve decompositions with two predictors. Each three-predictor decomposition has three computable shares, a total of twelve. Any two-predictor decomposition entails two computable shares, a total of twenty-four. Overall, thirty-six shares are available. This number is relatively large, but to keep things interpretable and simple, the reader should note that only the shares computed under the same decomposition, that is, two or three shares each, depending on the number of predictors, will be compared pairwise, to inspect simplex structure. How many of these pairwise comparisons do I have, for the four basic variables? For each three-predictor decomposition, I have three pairs of shares for comparisons, so twelve pairs altogether. For any two-predictor decomposition, there is only one pair of shares, again twelve comparison pairs in total. Overall, I have twenty-four such admissible pairs, which are the simplex structure pairs. Depending on the envisaged granularity of the analysis, I can choose among the simplex structure pairs and check whether the two shares of any of those pairs are in accordance with theory.

The simplex structure definition is mathematical, in population quantities, and may be violated in data. Thus, I allow for deviations of the requirements for a simplex structure and define a violation. Let $(M, \sqsubseteq, d)$, in k, l, denote a linear motivational structure. For any simplex structure pair in $(M, \sqsubseteq, d)$, if its two shares do not order in accordance with theory, that is, if they do not satisfy either property (8.2) or property (8.3), I call that pair a violation of the simplex structure (definition, property, or assumption). In particular, for the four motivations with no gradations of intermediate regulations ($k = l = 1$), a maximum of twenty-four violations are possible. In data, the typical procedure, I substitute population share parameters with sample share estimates, which are computed quadratic program minimizers. In the estimates, violations can occur. Depending on how many violations occur, I may define a measure of the plausibility of the simplex structure in the data. Descriptively, the following heuristic is reasonable. The more simplex structure violations appear, the more implausible this assumption may be deemed. Stated differently, the smaller the count of the observed violations of the simplex structure definition in a data set is, the stronger the assumption of the simplex structure is corroborated in this data set. Mathematically though, $(M, \sqsubseteq, d)$ cannot be a simplex structure, if I allow for one or more violations. But this is in order. Acceptably small numbers of violations are necessary in real data applications.

I apply the concepts of the preceding paragraphs to empirical data by Müller et al. (2007). The data comprise the learning motivation subscale scores of Austrian pupils in different school class subject areas. If need may be, the reader can reproduce the following plots using the related empirical data set of the package SDT. Figures 8.3 and 8.4 visualize the shares computed in the empirical data for autonomous and non-autonomous target variables, respectively. Two-predictor decomposition shares and three-predictor decomposition shares were computed with the functions internalization and simplex of the package SDT in R, respectively.

Figure 8.3 depicts the shares for the autonomous motivation variables intrinsic regulation InR and identified regulation IdR as the target variables of the convex decompositions represented by the numbered lines, in black and red colors, respectively. There are two three-predictor decompositions, the lines numbered 4, and six two-predictor decompositions, the lines numbered 1, 2, and 3, always with the autonomous variables as the target variables. The discrete x-axis has only three positions, represented by the vertically stacked numbers. These positions (vertical layers) stand for different motivation regulation types, depending on the target variable. For intrinsic regulation as target variable (all black lines), the positions 1, 2, and 3, from left to right, are identified regulation, introjected regulation, and external regulation. For identified regulation as target variable (all red lines), the positions are intrinsic regulation, introjected regulation, and external regulation, respectively. The numbers are points representing these position specific regulations and their shares computed under the corresponding decompositions. For example, the (ordinate of) point 4, in red at mid position, is the share of identified regulation (target variable) with introjected

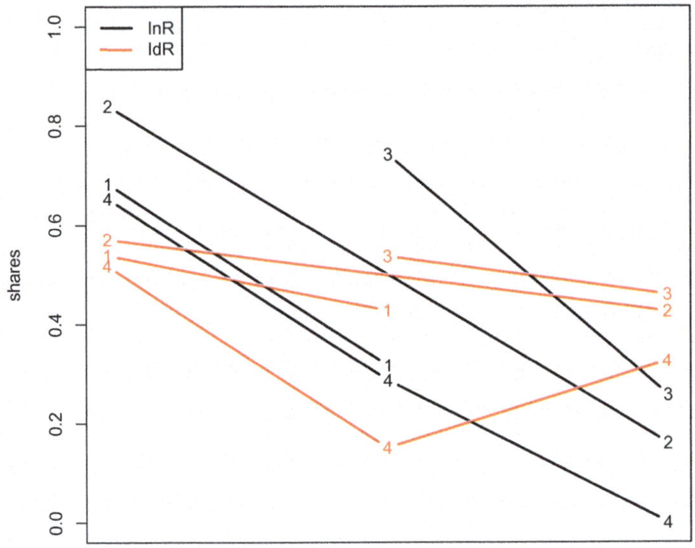

Fig. 8.3 Decomposition and autonomous motivation

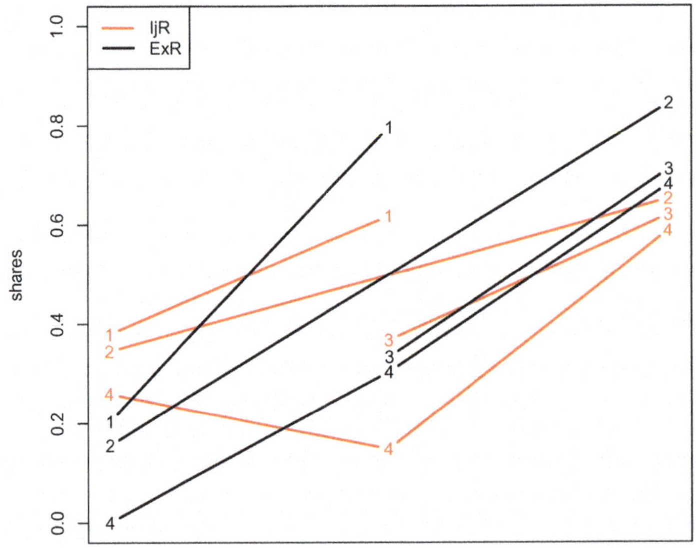

Fig. 8.4 Decomposition and non-autonomous motivation

regulation (one of the three predictors) of the decomposition into intrinsic regulation, introjected regulation, and external regulation (reference system). Or the point 1, in black at first position, depicts the share of intrinsic regulation with identified regulation of the decomposition into identified regulation and introjected regulation.

Yet another example, the point 2, in black at third position, is the share of intrinsic regulation with respect to external regulation in the reference system of identified regulation and external regulation.

In Fig. 8.3, for pole intrinsic regulation (black lines), all numbered lines are (strictly) monotonically decreasing (from left to right). This implies that all simplex structure pairs of points of those (black) lines are in accordance with theory, in $(M, \sqsubseteq, d)$. Why? Under that model, let $Y = InR$, $X_1 = IdR_m$, $X_2 = IjR_n$, and $X_3 = ExR$ for $1 \leq m \leq l$ and $1 \leq n \leq k$ be the observed intrinsic, identified, introjected, and external motivation variables of the data, respectively. Then, $d(Y, X_1) = m < d(Y, X_2) = l + n < d(Y, X_3) = l + k + 1$. Let π_1, π_2, and π_3 be placeholders for the two or three shares of each of the lines at positions 1, 2, and 3, respectively, if applicable. Then, a simplex structure pair is in accordance with theory iff $\pi_1 > \pi_2 > \pi_3$, if applicable. For example, the line numbered 1 has shares $\pi_1 > \pi_2$ (π_3 not applicable), and the line numbered 2 has shares $\pi_1 > \pi_3$ (π_2 not applicable). This answers the question why. In particular, there is no violation of the simplex structure definition among the simplex structure pairs for intrinsic regulation as the target variable. Overall, in the pole variable, the results are perfectly supportive of a simplex structure.

In intermediate identified regulation (red lines), the picture is not anymore only positive. I see a violation of the simplex structure definition, the pair of the shares of identified regulation with introjected regulation (4 at mid position) and identified regulation with external regulation (4 at third position) of the three-predictor decomposition. In respective order, these shares π_2 and π_3 at positions 2 and 3 are not in accordance with theory. I have $d(IdR_m, IjR_n) < d(IdR_m, ExR)$, but (considerably) $\pi_2 < \pi_3$ (not $\pi_2 > \pi_3$). This pair $\{\pi_2, \pi_3\}$ is the only violation of the simplex structure property in the plot. Thus, in Fig. 8.3, the count of violating simplex structure pairs is $v_a = 1$. The relative frequency as a frugal measure for the plausibility of a simplex structure is $s_a = v_a/12 = 0.083$, relative to autonomous motivation variables. Is intermediate identified regulation closer to contiguous intrinsic regulation or contiguous introjected regulation? Consider the two decompositions of identified regulation containing (both) the predictors intrinsic regulation and introjected regulation. Let π_1 and π_2 be the shares at positions 1 and 2 of the two-predictor decomposition (line numbered 1), and μ_1 and μ_2 the shares at positions 1 and 2 of the three-predictor decomposition (line numbered 4). The shares of identified regulation with respect to intrinsic regulation are larger than the corresponding shares with respect to introjected regulation. That is, $\pi_1 > \pi_2$ and $\mu_1 > \mu_2$. Thus, data-analytically (greater load), identified regulation seems to be closer to intrinsic regulation than to introjected regulation. I may also argue as follows. Under the assumption that simplex structure is valid in population shares in $(M, \sqsubseteq, d)$, if identified regulation is equally close or closer to introjected regulation $d(IdR_m, InR) \geq d(IdR_m, IjR_n)$, this implies that, in respective population shares, $\pi_1 \leq \pi_2$ and $\mu_1 \leq \mu_2$. By contraposition, $[\pi_1 > \pi_2$ or $\mu_1 > \mu_2]$ implies $d(IdR_m, InR) < d(IdR_m, IjR_n)$. If the estimates are sufficiently good proxies of the population parameters, under the assumption of the simplex structure, I can conclude from the premise of the above implication that identified regulation is theoretically, in population, closer to intrinsic regulation than to introjected regulation.

Figure 8.4 depicts the shares for the non-autonomous motivation variables introjected regulation IjR and external regulation ExR, the target variables of the decompositions represented by the red and black numbered lines, respectively. This plot is similarly read, see Fig. 8.3. For introjected regulation (red lines), the positions on the x-axis are intrinsic regulation, identified regulation, and external regulation. For external regulation (black lines), the positions are intrinsic regulation, identified regulation, and introjected regulation. For example, the point 3, in red at mid position, is the share of introjected regulation with identified regulation by decomposition into identified regulation and external regulation.

Figure 8.4 can be interpreted similarly. I only summarize. For the pole variable external regulation (black lines), all numbered lines are monotonically increasing. Thus, all simplex structure pairs of those lines are in accordance with theory, with no violations of simplex structure. In the other pole of external regulation also, the results are fully supportive of an underlying simplex structure among the motivation variables. In the intermediate variable introjected regulation (red lines), there is only one violation, which is the simplex structure pair of the shares at positions 1 and 2 of the three-predictor decomposition. The data suggest that intermediate introjected regulation compared to its direct neighbor regulations should be closer to external regulation than to identified regulation. Thus, I may assume that the two simplex structure pairs of the shares at positions 2 and 3 of the two-predictor decomposition and three-predictor decomposition are not violations. Relative to non-autonomous motivation variables, again, $v_{na} = 1$ and $s_{na} = v_{na}/12 = 0.083$. Pooled, $v = 2$ and $s = v/24 = 0.083$.

In sum, the findings corroborate the simplex structure assumption for the motivation variables of this empirical application. What is peculiar though, is the observation that the two violations of the simplex structure assumption in this example are dually expressible. They are symmetrical to each other. In-depth analyses of this and other applications and the method could study why.

8.5 A Systematic Empirical Application

The technique for internalization and simplex structure analyses was exemplified with an empirical study exhaustively in Ünlü and Dettweiler (2015, pp. 678–680, 682–686), in continuation of work by Dettweiler et al. (2015). The study compared science teaching in two different settings, a classical school class versus an expeditionary outdoor program. The empirical findings based on the discussed approach were interesting. Based on a sample of 84 German pupils, the authors found that the science teaching format did not influence the internalization of identified regulation. However, the internalization of introjected regulation differed between the two settings. It shifted more toward the external pole in the outdoor teaching format. In essence, the method of optimal shares could support the simplex structure in the

Table 8.3 Internalization of science learning motivation

Intermediate type	Extent of internalization	
	Internal share	External share
Identified regulation	0.7552667 0.7557941	0.2447333 0.2442059
Introjected regulation	0.2055134 0.1649183	0.7944866 0.8350817

This is Table 1 originally published in Ünlü and Dettweiler (2015, p. 683). Summarized are the findings of the empirical study on learning motivation in science about the degrees of internalization of identified regulation and introjected regulation in the two teaching settings of classical school class and expeditionary outdoor program, in first and second lines, respectively

study data on science learning motivation. For more details about this study, I refer the reader to the above-mentioned papers.

The results pertaining to internalization of the empirical study are reproduced in Table 8.3.

8.6 Discussion

A summary and further remarks and personal views are presented.

8.6.1 Summary

Self-determination theory (SDT) by Deci and Ryan (1985, 2000) is a popular motivation theory. Quantitative papers on SDT have reported mainly psychometric analyses, for example, factor analysis or structural equation modeling. SDT is a very practical theory, and thus, any mathematization of its concepts is an interesting and useful contribution, which adds to the quantitative literature on SDT. By providing the mathematical definitions of such important concepts as the simplex structure, this work is a step in that direction (see also Ünlü 2023a, b, c). A sound mathematical model, in population parameters, is a prerequisite for the development of corresponding statistical estimation or testing procedures, in sample estimates. With the exploratory data analysis in this chapter, I have tried to illustrate that the estimated shares, the computed quadratic program minimizers, may approximate well population shares, such that I could infer which motivations are likely theoretically closer to one another. More work in this regard is needed.

I summarize. In Sect. 8.1, I have reviewed SDT. In Sect. 8.2, I have motivated why the computational perspective on motivation, here SDT, is useful, by three or four exemplary important questions, which naturally arise from the formulation of SDT. In Sect. 8.3, to simplify presentation, I have discussed the simplest form of the convex decomposition model, in two predictor variables, and the corresponding quadratic program, for the computational analysis of internalization. A numerical example helped to illustrate the computation steps. In Sect. 8.4, I have considered the general form of the model and quadratic program. Importantly, I have defined linear motivational structure, including closeness as geodesic distance, mathematically. A linear motivational structure is a linearly ordered set of motivations, equipped with the canonical geodesic distance, making the linearly ordered motivation set also a metric space. This allowed to define simplex structure. Two example empirical data sets were analyzed for the simplex structure. Based on these examples, it seems that simplex structure is a relatively robust property, especially for poles of the motivation continuum as target variables. In contrast, intermediate regulations caused violations of the simplex structure and were less robust in this regard. Obviously so, for reasons that I outline in the next section, indeterminacy in closeness. In Sect. 8.5, I have only reviewed briefly the results of a systematic real application, which used the convex decomposition approach to investigate science learning motivation. Overall, the findings were promising and interpretable. That goes without saying, or probably saying it all too often, I used the R package SDT and reported code for computations.

The next section contains a few remarks, which are clarifications, suggestions, or simply ideas, that I find interesting. These are my personal views. For better orientation, the paragraphs are labeled. The paragraphs are, in their order of appearance, unfounded motivation aggregation, R implementation, indeterminacy in closeness, amotivation, and basic psychological needs.

8.6.2 Concluding Remarks

Unconfounded motivation aggregation. I briefly address question 3 of Sect. 8.2, related to the usefulness of the computational perspective on motivation. The convex decomposition technique has another computational application to SDT. The method can be used to adjust scoring protocols or rules for the unconfounded aggregation of motivation. This is a pertinent problem. Researchers in SDT are interested in scoring protocols, since such rules aggregate a person's individual subscale scores of a motivation test to yield an overall informative summary statistic, as a proxy for the self-determination inherent in the person. In SDT, the popular scoring protocol used for motivation aggregation is the relative autonomy index (RAI), also called the self-determination index (SDI). For example, see Grolnick and Ryan (1989), Ryan and Connell (1989), and Kusurkar et al. (2013). This index was criticized by other authors as well, Chemolli and Gagné (2014) and Wilson et al. (2012), and improved on by Ünlü (2016). Based on the quadratic program minimizers for the convex shares, an adjusted variant of the scoring rule RAI was derived. Why to correct the widely used RAI index? The general problem can be described as follows. If the inventory scores measured for the individual regulation types are aggregated in a scoring protocol formula, including regulations comprised of mixed internal motivation and external motivation, as it is the case with the RAI, that formula should contain separate, and of course proper, mathematical terms for the internal motivation and external motivation of an intermediate regulation. If necessary, this can include weighting the subscale scores according to the extents to which the intermediate regulations are internal and external, which is not the case with the RAI. Thus, this popular SDT index for quantifying self-determined behavior is a potentially confounded scoring protocol, which may be distorted or lack interpretability in some cases. However, see Ünlü (2016), improvements on it are possible, with an adjusted and properly weighted variant of the SDI. The weights used in the improved scoring protocol are the internal motivation and external motivation shares of any of the intermediate regulation types, numerically obtained by the described quadratic program. Thus, this alternative definition of the SDI is designed to computationally account for the mixture or confounding of internal motivation and external motivation amounts, inherent in the intermediate regulation types.

R implementation. In answer to question 4 of Sect. 8.2, I cannot emphasize enough the importance of the computing and graphics environment and language R (The R Core Team 2022) for the implementation and realization of the computational approach to SDT. R is the lingua franca, indispensable for statistics and data analysis, including applications in such disciplines as psychometrics, econometrics, chemometrics, or computational physics. The list is long. To have contributed the R package SDT, see Ünlü (2019), is only the first step in introducing and applying R to SDT. More work in this direction is required. I hope to have laid a basis for, or at least motivated, such a program. What is included in the package SDT? The purpose of the package, in its current version, is to provide the necessary functions that implement and realize the presented research program on the computational analysis of internalized and simplex-structured motivation. That is, it contains the functions for motivation internalization analysis, motivation simplex structure analysis, and for computing the confounded original and unconfounded adjusted self-determination scoring indices. An advantage of R is that in the ecosystem of contributed packages, you can find a plethora of methods that could be applied to motivation data, in addition to the functions of the package SDT (Ünlü and Yanagida 2011).

Indeterminacy in closeness. Without additional knowledge, in the general case, I cannot decide unequivocally whether a (two-sided) intermediate regulation is empirically closer to one or the other of the two contiguous regulations, in contrast to poles (below). That is, there is indeterminacy in closeness of intermediate regulations to their direct neighbor regulations. I guess necessarily, as the theory cannot be stationary in this regard, singling out a universal solution, due to the crucial assumption that intermediate regulations are defined to vary in their internalizations on the continuum, thus entailing, if not restricted by additional assumptions, variable locations of fairly arbitrary distances. For example, introjected regulation could be closer to identified regulation than to external regulation, or to external regulation than to identified regulation. Both situations may be empirically conceivable. That is, for intermediate regulations, I generally cannot know the empirically closer regulations. This cannot happen with the (one-sided) poles of the motivation continuum. For example, for external regulation, since it is the pole far left (or far right) of the chain of cumulative internalities, for any other two regulations, which may include the opposite pole, one of the two regulations is empirically closer to external regulation. This is also true in model, universally. Under *any* model $(M, \sqsubseteq, d)$, external regulation and intrinsic regulation are increasingly closer to $InR, IdR_l, \ldots, IdR_1, IjR_1, \ldots, IjR_k$ and $ExR, IjR_k, \ldots, IjR_1, IdR_1, \ldots, IdR_l$, respectively, with (strictly) monotonically decreasing geodesic distances in both cases. In particular, for poles, as compared to intermediate target variables, I can always check for accordance with theory of the computed shares without reliance on procedures to determine the theoretically closer regulations. The simple approach is to omit the indeterminate cases and only consider those where one can always know the empirically closer regulations (e.g., with poles as the target variables). Or, as I have suggested in this chapter, you can estimate that hidden information based on the computed shares of simplex structure pairs. To sum up, for the proposed technique of optimal shares, the theory of self-determination is universally unambiguous in poles, but not in intermediate regulations, in the above sense.

Amotivation. I give a remark about a more general interpretation of amotivation. In SDT (Table 8.1), intrinsic motivation and extrinsic motivation are assumed to be the only possible forms of motivation, and amotivation is meant to stand for no motivation. Thus, amotivation, as the state of absence of motivation of any (describable and in this case intrinsic and extrinsic) form, can be mathematized to be complementary to intrinsic motivation and extrinsic motivation. This complementarity of types generalizes as follows. Amotivation could comprise additional, more subtle motivations, or motivation cognate variables, that are different from the postulated two main types. That is, assume that intrinsic motivation and extrinsic motivation may not be the only true motivations possible, that there may be one or more other existent and scalable, or even existent but unscalable, forms of motivation that accrue in amotivation and that I have not discriminated so far, or basically cannot, respectively, or that there may be spurious forms of alleged motivation that were artificially measured inflating amotivation. Thus, I could interpret amotivation, especially if measured, to be the aggregate of all of those existent or spurious, scalable or unscalable, true or alleged motivations, other than the primary two intrinsic and extrinsic motivation types. In a sense, amotivation could be theorized to represent the type of all the unaccounted, motivation or motivation related, residuals, which in their totality may be assumed to be complementary to the essential types of intrinsic motivation and extrinsic motivation. In this case, I would assume and account for heterogeneity in amotivation, which seems to be supported empirically (e.g., Pelletier et al. 1999, 2013). The use or further differentiation of an auxiliary type of this sort, either extending or in addition to amotivation, could allow for a generalized motivation theory of also behaviors that may not be classified, or unambiguously categorized, into one of the common two or three motivation types of intrinsic motivation, extrinsic motivation, or amotivation. This could be useful for analyses of motivation in abnormal behaviors such as that of paranoid schizophrenia (cf., Gard et al. 2014; Vancampfort et al. 2013), or perhaps, if feasible, in extreme behaviors that, as possibly real events, seem to occur so randomly or infrequent that they cannot be adequately replicated and studied based on the commonly agreed scientific approach to experiments (e.g., parapsychology, Hyman 1986; Irwin and Watt 2007).

Basic psychological needs. SDT posits that satisfaction of the basic psychological needs for competence, autonomy, and relatedness facilitates internalized or intrinsic motivation. In SDT, this postulate is fundamental and has a law like character, comparable to physics. In analogy, for example within classical mechanics, force F derives on momentum p. The former is the time derivative of the latter, $F = f(p) = p'$. Similarly, typically in latencies, intrinsic motivation I may be specific, linear or nonlinear, mathematical functions f of the basic psychological needs B, $I = f(B)$, which would conform to the many empirical studies reported in SDT. If for a basic psychological need, say competence C, that true functional relationship f between intrinsic motivation I and C exists and is invertible with inverse f^{-1}, each of intrinsic motivation and competence could be uniquely obtained from knowledge of the other, $I = f(C)$ and $C = f^{-1}(I)$, respectively. In addition to knowledge about f and f^{-1}, if I could assume that competence is a different manifestation of intrinsic motivation and intrinsic motivation is a specific manifestation of competence, intrinsic motivation and competence could be interpreted to be equivalent, that is, different but interchangeable manifestations of a same abstract entity, analogous to the equivalence of energy and mass in special relativity (Born 2003), where energy E is a linear function f of mass m with proportionality constant the square of the speed of light c, $E = f(m) = c^2 m$ and $m = f^{-1}(E) = c^{-2}E$. Such an endeavor would most probably necessitate the development of far more advanced physical or technical devices and experiments for the measurement of mental or psychological properties. I believe that properties of the mind and psyche physically can exist, we seem to be lacking the instruments to measure them at a physical level, but I doubt their discoverability can be possible only through the current approach primarily based on questionnaires or psychological tests. This will not be realizable in the near future, most probably. Only time can tell.

Acknowledgment Bu çalışmamı sevdiğim büyükannem Hanım Ünlü ve sevgili amcam İmdat Ünlü'ye ithaf ediyorum. Desteğini esirgemeyen değerli arkadaşım Dr. med. Erol Koçdemir'e bir kez daha en içten şükranlarımı sunuyorum.

References

Bellg AJ (2003) Maintenance of health behavior change in preventive cardiology: internalization and self-regulation of new behaviors. Behav Modif 27:103–131

Block R, Niemiec C, Rigby S, Williams GC (2016) Virtual clinician need support, health-behavior change, and its maintenance. J Clin Lipidol 10:667

Born M (2003) Die Relativitätstheorie Einsteins [Einstein's theory of relativity]. Springer, Berlin

Boyd S, Vandenberghe L (2009) Convex optimization. Cambridge University Press, New York, NY

Chemolli E, Gagné M (2014) Evidence against the continuum structure underlying motivation measures derived from self-determination theory. Psychol Assess 26:575–585

Conesa PJ, Onandia-Hinchado I, Duñabeitia JA, Moreno MÁ (2022) Basic psychological needs in the classroom: a literature review in elementary and middle school students. Learn Motiv 79:101819

Davey BA, Priestley HA (2002) Introduction to lattices and order. Cambridge University Press, Cambridge

Deci EL (1975) Intrinsic motivation. Plenum, New York, NY

Deci EL, Ryan RM (1985) Intrinsic motivation and self-determination in human behavior. Plenum, New York, NY

Deci EL, Ryan RM (2000) The "what" and "why" of goal pursuits: human needs and the self-determination of behavior. Psychol Inq 11:227–268

Deci EL, Ryan RM (2008) Self-determination theory: a macrotheory of human motivation, development, and health. Can Psychol 49:182–185

Dettweiler U, Ünlü A, Lauterbach G, Becker C, Gschrey B (2015) Investigating the motivational behavior of pupils during outdoor science teaching within self-determination theory. Front Educ Psychol 6(125):1–16

Diestel R (2017) Graph theory. Springer, Berlin

Gagné M, Parker SK, Griffin MA, Dunlop PD, Knight C, Klonek FE, Parent-Rocheleau X (2022) Understanding and shaping the future of work with self-determination theory. Nat Rev Psychol

Gard DE, Sanchez AH, Starr J, Cooper S, Fisher M, Rowlands A, Vinogradov S (2014) Using self-determination theory to understand motivation deficits in schizophrenia: the 'why' of motivated behavior. Schizophr Res 156:217–222

Gillison FB, Rouse P, Standage M, Sebire SJ, Ryan RM (2019) A meta-analysis of techniques to promote motivation for health behaviour change from a self-determination theory perspective. Health Psychol Rev 13:110–130

Goldfarb D, Idnani A (1983) A numerically stable dual method for solving strictly convex quadratic programs. Math Program 27:1–33

Gorin AA, Powers TA, Koestner R, Wing RR, Raynor HA (2014) Autonomy support, self-regulation, and weight loss. Health Psychol 33:332–339

Grolnick WS, Ryan RM (1989) Parent styles associated with children's self-regulation and competence in school. J Educ Psychol 81:143–154

Halvari AEM, Halvari H, Williams GC, Deci EL (2017) Predicting dental attendance from dental hygienists' autonomy support and patients' autonomous motivation: a randomised clinical trial. Psychol Health 32:127–144

Hyman R (1986) Parapsychological research: a tutorial review and critical appraisal. Proc IEEE 74:823–849

Irwin HJ, Watt CA (2007) An introduction to parapsychology. McFarland & Company, Jefferson, NC

Keenan J, Rahman R, Hudson J (2021) Exploring the acceptance of telehealth within palliative care: a self-determination theory perspective. Health Technol 11:575–584

Kusurkar RA, Croiset G (2015) Autonomy support for autonomous motivation in medical education. Med Educ Online 20:27951

Kusurkar RA, Croiset G, Mann KV, Custers E (2012) Have motivation theories guided the development and reform of medical education curricula? A review of the literature. Acad Med 87:735–743

Kusurkar RA, Ten Cate TJ, Vos CMP, Westers P, Croiset G (2013) How motivation affects academic performance: a structural equation modelling analysis. Adv Health Sci Educ 18:57–69

Morbée S, Vermote B, Waterschoot J, Dieleman L, Soenens B, Van den Bergh O, Ryan RM, Vanhalst J, De Muynck G-J, Vansteenkiste M (2021) Adherence to COVID-19 measures: the critical role of autonomous motivation on a short- and long-term basis. Motiv Sci 7:487–496

Müller FH, Hanfstingl B, Andreitz I (2007) Skalen zur motivationalen Regulation beim Lernen von Schülerinnen und Schülern: adaptierte und ergänzte Version des Academic Self-Regulation

Questionnaire (SRQ-A) nach Ryan & Connell [Scales of motivational regulation for student learning: adapted and supplemented version of the academic self-regulation questionnaire (SRQ-A) by Ryan & Connell]. In: Institut für Unterrichts- und Schulentwicklung (ed) Wissenschaftliche Beiträge [Scientific contributions]. Alpen-Adria-Universität, Klagenfurt, Austria, pp 1–17

Nelson LE, Wilton L, Whitfield DL, Williams GC, Mayer KH, Komárek A, Boyd DT, Beauchamp G, Fields SD, Wheeler DP (2022) Client-centered care coordination (C4™) for HIV/STI prevention: a theoretical, conceptual, and methodological overview—HIV prevention trials network (HPTN) 073. Sex Res Soc Policy

Ng JYY, Ntoumanis N, Thøgersen-Ntoumani C, Deci EL, Ryan RM, Duda JL, Williams GC (2012) Self-determination theory applied to health contexts: a meta-analysis. Perspect Psychol Sci 7:325–340

Nicholas JC, Ntoumanis N, Smith BJ, Quested E, Stamatakis E, Thøgersen-Ntoumani C (2021) Development and feasibility of a mobile phone application designed to support physically inactive employees to increase walking. BMC Med Inform Decis Mak 21:23

Ntoumanis N, Ng JYY, Prestwich A, Quested E, Hancox JE, Thøgersen-Ntoumani C, Deci EL, Ryan RM, Lonsdale C, Williams GC (2021) A meta-analysis of self-determination theory-informed intervention studies in the health domain: effects on motivation, health behavior, physical, and psychological health. Health Psychol Rev 15:214–244

Orsini C, Binnie VI, Wilson SL (2016) Determinants and outcomes of motivation in health professions education: a systematic review based on self-determination theory. J Educ Eval Health Prof 13:19

Pelletier LG, Dion S, Tuson K, Green-Demers I (1999) Why do people fail to adopt environmental protective behaviors? Toward a taxonomy of environmental amotivation. J Appl Soc Psychol 29:2481–2504

Pelletier LG, Rocchi MA, Vallerand RJ, Deci EL, Ryan RM (2013) Validation of the revised sport motivation scale (SMS-II). Psychol Sport Exerc 14:329–341

Rudin W (1976) Principles of mathematical analysis. McGraw-Hill, New York, NY

Ryan RM (ed) (2019) The Oxford handbook of human motivation. Oxford University Press, New York, NY

Ryan RM, Connell JP (1989) Perceived locus of causality and internalization: examining reasons for acting in two domains. J Pers Soc Psychol 57:749–761

Ryan RM, Deci EL (2000) Self-determination theory and the facilitation of intrinsic motivation, social development, and well-being. Am Psychol 55:68–78

Ryan RM, Deci EL (2002) Overview of self-determination theory: an organismic dialectical perspective. In: Deci EL, Ryan RM (eds) Handbook of self-determination research. University of Rochester Press, Rochester, NY, pp 3–33

Ryan RM, Deci EL (2017) Self-determination theory: basic psychological needs in motivation, development, and wellness. Guilford Publishing, New York, NY

Ryan RM, Deci EL (2020) Intrinsic and extrinsic motivation from a self-determination theory perspective: definitions, theory, practices, and future directions. Contemp Educ Psychol 61:101860

Ryan RM, Patrick H, Deci EL, Williams GC (2008) Facilitating health behavior change and its maintenance: interventions based on self-determination theory. Eur Health Psychol 10:2–5

Silva MN, Marques MM, Teixeira PJ (2014) Testing theory in practice: the example of self-determination theory-based interventions. Eur Health Psychol 16:171–180

Smit ES, Dima AL, Immerzeel SAM, Van den Putte B, Williams GC (2017) The virtual care climate questionnaire: development and validation of a questionnaire measuring perceived support for autonomy in a virtual care setting. J Med Internet Res 19:e155

Teixeira PJ, Marques MM, Silva MN, Brunet J, Duda JL, Haerens L, La Guardia J, Lindwall M, Lonsdale C, Markland D, Michie S, Moller AC, Ntoumanis N, Patrick H, Reeve J, Ryan RM, Sebire SJ, Standage M, Vansteenkiste M, Weinstein M, Weman-Josefsson N, Williams

K, Hagger MS (2020) A classification of motivation and behavior change techniques used in self-determination theory-based interventions in health contexts. Motiv Sci 6:438–455
The R Core Team (2022) R: a language and environment for statistical computing. R Foundation for Statistical Computing, Vienna, Austria
Ünlü A (2016) Adjusting potentially confounded scoring protocols for motivation aggregation in organismic integration theory: an exemplification with the relative autonomy or self-determination index. Front Educ Psychol 7(272):1–4
Ünlü A (2019) Self-determination motivation theory in R: the software package SDT. Arch Psychol 3:1–19
Ünlü A (2023a) Qualitative motivation with sets and relations. Front Psychol 13:993660
Ünlü A (2023b) Mathematical self-determination theory I: real representation. J Math Psychol 116:102792
Ünlü A (2023c) Mathematical self-determination theory II: affine space representation. J Math Psychol 116:102793
Ünlü A, Dettweiler U (2015) Motivation internalization and simplex structure in self-determination theory. Psychol Rep 117:675–691
Ünlü A, Yanagida T (2011) R you ready for R?: the CRAN psychometrics task view. Br J Math Stat Psychol 64:182–186
Vancampfort D, De Hert M, Vansteenkiste M, De Hert A, Scheewe TW, Soundy A, Stubbs B, Probst M (2013) The importance of self-determined motivation towards physical activity in patients with schizophrenia. Psychiatry Res 210:812–818
Vlachopoulos SP, Letsiou M, Palaiologou A, Leptokaridou ET, Gigoudi MA (2010) Assessing multidimensional exercise amotivation among adults and older individuals: the amotivation toward exercise scale–2. Eur J Psychol Assess 26:248–255
Williams GC, Lynch M, Glasgow RE (2007) Computer-assisted intervention improves patient-centered diabetes care by increasing autonomy support. Health Psychol 26:728–734
Wilson PM, Sabiston CM, Mack DE, Blanchard CM (2012) On the nature and function of scoring protocols used in exercise motivation research: an empirical study of the behavioral regulation in exercise questionnaire. Psychol Sport Exerc 13:614–622
Wuyts D, Van Hecke A, Lemaire V, Vandepoel I, Duprez V (2021) Development and validation of INTENSS, a need-supportive training for nurses to support patients' self-management. Nurse Educ Today

Ali Ünlü is interested in mathematical and statistical methods in the modeling, measurement, and visualization of information in the quantitative behavioral sciences. His research mainly focuses on latent variable and knowledge structure models. Having earned a first degree in mathematics, he completed doctoral studies in psychometrics and mathematical psychology. Professor Ünlü worked as Visiting Professor and Junior Specialist. He assumed the position of Assistant Professor in computer-oriented statistics and data analysis and acquired postdoctoral teaching qualification known as habilitation. He became Associate Professor for statistical methods in the social and educational sciences. Professor Ünlü has been Endowed Chair and Full Professor of empirical methods in education research.

Chapter 9
Recent Developments of Multiple Myeloma: Analysis and Analytical Modeling Using Real Data

Chris P. Tsokos and Lohuwa Mamudu

Abstract Multiple myeloma cancer (MMC), also known as Kahler disease, myelomatosis, and plasma cell myeloma, is a devastating type of cancer that still remains incurable. In this chapter, we reviewed and identified some of the very recent research developments and findings from real data of patients diagnosed with MMC which we believe are significant to the subject matter of MMC survivorship. Our recent findings involved developing a parametric approach to survival analysis. We conducted parametric and nonparametric analyses of the survival times of patients diagnosed with MMC. We included in our analysis the applicability of the Cox-proportional hazard analytical model that is being driven by significant risk factors that significantly contribute to the survival of MMC and compared it with our parametric findings and other research findings on survival analysis of MMC. We concluded our studies with a very recent and effective analytical model that was developed using real data that very accurately predicts the survival time of MMC patients. This parametric analytical modeling approach is driven by the significant risk factors and interactions that are associated with the survival time of MMC patients. In addition, we identified new risk factors associated with the survival of MMC. Our recent findings about the survival time of MMC and the new approach to survival analysis provide a novel

---

This book chapter was written by quoting Lohuwa Mamudu's doctoral thesis titled "Driven Analytical Modeling of Multiple Myeloma Cancer, U.S. Crop Production and Monitoring Process".

Mamudu, Lohuwa, "Data-Driven Analytical Modeling of Multiple Myeloma Cancer, U.S. Crop Production and Monitoring Process" (2021). USF Tampa Graduate Theses and Dissertations. https://digitalcommons.usf.edu/etd/9698.

C. P. Tsokos
Distinguished University Professor Department of Mathematics and Statistics, University of South Florida, Tampa, FL, USA
e-mail: ctsokos@usf.edu

L. Mamudu (✉)
Department of Public Health, California State University, Fullerton, CA, USA
e-mail: lohuwam@fullerton.edu

E. Çetin and H. Özen (eds.), *Healthcare Policy, Innovation and Digitalization*, Accounting, Finance, Sustainability, Governance & Fraud: Theory and Application,
https://doi.org/10.1007/978-981-99-5964-8_9

strategy and ways to investigate the survival of MMC and other types of cancers. The findings from our studies can help improve the therapeutic/treatment strategy of MMC disease, hence, improvement in MMC survival.

Keywords Multiple myeloma cancer · Cox-PH model · Statistical modeling · Cancer survival analysis · Cancer therapeutic · Parametric and nonparametric probability estimation

9.1 Introduction

Multiple myeloma cancer (MMC), also known as Kahler disease, myelomatosis, and plasma cell myeloma, is a type of cancer that starts from malignant plasma cells (specifically the white blood cell) (Rabb et al. 2009). As part of the human immune system are antibodies produced by the plasma cell which fight against germs and other substances harmful to the human body. When the plasma cell becomes abnormal, called the myeloma cell, it causes myeloma (Bethesda 2020). When the myeloma cells increase, it accumulates in the bone marrow and overcrowds the active blood cells, and with time may destroy the solid part of the bone. Hence, the collection of several myeloma cells in the bones causes MMC (Ferri 2013; National Cancer Institute 2017).

Abnormal antibodies are produced by the abnormal plasma cells causing kidney problems and highly thick blood (American Cancer Society 2020). MMC has no specific causes. However, some research has found obesity, radiation exposure, family history, and certain chemicals as associated with the cause of MMC (Durie et al. 1980; Roberts et al. 2010; Van de Donk et al. 2016a, b; World Health Organization 2014a). There have been some treatment recommendations for MMC focused on decreasing the clonal plasma cell population and consequently decreasing the symptoms of the disease (Roberts et al. 2010). With a preferred treatment like high-dose chemotherapy, commonly with bortezomib-based regimens, and lenalidomide-dexamethasone followed by autologous hematopoietic stem-cell transplantation (ASCT), the transplantation of a person's stem cell has been recommended for MM patients under 65 years (Dutta et al. 2017). In 2017, a meta-analysis performed has shown that post-ASCT maintenance therapy with lenalidomide has improved the progression-free survival and overall survival in persons at standard risk (Van de Donk et al. 2016a; b). Whereas in 2012, it was found from a clinical trial that intermediate and high-risk disease patients benefit from a bortezomib-based maintenance regimen (Korde et al. 2011).

Statistically, approximately 30,000 new patients are diagnosed with MMC in the United States (U.S.) every year, making it the second most common hematologic malignancy in the U.S. (Kyle and Rajkumar 2008). In 2019, a report by the Surveillance, Epidemiology, and End Results (SEER) Cancer Institute reported that of all new cancer cases in the U.S, MMC constitutes 1.8% and ranks among the top 14 cancer diseases (National Cancer Institute 2017). A further projection by SEER indicates 32,110 estimated new cases of MMC, and an estimated 12,960 MM patients are expected to die. There is a sufficient increase compared with the 24,050 estimated new MMC cases reported in 2014 (McCarthy et al. 2017). The identified risk factors of MMC are reported to be common among the black race, families with MMC history, and being a male (National Cancer Institute 2017; Sonneveld et al. 2012). SEER cancer institute reported from 2012 to 2016 that 63.1% of all races and sexes of MM cases are aged 65 or greater.

Though MMC disease remains incurable, most research into MMC focuses on how to improve the survival times of patients diagnosed with MMC. The Kaplan–Meier (KM) method has been popularly used for analyzing cancer survivorship data in recent times due to the simplicity of its usage. It is often used to compare the survival difference of observations/groups based on the log-rank test of the null hypothesis that there is no difference. KM is mostly used for longitudinal studies like a cohort study (Siegel et al. 2021); an example is the present chapter (i.e., the survival time of patients diagnosed with MMC). Brain et al. (Krall et al. 1975) used Kaplan–Meier to test whether there was a significant difference in the overall survival duration between the categories of risk factors based on the generalized Wilcoxon test and the log-rank test. They found a significant difference in the survival duration between MMC patients with LI% $< 1\%$ (i.e., low percentage labeling index) and LI% $\geq 1\%$ (i.e., high percentage labeling index). Also, there was a significant difference in the survival duration for MMC patients with the number of DNA synthesizing (S) values $< 1.0 \times 1011$ and S values $\geq 1.0 \times 1011$. Shaji K. Kumar et al. (Harley 1971) used the Kaplan–Meier to test for the significant difference in the overall survival from the time of post-transplantation relapse between MMC groups treated subsequently with one or more of the newer drugs (thalidomide, bortezomib, or lenalidomide) and those not exposed to the newer drug, and they found a significant difference between the two groups. It is well known that parametric analysis is more powerful in decision analysis than its nonparametric counterpart. Feigl and Zelen (1965, p. 835) and other authors have pointed out that assuming the exponential distribution works well for studying the survival of cancer-related cases (Kaplan and Meier 1958). However, almost every data given on any cancer survival problem may have an associated well-defined probability distribution. Hence, assuming an exponential distribution for a given cancer survival case without any further investigation is a serious mistake that will lead to making incorrect decisions.

In this chapter, we summarize some of the most recent research findings by the authors on MMC (Mamudu et al. 2020; Mamudu and Tsokos 2020a, b; Mamudu and Tsokos 2021). This chapter has been organized into four sections. In Sect. 9.2, we assess the parametric and nonparametric analysis of the survival times of patients with MMC (Mamudu and Tsokos 2020a; b). In this section, we assessed the survival probability of 48 patients diagnosed with MMC based on parametric and nonparametric techniques. We performed parametric survival analysis and found a well-defined probability distribution of the survival time follows a three-parameter log-normal. We then estimated the survival probability and compared it with the commonly used nonparametric Kaplan–Meier survival analysis of the survival times. The comparison of the survival probability estimates of the two methods revealed a better survival probability estimate by the parametric method than the Kaplan–Meier. The parametric survival analysis is more robust and efficient because it is based on a well-defined parametric probabilistic distribution, hence preferred over the nonparametric Kaplan–Meier.

In Sect. 9.3, we utilized the semi-parametric Cox-proportional hazard model to examine the survival probability considering sixteen risk factors presumed to be associated with the survival times of patients with MMC (Mamudu et al. 2020). We carefully and rigorously assessed the risk factors using the stepwise selection technique, selecting the final model with the least AIC. This resulted in seven risk factors and one interaction term that were found to be statistically significantly associated with the survival times of patients with MMC. The results were reported based on hazard ratios (HR) and confidence intervals (CI). They included blood urea nitrogen (BUN)/serum creatinine (HR = 7.45; CI = 2.22–25.06), white blood cells (WBC) (HR = 6.54; CI = 1.43–29.75), absence of Bence Jone protein in the urine (BJPU) (HR = 4.84; CI = 2.13–1097), presence of fractures (HR = 2.35; CI = 1.05–5.24), proteinuria (HR = 1.11, CI = 1.05–118), and the interaction of infections and serum calcium (HR = 1.12; CI = 1.04–1.21), all of which had increased risk of association with the survival time of patient diagnosed with MMC; but females (HR = 0.44; CI = 0.21–0.95) and normal platelets (HR = 0.2; CI = 0.08–0.54) had decreased risk of association with the length of survival. The proposed Cox-PH model was well-validated and satisfied the key assumptions. The identified risk factors were ranked according to the prognostic effect on the survival times based. Blood urea nitrogen (BUN)/serum creatinine had the greatest prognostic effect on the survival time of MMC patients, followed by white blood cells (WBC), and normal platelet was found to be the minimum. This section of the chapter offers prognostic and therapeutic significance for further enhancement in the treatment strategy of MMC disease.

In Sect. 9.4.of this chapter, we proposed a data-driven statistical predictive model for the survival times of patients diagnosed with MMC (Mamudu and Tsokos 2020a; b). The proposed model identified nine attributable risk factors out of 16 and one interaction term to be statistically significantly contributing to the survival times of MMC patients. The absence of Bence Jone protein in urine contributed the highest (30.57%) to the survival time, blood urea nitrogen (BUN)/serum creatinine contributed 23.48%, presence of infections had 10.18% contribution, percent myeloid cells in peripheral blood (10.18%), presence of fractures (7.01%), serum calcium (7.56%), female (3.76%), normal platelets (1.26%), and age (0.98%), and the interaction of white blood cells and total serum protein contributed 4.34% to the survival duration of MMC patients. This finding was based on a well-structured and well-validated model that satisfies all the assumptions and has a very high prediction accuracy of 93%. Thus, it passes the goodness-of-fit test and the qualities of a good statistical model. Our model was evaluated by comparing it with other existing models of survival analysis of MMC. It proved to be more robust as 87.14% variation in the survival times was explained by the identified risk factors. The section offers an improved strategy for the therapeutic/treatment process of MMC.

Finally, the Cox-proportional hazard (Cox-PH) model has been a popularly used method for survival analysis of cancer data given the survival times as a function of covariates or risk factors. However, it is very seldom to see the assumptions for the application of the Cox-PH model satisfied in most of the research studies, raising questions about the effectiveness, robustness, and accuracy of the model in predicting the proportion of survival times. This is because the necessary assumptions in most cases are difficult to satisfy, as well as the assessment of interaction among covariates. To further improve the therapeutic/treatment strategy for cancer diseases, in Sect. 9.5, we proposed a new approach to survival analysis using MMC data (Mamudu and Tsokos 2021). We first developed a data-driven nonlinear statistical model that predicts the survival times with 93% accuracy. We then performed a parametric analysis on the predicted survival times to obtain the survival function which is used in estimating the proportion of survival times. Results: The new proposed approach for survival analysis has proved to be more robust and gives better estimates of the proportion of survival than the Cox-PH model. Also, satisfying the proposed model assumptions and finding interactions among risk factors is less difficult compared to the Cox-PH model. The proposed model can predict the real values of the survival times, and the identified risk factors are ranked according to the percent of contribution to the survival time. The new proposed nonlinear statistical model approach for survival analysis of cancer diseases is very efficient and provides an improved and innovative strategy for cancer therapeutic/treatment.

9.2 Parametric and Nonparametric Analysis of the Survival Times of Patients with Multiple Myeloma Cancer (MMC)

In this section of this chapter, we investigate the survival times of patients diagnosed with MMC by comparing the estimation of survival probability between parametric and nonparametric analysis.

9.2.1 Data Description

The data for this study was obtained from West Virginia University Medical Center provided by Harley (Harley 1971; Krall et al. 1975). It originally constituted the survival times of 72 MMC patients diagnosed and treated with alkylating agents (Krall et al. 1975). We used 65 out of 72 patients who provided complete data for 16 risk factors. The remaining 7 patients were eliminated due to missing data in at least one of the 16 risk factors. Given that a patient is diagnosed with myeloma, the 16 risk factors were recorded, and the length of survival was noted (called the survival time from diagnosis to the nearest month). Of the 65 patients, 48 and 17 were dead

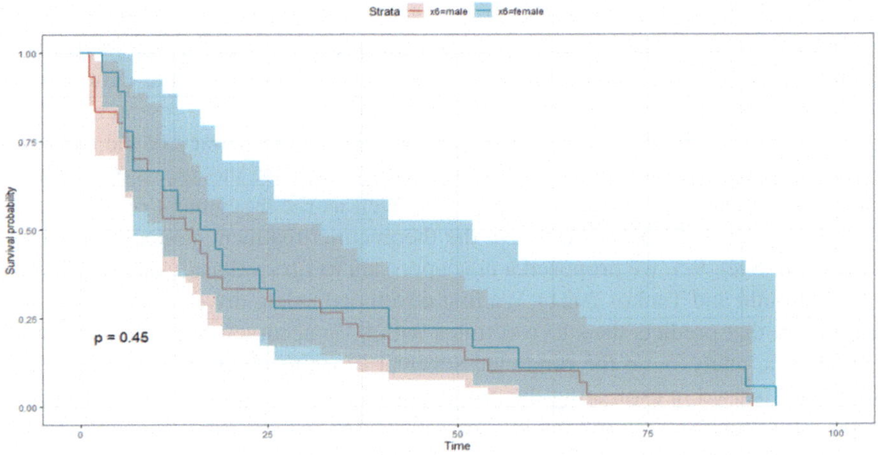

Fig. 9.1 Log-rank test for difference in survival time of gender

and alive, respectively. In the present study, we assessed the 16 risk factors of the complete data of 48 patients' survival times.

We first performed the parametric analysis of the survival times of patients to assess the difference in the survival times between males and females. Given that we have small data of only 48 patients, we used the log-rank test (Richard Peto 1972) to compare the difference in survival times of males and females. Figure 9.1 shows a log-rank test with a large p-value = 0.45, indicating the null hypothesis (i.e., $H_0 : \mu_M = \mu_F$) was not rejected, i.e., there was no difference in the survival times of males and females. As a result, we proceeded with parametric analysis without considering stratification based on gender.

9.2.1.1 Descriptive Statistics of the Survival Times of Multiple Myeloma Cancer (MMC)

This section investigated the distribution of the survival time of MMC as shown in Fig. 9.2. The distribution of the survival time of MMC was right-skewed. Table 9.1 presents the descriptive statistics of the survival time of patients diagnosed with MMC. The spread of the distribution of the survival times was measure based on the kurtosis. A positive kurtosis value implies a leptokurtic behavior of the distribution, and a negative value implies a platykurtic behavior of the distribution. A kurtosis value of zero implies the distribution behavior is mesokurtic (Sharma and Bhandari 2015). The kurtosis value for the survival times of MMC patients of 0.78 indicates the leptokurtic nature of the survival times of MMC patients.

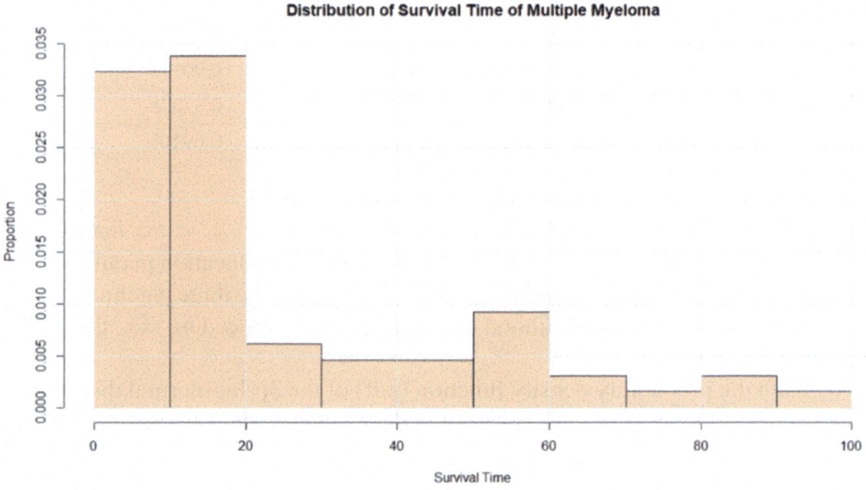

Fig. 9.2 Histogram showing the distribution of survival time (to the nearest month) of multiple myeloma cancer

Table 9.1 Descriptive statistics of survival time (to the nearest month) of multiple myeloma cancer

Mean	Median	Std Err	Std Dev	Kurtosis	Skewness
24.43	15.50	3.56	24.65	0.78	1.33

9.2.1.2 Three-parameter (3P) Log-normal Probability Estimation of the Survival Times of Patients with Multiple Myeloma Cancer

From Fig. 9.1 and Table 9.2, we find that the probability distribution of the survival time of MMC follows the three-parameter log-normal distribution. A formal statistical goodness-of-fit tests in Table 9.2 further showed that the distribution of the survival times of the MMC patients followed 3p-log-normal probability distribution, given by the Kolmogorov–Smirnov, Anderson–Darling, and Chi-square test. Each of the tests resulted in a large p-value, which indicate that we do not reject the null hypothesis, H_0: the probability distribution follows the 3p-log-normal. In the present section, we find the probability density function (pdf) of the 3p-log-normal distribution and estimated the parameters. That is, for a random survival time of MMC, t, pdf of the 3p-log-normal probability distribution can be expressed as,

$$f\left(t;\gamma,\mu,\sigma^2\right)=\begin{cases}0 & \text{if } t\leq\gamma\\ \left(2\pi\sigma^2\right)^{-\frac{1}{2}}(t_i-\gamma)^{-1}\exp\left(-\frac{1}{2}\left(\frac{\ln(t_i-\gamma)-\mu}{\sigma}\right)^2\right), & \text{if } t>\gamma\end{cases} \tag{9.2.1}$$

Table 9.2 Goodness-of-fit test of the 3P-log-normal distribution of the survival time

Type of test	p-value
Kolmogorov–Smirnov	0.90171
Anderson–Darling	0.37878
Chi-squared	0.69163

where $\sigma > 0$ is the continuous shape parameter, $-\infty < \mu < \infty$ denotes the continuous scale parameter, and γ represents the continuous location parameter. For $\gamma \equiv 0$ yields the 2p-log-normal distribution. We estimated the three parameters σ, μ, and γ, using the maximum likelihood estimation (MLE) procedure (see the results given in Table 9.3).

To obtain the probability density function (pdf) of the 3p-log-normal distribution of the survival time of MMC, we substituted the estimates into Eq. (9.2.1) given by,

$$f(t) = \begin{cases} 0, & \text{if } t \leq -0.17824 \\ 0.38253(t_i + 0.17824)^{-1} \\ \exp\left(-\frac{1}{2}\left(\frac{\ln(t_i - 0.17824) - 2.7015}{1.0429}\right)^2\right), & \text{if } t > -0.17824 \end{cases}$$

The above findings show that the survival times of MMC patients given by

$$\boldsymbol{F_T}\left(\boldsymbol{t}; \boldsymbol{\gamma}, \boldsymbol{\mu}, \boldsymbol{\sigma}^2\right) = \boldsymbol{\Phi}\left(\frac{\ln(\boldsymbol{t_i} + 0.17824) - 2.7015}{1.0429}\right). \tag{9.2.2}$$

where $\Phi(.)$ denotes the standard normal CDF for a given survival time t_i. Figure 9.3 shows that the CDF plot of the survival times of MMC determines the probability that a given random observation (survival time t) would be less than or equal to some value T; thus, $P(t \leq T)$. For instance, we can calculate the probability that a patient with MMC survives up to time $t = 16$ months is approximately 0.53.

Further, we obtained the survival function or reliability of the survival time t of MMC patients, given the CDF of the survival times of MMC patients in Eq. (9.2.2), expressed as

$$\begin{aligned} \hat{S}\left(t_i; \gamma, \mu, \sigma^2\right) &= 1 - F_T\left(t; \gamma, \mu, \sigma^2\right) \\ &= 1 - \Phi\left(\frac{\ln(t_i + 0.17824) - 2.7015}{1.0429}\right) \end{aligned}$$

Table 9.3 Parameter estimates for the three-parameter log-normal probability distribution of the survival time of multiple myeloma cancer

Location ($\hat{\gamma}$)	Scale ($\hat{\mu}$)	Shape ($\hat{\sigma}$)
−0.17824	2.7015	1.0429

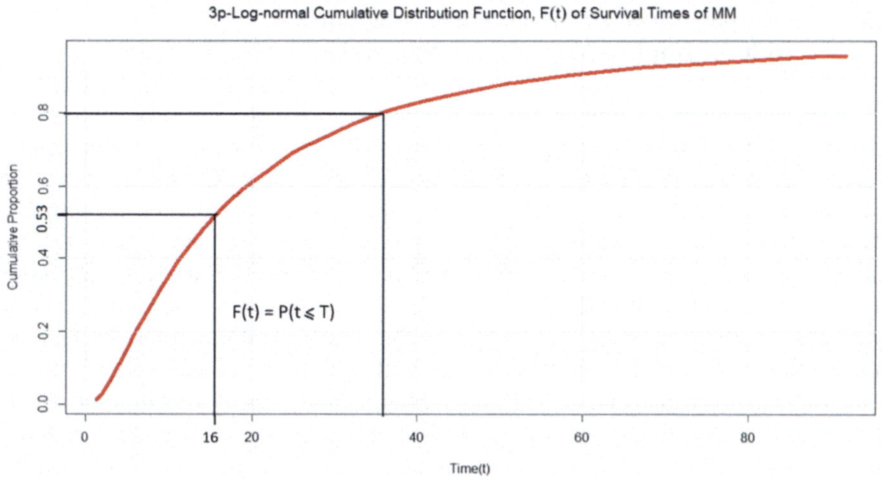

Fig. 9.3 Cumulative distribution function plot for the survival time of MMC

The above survival function can be used to assess the probability that a patient diagnosed with MMC survives beyond time T; thus, $P(t > T)$. Figure 9.4 displays the estimated survival function $\hat{S}(t)$ of the survival times of MMC.

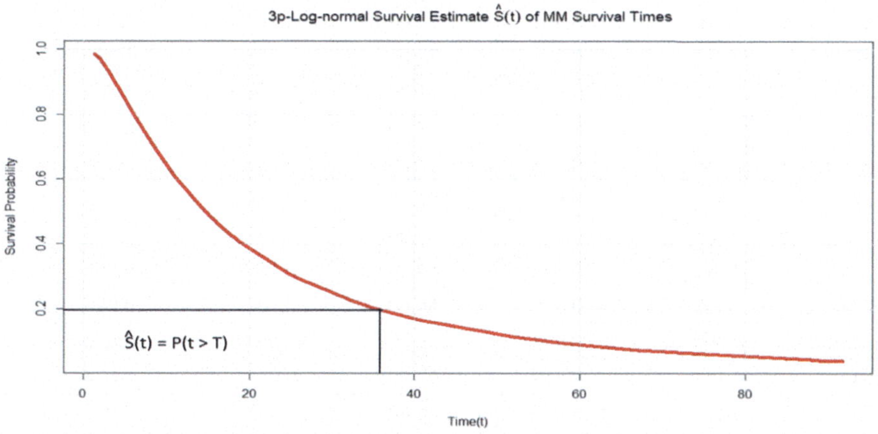

Fig. 9.4 Survival estimate for the survival time of MMC

9.2.1.3 Comparison of Three-parameter Log-normal with the Kaplan–Meier Estimation of the Survival Function

The parametric analysis showed that the probability distribution of survival times of patients with MMC followed the three-parameter log-normal. We further conducted a Kaplan–Meier (KM) nonparametric analysis to estimate the probability of surviving from the MMC. We then compare the survival estimate of the 3p-log-normal with that of the Kaplan–Meier survival estimate of the survival times of the MMC patients. The survival function of the two probabilistic methods estimated the survival probability of a patient diagnosed with MMC beyond a given time. Table 9.4 presents the survival times along with the survival probabilities of the two survival functions. We can see that the 3p-log-normal survival function generally gives a better and higher estimate of the survival times than that of the Kaplan–Meier. Though there were times in which the KM estimates higher probabilities, the 3p-log-normal survival function from the parametric probability distribution had accurate estimation of the survival proportion than the Kaplan–Meier. Noticeably, parametric methods are more powerful, robust, and efficient than nonparametric methods.

Table 9.4 Kaplan–Meier ($\hat{S}_{KM}(t)$) versus parametric (3P-log-normal, $\hat{S}_P(t)$) survival function estimate of the survival times

t_i	$\hat{S}_{KM}(t)$	$\hat{S}_P(t)$	t_i	$\hat{S}_{KM}(t)$	$\hat{S}_P(t)$
1.25	0.958	0.988	25.00	0.313	0.308
2.00	0.896	0.967	26.00	0.292	0.295
3.00	0.875	0.931	32.00	0.271	0.230
5.00	0.833	0.845	35.00	0.250	0.205
6.00	0.750	0.801	37.00	0.229	0.190
7.00	0.688	0.758	41.00	0.188	0.164
9.00	0.667	0.679	51.00	0.167	0.118
11.00	0.563	0.609	52.00	0.146	0.115
13.00	0.542	0.547	54.00	0.125	0.108
14.00	0.521	0.519	58.00	0.104	0.098
15.00	0.500	0.493	66.00	0.083	0.076
16.00	0.458	0.469	67.00	0.063	0.074
17.00	0.417	0.446	88.00	0.042	0.044
18.00	0.396	0.424	89.00	0.021	0.043
19.00	0.354	0.404	92.00	0.000	0.040
24.00	0.333	0.321			

9.2.2 Discussion

Given the danger posed by MMC in recent years, it is important to assess the prognosis influence of risk factors on survival and ways to enhance the therapeutic/treatment of MMC. While MMC currently remains incurable, some improvement in the treatment process has been observed. Some treatment progress includes the introduction of therapeutic agents such as thalidomide, lenalidomide, and bortezomib, and high-dose chemotherapy and stem-cell rescue (ASCT). Research to improve MMC patients' survival has resulted in adopting several different research approaches and methodologies.

In this section, we found that the survival times of both males and females diagnosed with MMC are not different, so we performed the analysis using the combined data of the males and females. We also identified a well-defined probability distribution that characterizes behavior of the survival times of the 48 patients diagnosed with MMC which was used to estimate the survival function. We further estimated the proportion of survival time adopting KM technique and then compare and contrast its relevance and efficiency with the 3p-log-normal analysis of survival probability estimation of patients diagnosed with MMC beyond a given survival time.

Given that we found a well-defined probability distribution that characterizes the survival time of the 48 patients diagnosed with MMC to be 3p-log-normal probability distribution refutes the suggestion of assuming exponential distribution (as suggested by Feigl and Zelen (1965, p. 835) and other authors) or using the nonparametric Kaplan–Meier to investigate most cancer survivorship studies.

Further, we found that the 3p-log-normal survival often estimated higher survival probability than the KM survival function, given in Table 9.4. KM is the most commonly used technique for cancer survival analysis. However, parametric analyses have been found to be more robust and efficient than the nonparametric analyses of the same subject matter. One key demerit of KM is that it cannot be used to estimate the rate at which patients die or survive with MMC (i.e., the hazard function). On the other hand, the hazard function can be easily calculated after finding the parametric probability distribution by dividing the probability density function, pdf, by the survival function. This supports our finding of the parametric 3p-log-normal probability distribution to be better in estimating the survival probability of the patients diagnosed with MMC than the Kaplan–Meier. The KM technique has often been used to assess the differences in the proportion of survival between two or more entities or groups usually using the log-rank test. However, by obtaining the best parametric probability distribution that characterizes the survival times, we can find the survival function and estimate the survival rate and compare the results of two or more entities with a high degree of accuracy. Although, all argument tilted in favor of parametric survival analysis, we still recommend the use of nonparametric approach to survival analysis if there is no parametric distribution form for the given cancer data.

9.3 Survival Analysis of Multiple Myeloma Cancer (MMC) Using the Cox-proportional Hazard Model

In this section of this chapter on MMC, we developed a Cox-proportional hazard (Cox-PH) model to assess risk factors influencing the length of survival of patients diagnosed with MMC.

9.3.1 Review of the Cox-proportional Hazard Model

There are two major important characteristics to be considered when conducting survival analysis, i.e., time and event. In survival analysis, the survival time is defined as time an event occurred (e.g., die, survive, etc.). For instance, the length of time a patient lived with MMC. Several factors/risks can influence the survival time. These factors are often termed as covariates by most survival analysis literature. The Cox-proportional hazard model, commonly referred to as the Cox model, was introduced by Cox (1972) which is widely recommended for modeling of the relationship between the survival time and covariates in survival analysis. A good introductory review is given by Kleinbaum (Kleinbaum and Klein 2012), and an extensive is given by Kalbfleisch and Prentice (Kalbfleisch and Prentice 2011).

An important feature of the Cox model is the hazard function, which measures the rate/risk of death or survival at time t. The hazard function is defined as follows; given a random survival time T of survival time with cumulative density function $F_T(t)$, given by

$$F_T(t) = P(T \leq t) = \int_0^t f_T(t)dt$$

where $F_T(t)$ is the probability of failure or survival by time t and $f_T(t) = dF_T(t)/dt$ is the probability density/distribution function. Giving $F_T(t)$, the survival function is defined as

$$S_{T(t)} = P(T > t) = 1 - P(T \leq t) = 1 - F_{T(t)}$$

The hazard function measuring the risk of instantaneous death or survival at time t can be defined as

$$\begin{aligned} h(t) &= \lim_{\partial t \to 0} \frac{F_T(t + \partial t) + F_T(t)}{\partial t . S_T(t)} \\ &= \lim_{\partial t \to 0} \frac{P(t < T \leq t + \partial t)}{\partial t . S_T(t)} \end{aligned}$$

$$= \lim_{\partial t \to 0} \frac{P(t < T \leq t + \partial t | T > t)}{\partial t}$$

$$= \frac{f_{T(t)}}{S_{T(t)}} \tag{9.3.1}$$

We can obtain the cumulative hazard function from the hazard function (Eq. 9.3.1), which is expressed as.

$$H(t) = \int_0^t h(s)ds.$$

The close form of the above integral function can be expressed as $H(t) = -lnS(t) = -lnR(t)$. We estimated the Cox model with interacting covariates which is expressed using the hazard functions as follow:

$$h_i(t) = \alpha_{i(t)}\exp\left(\sum_{i=1}^{k} \beta_i X_i + \sum_{i \neq j=1}^{k} \rho_{ij} X_i X_j\right),$$

and

$$\ln\left(\frac{h_i(t)}{\alpha_i(t)}\right) = \sum_{i=1}^{k} \beta_i X_i + \sum_{i \neq j=1}^{k} \rho_{ij} X_i X_j. \tag{9.3.2}$$

where t denotes the survival time, $h_i(t)$ is hazard function obtained by the set of k covariates, β_i measures the influence of the covariates X_i on $h_i(t)$, ρ_{ij} is the coefficient measuring the effect of interacting covariates $X_i X_j$ on $h_i(t)$, and $\alpha(t)$ represent the baseline value of $h_i(t)$ if all X_i and $X_i X_j$ equals zero. The above Cox model is a representation of multiple linear regression of the logarithmic form of the hazard on X_i's and $X_i X_j$'s, with $\alpha(t)$ denoting the intercept that changes with time t. A key assumption of the Cox model is the proportional hazard (PH) assumption. The PH assumption explains that the hazard function of data observations (or patients) should be proportional and independent of time t (Brody 2016). For example, in the case of two patients i and i' having varying values of covariates; their corresponding hazard functions for ith patient is defined as

$$\eta_i(t) = \alpha(t)\exp\left(\sum_{i=1}^{k} \beta_i X_i + \sum_{i \neq j=1}^{k} \rho_{ij} X_i X_j\right),$$

and the corresponding hazard functions for ith is

$$\eta_i{'}(t) = \alpha(t)\exp\left(\sum_{i'=1}^{k} \beta_{i'} X_{i'} + \sum_{i' \neq j'=1}^{k} \rho_{i'j'} X_{i'} X_{j'}\right).$$

The hazard ratio between the two patients is

$$\frac{\eta_i(t)}{\eta_i'(t)} = \frac{\alpha(t)\exp\left(\sum_{i=1}^{k}\beta_i X_i + \sum_{i\neq j=1}^{k}\rho_{ij}X_iX_j\right)}{\alpha(t)\exp\left(\sum_{i'=1}^{k}\beta_{i'} X_{i'} + \sum_{i'\neq j'=1}^{k}\rho_{i'j'}X_{i'}X_{j'}\right)}$$
$$= \frac{\exp\left(\sum_{i=1}^{k}\beta_i X_i + \sum_{i\neq j=1}^{k}\rho_{ij}X_iX_j\right)}{\exp\left(\sum_{i'=1}^{k}\beta_{i'} X_{i'} + \sum_{i'\neq j'=1}^{k}\rho_{i'j'}X_{i'}X_{j'}\right)} = \exp(coef), \quad (9.3.3)$$

which is independent of time t. The result of the above hazard ratio (HR) implies that the Cox model is a proportional hazards model. The HR measures the relative hazards between observations/groups (Case et al. 2002). It is interpreted in the following three ways: (1) $HR = 1$ implies that there is no hazard or relative risk effect. Thus, the covariates have no relationship with the event probability, hence, no influence on the length of survival. (2) $HR > 1$ (i.e., equivalently $\beta_i > 0$) implies an increase in hazard or risk, i.e., the given covariate has a positive impact on the event probability, hence, negatively associated with the length of survival (bad prognostic factor). (3) $HR < 1$ (i.e., equivalently $\beta_i < 0$) implies a decrease in hazard or risk, i.e., the given covariate has negative association with the probability of the event, hence, has positive impact on the length of survival (good prognostic factor). An extensive review of the hazard ratio can be found in a study by Case et al. (2002). We compute the baseline hazard function by performing the following computation:

$$\hat{\alpha}(t) = \sum_{t_i \le t} \hat{h}(t_i),$$

for

$$\hat{h}(t_i) = \frac{d_i}{\sum_{i\in R(t_i)}\exp(X_i'\hat{\beta})},$$

where $t_1 < t_2 < \ldots < t_n$ represents the distinct event times, d_i represents the number of events at t_i, and $R(t_i)$ is the hazard or risk set at t_i which represent individuals still susceptible or at risk of the event at t_i. The baseline hazard function can assume any functional form of the covariates. We discussed in detail the major assumptions of the Cox-PH model in Sect. 9.3.1.1.

9.3.1.1 Cox-proportional Hazards (PH) Model Assumptions

For any given Cox-PH model, it should satisfy the following three key assumptions. Not satisfying the assumptions may result in making wrong decision about the subject matter.

A. **Proportional hazard (PH) assumption**. A formal statistical test is needed to evaluate the PH assumption of the Cox-PH model. The PH assumption is said to be satisfied if the covariates and the global tests are not statistically significant (p-value > 0.05). We can also assess for the presence of PH assumption by investigating the plot of scaled Schoenfeld residuals against the transformed time. The Schoenfeld residuals are independent of time, and hence, a non-random pattern against time is evidence of a violation of the PH assumption. We compute the Schoenfeld residuals with one per observation per covariate, which is expressed as

$$r_{ik} = X_{ik} - \widehat{\overline{X}}_{w_i k}(\beta, t_i),$$

where X_{ik} is the value of the kth covariate for ith observation. $\overline{X}_{w_i k}(\beta, t_i)$ is the weighted mean values of covariates at risk at the given event time, t_i, denoted by $R(t_i)$, and is given by

$$\widehat{\overline{X}}_{w_i k}(\beta, t_i) = \sum_{j \in R(t_i)} X_{ik} w_i(\beta, t_i).$$

The weight function, $w_i(\beta, t_i)$ for ith observation at risk, $R(t_i)$ is the probability that observation i fails at time t_i, which is defined by

$$w_i(\beta, t_i) = \frac{\exp\left(\beta^T X_i\right)}{\sum_{l \in R(t_i)} \exp(\beta^T X_l)}.$$

A positive value of r_{ik} represents an X value higher than expected at that death time. For a binary outcome (0,1) variable, Schoenfeld residuals will be between -1 and 1. In that situation,

$$r_{ik} = \begin{cases} 0 - \widehat{\overline{X}}_{w_i k}, & \text{for } X = 0 \\ 1 - \widehat{\overline{X}}_{w_i k}, & \text{for } X = 1. \end{cases}$$

B. **Linear functional form of continuous covariates**. Another assumption is that the functional form of the covariates is linear. A study by Therneau et al. (1990) outlined that this assumption can be assessed by visualizing the plot of Martingale residuals against the continuous covariates with fitted lowess (i.e., locally weighted smoothing) line function. Thus, if a trend or pattern in the plot is observed, imply an evidence violation of the linear functional form of the covariates. Martingale residual is defined by

$$\hat{M}_i = \delta_i - \hat{\Gamma}_0(t_i)\exp\left(\hat{\beta}_1 X_{1i} + \cdots + \hat{\beta}_k X_{ki}\right),$$

where δ_i represents the event indicator for ith observation, $\hat{\Gamma}_0(ti)$ is the estimated cumulative hazard at the final follow-up time for ith observation, and $\exp\left(\hat{\beta}_1 X_{1i} + ...\right)$ denotes the estimated coefficients applied to the observed covariate for the ith observation. The distribution of Martingale residuals, $\hat{M}_i$, is skewed distribution. The $\hat{M}_i$ values are.

$$\hat{M}_i = \begin{cases} 1, & \text{for minimum possible values} \\ -\infty, & \text{for minimum possible values} \end{cases}$$

If $\hat{M}_i$ is positive, it implies individuals have very short time to die; and if $\hat{M}_i$ is negative, it implies individuals lived too long. Most often a transformation of $\hat{M}_i$ to obtain approximate symmetric distribution may be essential in the model development process. Such a transformation is motivated by deviance residuals defined below.

C. **Examining influential observations (or outliers).** To examine influential observations, we visualized the $dfbeta$ values. The $dfbeta$ values estimate the influence of ith − case (or observation) on the estimated regression coefficients. A very high value of $dfbeta$ should be carefully investigated. We can also assess the influential observations which is by assessing the deviance residuals (symmetric/normalized transformation of the Martingale residuals) plot. The deviance residual is defined by

$$d_i = \sin\left(\hat{M}_i\right)\sqrt{2}\sqrt{-\hat{M}_i - \delta_i \log\left(\delta_i - \hat{M}_i\right)}$$

Note that $\mathrm{d_i} = 0$ is only when $\hat{M}_i = 0.$. The square root shrinks the large negative martingale residuals, while the logarithm transformation expands those residuals that are close to zero (i.e., as $n \to \infty$). The distribution of the residuals should be roughly normal about zero mean and standard deviation of one. A very large/small/distant deviance residual values is evidence of presence of influential observations or outliers. The values of the deviance residual values can be compared with the expected value of survival time. A positive value implies individuals demised too soon, and negative value implies individuals lived too long. The above-stated assumptions will be assessed and evaluated during the development of the Cox-PH model for the survival of patients diagnosed with MMC.

9.3.1.2 Proposed Cox-PH Model for the Survival Times of Patients with Multiple Myeloma Cancer

In this section, we fitted the Cox-PH model to the survival times for all the 16 covariates X_i, including their two-way interactions. We utilized the stepwise model selection method to select the final model with the least Akaike information criterion

(AIC = 2ln(L) + 2 k, where L denotes the value of the maximum likelihood function of the model and k represents the estimated model parameters) (Akaike 1974). Using AIC, we estimated the relative amount of information missing in the model, with the smaller the AIC value the better the quality of the model. It reduces the danger posed by overfitting or under-fitting the model.

The stepwise model variable selection procedure allowed to determine the significant covariates for Cox-PH models. It uses iterations between forwarding and backward steps during the model selection process. The individual covariates and interactions are included in the "variable list" for selection. The significance levels for entry (α_{entry}) and stay (α_{stay}) was set at a conservative value 0.15 or larger. We obtained the best Cox-PH model by manually removing the covariates with $p - value > 0.05$ one at a time until the remaining model coefficients are statistically significant (p-value < 0.05) at the chosen level of significance, $\alpha = 0.05.$. The final model has the least AIC value. Our final proposed model that significantly contributes to the probabilistic survival time of patients diagnosed with MMC includes seven significant contributable covariates (risk factors) and one interaction: given by

$$\ln\left(\frac{h_i(t)}{\alpha_i(t)}\right) = 2.009X_1 - 1.608X_{3normal} - 0.815X_{6female} + 1.878X_7 + 0.854X_{8present} + 0.108X_{12} + 1.576X_{13none} + 0.114X_{4present} * X_{16}$$

Table 9.5 presents the statistical significance of each of the identified risk factors and interaction (coefficients) in Eq. (3.4). All the risk factors are statistically significant, with "three stars ***" indicating a very highly statistically significant risk factor. A positive coefficient ($\hat{\beta} > 0$) means a higher hazard rate, and thus a bad prognostic factor to survival. By contrast, a negative coefficient ($\hat{\beta} < 0$) means a lower hazard rate, and thus a good prognostic factor to survival. For instance, $\hat{\beta}_6 = -0.815$ representing gender means female MMC patients were good prognostic of the survival time; thus, females have a lower relative risk to die (higher survival rates) from MMC than males. The $\exp(\hat{\beta})$ represents hazard ratio which measures the extent or magnitude effect of the risk factor on survival. Thus, $\exp(-0.815) = 0.443 < 1$ for gender means being a female has relatively decreased risk of dying with MMC than being a male. We further rank the risk factors in order of their magnitude impact on the survival of MMC. The ranking of the significant attributable risk factors showed that blood urea nitrogen (BUN)/serum creatinine (X_1) is the greatest prognostic factor to the survival of MMC, followed by white blood cells (WBC) (X_7), and platelets (X_3) which is the least prognostic factor. Table 9.6 presents three different tests for the overall significance of the proposed Cox model, i.e., the likelihood-ratio test, the Wald test, and score log-rank statistics.

The three tests are asymptotically equivalent and give similar results for large samples. However, for small samples like in our case, the likelihood-ratio test is robust and generally preferred. The global statistical significance test demonstrates that the proposed Cox-PH model in Eq. (3.4) is highly statistically significant. Table 9.7

Table 9.5 Ranking of the significant contributing covariates (risk factors) based on prognostic effect on the survival time using the hazard ratios

Rank	Covariates	Coeff ($\hat{\beta}$)	HR ($\exp(\hat{\beta})$)	S.E. ($\hat{\beta}$)	Pr(>|z|)	95% CI of HR	
1	X_1	2.008	7.454	0.619	1.165 e^{-3**}	2.217	25.056
2	X_7	1.878	6.543	0.773	1.505 e^{-2*}	1.439	29.745
3	X_{13}	1.576	4.835	0.418	1.63 e^{-4***}	2.131	10.972
2	X_8	0.854	2.349	0.409	3.693 e^{-2*}	1.053	5.243
5	$X_4 * X_{16}$	0.113	1.121	0.040	4.873 e^{-3**}	1.035	1.213
6	X_{12}	0.108	1.114	011	3.84 e^{-4***}	1.049	1.183
7	X_6	−0.815	0.443	0.443	3.711 e^{-2*}	0.206	0.952
8	X_3	−1.608	0.200	0.502	1.355 e^{-3**}	0.075	0.536

Table 9.6 Global statistical significance of the model

Type of test	Type of statistic	df	p-value
Likelihood-ratio test	32.6	8	7 e^{-05}
Wald test	30.38	8	2 e^{-04}
Score (log-rank) test	32.49	8	8 e^{-05}

presents the baseline hazard function $\hat{\alpha}(t)$ for the first ten observations. Figure 9.5 displays the order of prognostic effect of the risk factors. We can see that blood urea nitrogen (X_1) is the greatest prognostic factor of the survival time of MMC patients, followed by white blood cells (X_7), and platelets (X_3) which is the least prognostic factor.

The proposed model in Eq. (3.4) was verified and validated to satisfy all the model assumptions outlined in Sect. 9.3.1.1. Given that a patient is diagnosed with MMC, we put into the model the seven contributing individual risk factors and the interacting factor to estimate the probability of survival beyond a given survival time (death time) along with 95% confidence interval, shown in Fig. 9.6.

Table 9.7 First ten baseline hazard estimates

Obs	Baseline hazard	Time
1	$2.498\ e^{-06}$	1.25
2	$1.463\ e^{-05}$	2.00
3	$1.951\ e^{-05}$	3.00
4	$3.162\ e^{-05}$	5.00
5	$6.104\ e^{-05}$	6.00
6	$8.818\ e^{-05}$	7.00
7	$9.806\ e^{-05}$	9.00
8	$1.549\ e^{-04}$	11.00
9	$1.676\ e^{-04}$	13.00
10	$1.818\ e^{-04}$	14.00

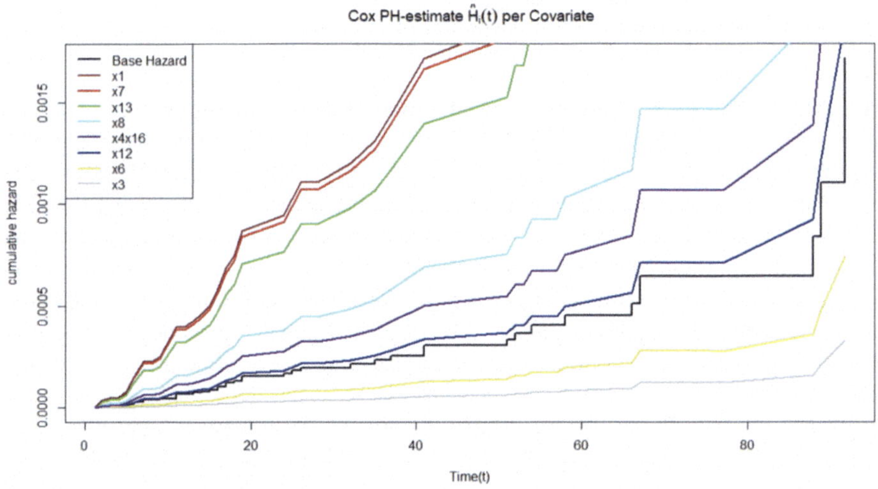

Fig. 9.5 Ranking of prognostic effect of risk factors on the patient's survival

9.4 Discussion

In the present section of this chapter, we conducted the Cox-PH model analysis of the survival times of patients diagnosed with MMC without considering stratification of the data, given that the survival times between males and females were not statistically significantly different (See Fig. 9.1 in Sect. 9.2). We then assessed the risk factors influencing the estimated proportion of the survival time as a function of covariates utilizing the Cox-proportional hazard regression model. Our final proposed Cox-PH model given by Eq. (3.4) identified seven significant risk factors and one interaction term to influencing the survival duration of MMC patients. They included blood urea nitrogen (BUN)/serum creatinine, white blood cells (WBC), Bence Jone protein in the

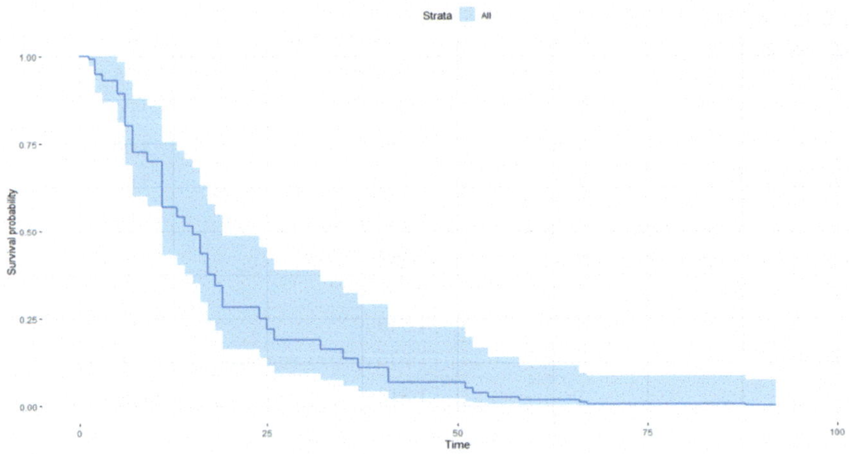

Fig. 9.6 Survival estimate $\hat{S}(t)$ from the proposed Cox-PH model

urine (BJPU), fractures, proteinuria, gender, platelets, and the interaction infections and serum calcium. An interesting and highly important observation from the model was that the two interacting risk factors did not individually significantly contribute to the survival probability, which is a very useful finding. Most studies using Cox-PH models fail to include interaction risk factors because they are difficult to investigate. However, not including interaction(s) in the Cox-PH model given that they exist and may impact the survival time of the patient can result in the wrong conclusion about the survival time, which can minimize the statistical power, relevance, and quality of the Cox-PH model, and consequently endangering the treatment process of the MMC disease.

Further, the Cox-PH model we proposed satisfied all the key assumptions of the Cox-PH model discussed in Sect. 9.3.1.1. If the Cox-PH models assumptions are not validated, we cannot justify the quality and validity of the findings. Our proposed Cox-PH model is of high quality because the underlying key assumptions are satisfied and well-validated. The Cox-PH model we proposed had the least AIC based on the stepwise model selection method and had very small variance inflation factor (VIF). The identified risk factors were ranked based on the prognostic effect on the survival time using the hazard ratio. We found all the identified risk factors, except the female gender and normal platelet to be highly prognostic factors (negatively associated with the survival time).

We found five new risk factors associated with the survival of patients with MMC. They include the individual risk factors platelets, gender, white blood cells, and fractures; and the interaction term, infections and serum calcium. These newly identified risk factors are not found in the findings among other studies (Durie et al. 1980; Durie and Salmon 1975; Krall et al. 1975; Merlini et al. 1980), which developed statistical models to determine the association of some risk factors to the survival time of MMC. The risk factor, serum calcium, was individually found to significantly contribute to the survival time by Merlini et al. (1980). However, we found it to be significant as it interacts with infections, and it negatively relates to the survival time, given by the hazard ratio (HR > 1). However, we believe our model is more genuine and accurate, given the fact that the identified significant contributing risk factors were carefully assessed and selected based on the AIC of stepwise model selection technique, and validated to satisfy all the key model assumptions.

9.5 Data-Driven Statistical Modeling and Analysis of the Survival Times of Multiple Myeloma Cancer (MMC)

Given the complexities in the development and limitations of the Cox-PH model for the survival times of MMC, we proposed a more robust data-driven statistical model that allows us to predict the real value of survival time of patients with MMC.

9.5.1 The Proposed Statistical Model

After conducting rigorous statistical analysis, utilizing the backward elimination model selection technique, the resulting final model had the least Akaike information criterion (AIC = 2ln(L) + 2 k and satisfies all assumptions with $R^2 = 0.8741$, which includes all the attributable risk factors and interaction that significantly contributes to the survival time of patients with MMC which was given by

$$\begin{aligned}\log(\hat{t}) &= -4.037 - 1.167X_1 + 0.267X_{3normal} - 0.977X_{4present} \\ &\quad + 0.016X_5 + 0.504X_{6female} - 0.581X_{8present} + 0.020X_{11} \\ &\quad - 1.209X_{13none} + 4.011X_{16'} - 0.228X_7 * X_{14'}\end{aligned}$$

Thus, there are nine attributable risk factors, namely, Bence Jone protein in urine (X_{13}), blood urea nitrogen (BUN)/serum creatinine (X_1), infections (X_4), % myeloid cells in peripheral blood (X_{11}), fractures (X_8), serum calcium (X_{16}), gender (X_6), platelets (X_3), and age (X_5), and one interaction term, namely, white blood cells (X_7) and total serum protein (X_{14}) that significantly contribute to the survival of MMC patients. The following remaining five risk factors do not contribute to the survival time of MMC patients at diagnostic: hemoglobin, plasma cells in bone marrow, lymphocytes in peripheral blood, proteinuria, and serum-globin (gm%). Because the estimated survival time t' and the attributable risk factors $X_{i'}$ are based on the log-transform data, we utilized the anti-logarithmic to transform back to the original values. The backward transformation of the response and attributable risk factors X_{14} and X_{16} can be expressed as

$$X_{i'} = \begin{cases} -\log(-X_i + 1), & \text{if } x < 0 \\ \log(X_i + 1), & \text{otherwise} \end{cases}$$

$$\Rightarrow X_i = \begin{cases} 1 - e^{-X_i'} & \text{if } x < 0, \\ -1 + e^{X_i'} & \text{otherwise, for } i = 14, 16. \end{cases}$$

9.5.1.1 Bootstrapping with the Proposed Statistical Regression Model

To further improve the efficiency of the proposed statistical model, we utilized the bootstrapping resampling method by Efron and Tibshirani (1993). Bootstrapping is a general approach to statistical inference that allows assigning measures of accuracy (defined in terms of bias, variance, confidence intervals, prediction error, or some other such measure) to sample estimates based on building a sampling distribution for a statistic by resampling from the actual data that we analyzed in the present chapter (Efron and Tibshirani 1993). We applied the bootstrap sampling to resampled with replacement the data used to build the proposed analytical model given by Eq. (4.1); increasing the sample size by 300. Thus, the bootstrap modeling involves the following algorithm.

- We bootstrap sample r = 300 drawn from n = 48 observations with replacement
- In bootstrapping statistical model:
  - We estimate the model coefficients α, β_i, and ρ_{ij} for the original sample $n = 48$, and calculate the fitted value and residual for each observation.
  - Select bootstrap samples of the residuals, $\in_b^*$, and from these, calculate bootstrapped t values, $t_b^* = \left[t_{b1}^*, t_{b2}^*, \ldots, t_{bn}^*\right]$, where $t_{bi}^* = \hat{t}_i + \in_{bi}^*$.
  - The bootstrap estimates are calculated by least-squares regression, then $b_b^* = \left(X'X\right)^{-1} X' t_b^*$ for $b = 1, \ldots, r$.

The bootstrap modeling asymptotically increased the level of significance of the coefficient estimates, making them equally highly significant, and increased both the R^2 and $R^2_{adjusted}$ to 91.16% and 90.85%, respectively. The modified version of the model in Eq. (4.2) based on the bootstrapping resampling method is given by

$$\begin{aligned} \log\left(\hat{t}_{Bootstrap}\right) = {} & -4.377 - 1.097X_1 + 0.332X_{3normal} - 0.949X_{4present} \\ & + 0.016X_5 + 0.562X_{6female} - 0.586X_{8present} + 0.022X_{11} \\ & - 1.268X_{13none} + 4.151X_{16'} - 0.252X_7 * X_{14'} \end{aligned}$$

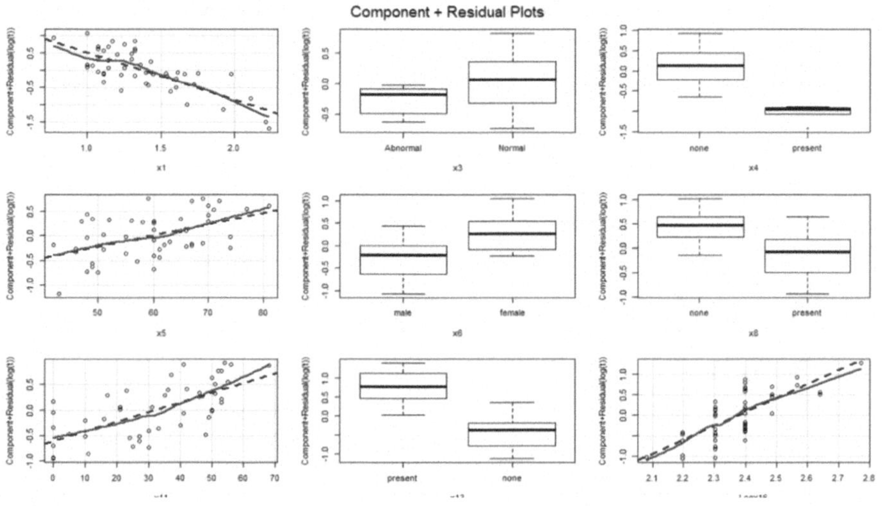

Fig. 9.7 Evaluation of linearity of the proposed statistical model

9.5.2 Validation of the Proposed Statistical Model

Before validating the proposed model, we need to be sure that all assumptions that underline our proposed model are satisfied. We tested for linearity by showing the linearity plot (sometimes referred to as the partial residual plot) of the response variable and the significant attributable risk factors as shown in Fig. 9.7. We can see that there is a well-established linear relationship between the response variable and the continuous attributable risk factors (shown by the blue and pink lines). Therefore, the linearity assumption which was initially a problem we encountered has been rectified in our final proposed statistical model.

To verify that the proposed statistical model satisfies the multivariate normal probability distribution assumption, we used the normal $Q - Q$ plot shown in Fig. 9.8. We see that the residuals are normally distributed with no major outlier and all the points in the plot fall within the 95% confidence bound. The evidence of normality is supported by Shapiro–Wilk's test of the normal probability distribution (a formal test), given a high p-value of 0.818. The plot of the distribution of studentized residuals in the second panel of Fig. 9.8 is further evidence that the proposed model's normality assumption is valid.

We performed residual analysis to assess the model residuals and constant variance. Figure 9.9 depicts the residual plot of the proposed model. Thus, we can conclude that there is no problem with homoscedasticity. Our proposed statistical model perfectly satisfies the assumption of constant variance, indicated by the randomly scattered points about the zero line with no major outliers. A formal test for homoscedasticity revealed a p-value of 0.506, which strongly supports that the homoscedasticity of our proposed model is valid. The mean absolute value

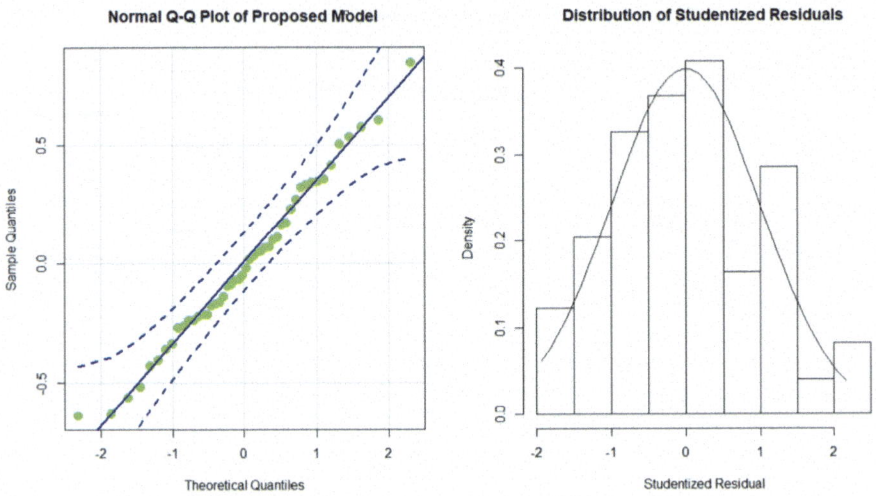

Fig. 9.8 Test for multivariate normal probability distribution

of the residuals, $|\bar{r}| = \sum_i^n e_i$ is $4.779 \times e^{-2}$, close to zero, and the variance $var(r) = 1/(n-1)\sum_i^n (r_i - \bar{r})^2$ is 0.636. The proposed statistical model has a very small root mean square error ($RMSE = \sqrt{\sum_i^n (y_i - \hat{y}_i)/n}$) of 0.384.

Multicollinearity is a major problem in statistical modeling that must be addressed. It can distort the precision of the estimated coefficients leading to overfitting and misinterpretation of the results of the model. All the estimates of the parameters in our proposed model have a very small variance inflation factor, $VIF < 3$, indicating

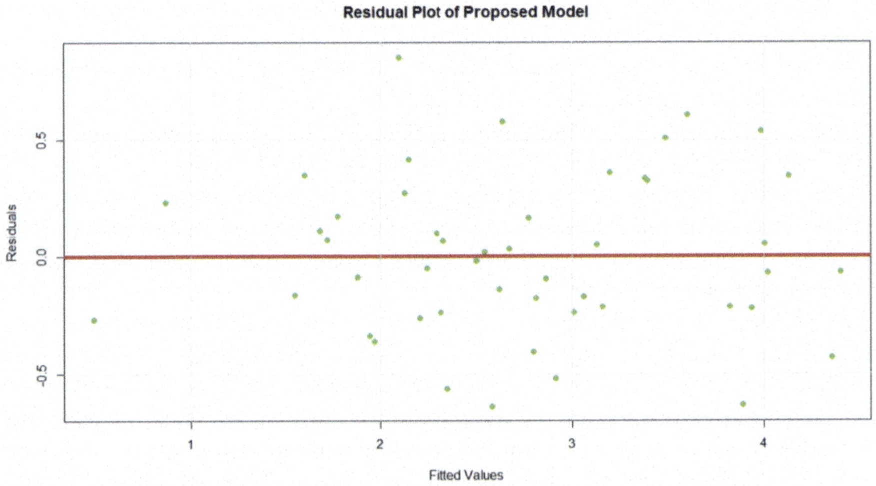

Fig. 9.9 Residual plot of the proposed statistical model

Table 9.8 Comparison of prediction of the survival time of multiple myeloma cancer

Log (t)	Original model	Trained model
0.2231	0.4917	0.6379
1.0986	0.8676	1.3378
1.6094	1.9708	2.3358
1.9459	1.5979	1.1624
2.3979	2.1245	2.0809
2.7726	3.0083	2.9017
3.1781	3.1269	2.6158
3.7136	3.3900	3.1649
3.9889	3.4851	3.2839
1.3863	1.5493	2.1449

that there is no problem with multicollinearity. Also, we expect the model residuals to be independent and uncorrelated. We tested for the presence of autocorrelation among errors in the proposed model using the Durbin–Watson of testing the null hypothesis H_0, no autocorrelation is present. Accepting the hypothesis with a large p-value of 0.624 indicated that there is no autocorrelation among residuals in our proposed model.

To validate the prediction accuracy of our proposed statistical model, we trained 80% of the data to build our model and tested on the remaining 20% test data. The prediction of the original model and the trained model using the test data is given in Table 9.8. We checked the accuracy of the predictions by finding the correlation coefficient r, and the corresponding R^2(square of r) between the actual and the predicted values. This resulted in R^2 of 0.943, a very high prediction accuracy. The comparison of the logarithmic survival times with the two models (i.e., the model developed using all the 48 patients and the 80% trained model) prediction on the test data resulted in R^2 of 0.943 and 0.930, respectively, attesting to the high prediction accuracy of our proposed model.

9.6 Ranking of the Contribution of Attributes/Risk Factors of the Survival Times of Multiple Myeloma Cancer

In this section, we ranked the individual significant risk factors and the interaction based on their contribution to the survival time of MMC patients using the percentage of R^2. Table 9.9 presents the rank of each of the identified significant risk factors and the interaction term. Bence Jone protein in urine is ranked first, followed by blood urea nitrogen (BUN), and the interaction term is ranked eighth, and age has the least contribution to the survival time of patients diagnosed with MMC among the significant attributable risk factors. A detailed discussion of the rankings will continue in the next section.

Table 9.9 Rank of contribution of attributing risk factors to survival time

Rank	Variable	Description	R^2	%Contribution
1	X_{13}	Bence Jone protein in urine at diagnosis; 1 = present, 2 = none	0.2672	30.57
2	X_1	Log BUN at diagnosis	0.2052	23.48
3	X_4	Infections at diagnosis; 0 = none, 1 = present	0.0949	10.86
4	X_{11}	% myeloid cells peripheral blood at diagnosis	0.089	10.18
5	$X_{16'}$	Serum calcium (mgm%) at diagnosis	0.0661	7.56
6	X_8	Fractures at diagnosis; 0 = none, 1 = present	0.0613	7.01
7	$X_7 * X_{14'}$	Lo WBC at diagnosis and total serum protein at diagnosis	0.0379	4.34
8	X_6	Gender; 1 = male, 2 = female	0.0329	3.76
9	X_3	Platelets at diagnosis; 0 = abnormal, 1 = normal	0.011	1.26
10	X_5	Age at diagnosis (complete years)	0.0086	0.98
Total			0.8741	100

9.7 Discussion

The assessment of the survival time of patients diagnosed with MMC is an essential for improving the prognosis and therapeutic/treatment strategy of MMC. This section of this chapter developed data-driven statistical model to predict the survival time from diagnosis to the nearest month of MMC patients' deaths. The following objectives were accomplished: (1) we found the single risk factors that significantly contribute to the survival time of MMC. (2) We found the interacting risk factors that significantly contribute to the survival time of MMC. (3) We estimated the percentage of contributions by each significant risk factor and interaction to the length of survival of MMC patients. It was imperative to examine whether there was a difference in the survival times between males and females, which we found no difference. With that, we proceeded with developing the model with the entire data with considering stratification. We developed the statistical model with 16 predictors (risk factors) reported to be contributing to the survival of MMC, but we only found nine (9) individual risk factors and one interaction term. Most of the risk factors in our data have been reported to be important by several studies (Alexanian et al. 1975; Costa et al. 1969; Durie et al. 1980; Durie and Salmon 1975; Kiang et al. 1973; Kyle et al. 1976; Merlini et al. 1980). The proposed model with all identified significant risk factors that accurately predict the survival time is given in a transformed form by Eq. (9.3.2). We back-transformed the model using anti-logarithm to obtain the original values of the survival time. Further, we validated the goodness-of-fit of the model as follows: (1) satisfied all the assumptions of a high-quality statistical regression model, (2) satisfied the residual test, i.e. $\epsilon_i \sim N(0, 1)$, (3) the model had a very high R^2 of 87.41%; and was bootstrapped to further increased the R^2 to 91.16%, and (4) it had a very high prediction accuracy of about 94% based on 80% training data and 20% test data.

The justification of the relevance of the proposed statistical model compared to other existing models or findings was assessed and evaluated. We found nine risk factors and one interaction term to be significantly contributing to the survival time of patients with MMC, given in Table 9.9. We can predict the survival time of a patient with MMC with at least 94% accuracy given a set of values of the identified risk factors. We found serum calcium, blood urea nitrogen (BUN)/serum creatinine, and Bence Jone protein in urine (BJPU) to be significantly contributing to the survival time, which is consistent with other studies (Durie and Salmon 1975; Merlini et al. 1980). BJPU was ranked as the highest contributor to the survival time, followed by BUN, and serum calcium was ranked sixth; see Table 9.9. Both BUN and serum calcium were identified to be a significant contributor to the survival time in the IgG myeloma group and BJ myeloma group, which is consistent with a finding reported by Merlini et al. (1980). We expected the percentage of bone marrow plasma cells (%BMPC) to significantly contribute to the survival time, but that was not the case in our findings; an observation difficult to explain, hence, require further studies. Giampaolo Merlini et al. found %BMPC not correlated with survival in the IgA myeloma group, parallel to our finding. We found age (ranked 10) and gender (ranked 8) to significantly influence the survival time. Our findings contradicts the finding by Merlini et al.(1980) who reported age and gender to have no major correlation with the survival of MMC. Our findings are consistent with that reported by the national cancer institute for Surveillance, Epidemiology, and Ends Results (SEER cancer) (Bethesda 2020), as they reported age and sex as important risk factors to MMC. This suggests that age and gender are important attributable risk factors to survival of MMC, as indicated by our findings.

Other risk factors we found to be significantly influencing to the survival time of MMC, but are not found in other studies included infections (ranked 3), percentage myeloid cells in peripheral blood (ranked 4), fractures (ranked 5), platelets (ranked 9), and an interaction between white blood cells (WBC) and total serum protein (ranked 7), all at diagnosis. With our proposed model, we can examine the influence of a given risk factor on the survival time, while adjusting for other risk factors. For instance, if the values of all the other risk factors remain the same for a particular patient with MMC, then we can predict that an increase in Bence Jone protein in urine would decrease the survival time of the patient, and vice Versa. This observation can be very helpful in aiding and improving the treatment process of MMC. Also, the fact that Bence Jone protein in urine was ranked to be the highest contributor to MMC survival, implies that an MMC patient with an increased Bence Jone protein in urine can be at high risk for life-threatening situation and would require immediate and critical treatment attention. WBC and total serum protein were not individually found to be significantly contributing to survival time. However, having both risk factors present at the same time of diagnosis with MMC significantly influence the survival time. Such a novel finding can be useful to physicians in the treatment of MMC.

9.8 A New Statistical Modeling Approach for Survival Analysis of Cancer Patients—Multiple Myeloma Cancer (MMC)

In Sects. 9.2 and 9.3 of this chapter, we conducted parametric, nonparametric, and semi-parametric analysis of the survival times of 48 MMC patients (Mamudu and Tsokos 2020a, 2020b). For the parametric analysis, we obtained the survival function of MMC survival times from three-parameter log-normal distribution. The survival function of the nonparametric analysis was obtained Kaplan–Meier model, which was used to estimate the probability of survival times. Cox-proportional hazard was used to obtain the survival function of the survival times for semi-parametric analysis (Mamudu et al. 2020). The parametric survival function was compared to the nonparametric analysis of the survival time, and we justified that the parametric method was more robust and efficient. Regardless, these two methods do not take into consideration the risk factors contributing to the survival time. As a result, the Cox-PH model was found to be more relevant and robust for estimating the proportion of patients' survival because it considers the additional useful information provided by the risk factors (covariates) contributing to the survival times. However, the Cox-PH also has some flaws, i.e., satisfying the assumptions needed to apply the Cox-PH model can be very difficult, as well as the assessing interaction among the risk factors. As a result, most research studies use the Cox-PH model without satisfying the underlying assumptions or assessing risk factors interactions. This makes it difficult to justify the accuracy of conclusions made from using the Cox-PH model.

Further, we developed a data-driven nonlinear statistical model of the 48 patients diagnosed with MMC in Sect. 9.4. (Mamudu and Tsokos 2020a, 2020b). We obtained a very accurate and high model coefficient of determination, $R^2 = 0.8741$ along with $R^2_{adj} = 0.8401$, which defines the amount of variation in the survival times explained by the significantly identified risk factors. We bootstrapped 300 samples of the survival times from the 48 original survival times with replacement, which resulted in asymptotic significance of the coefficients or parameters of the risk factors and an increase $R^2 = 0.9116$ along with $R^2_{adj} = 0.9085$. The model had 93% predictive accuracy from the test data. Similar to the Cox-PH model, the nonlinear statistical model considers the additional contributions from risk factors influencing the survival times. The Cox-PH model has been popularly utilized in analyzing survival data. While the Cox-PH model is used to estimate the proportion of patients' survival after a given time, the nonlinear statistical model can predict the survival

Table 9.10 Rank of significant attributable risk factors by the Cox-PH model and the nonlinear statistical model

Rank	Cox-PH	Nonlinear statistical model
1	Blood urea nitrogen (BUN)	Bence Jone protein in urine, X_{13}
2	White Blood Cells (WBC)	Blood urea nitrogen (BUN), X_1
3	Bence Jone protein in the urine	Infections, X_4
4	Fractures	% Myeloid cells in peripheral blood, X_{11}
5	Infections & serum calcium	Serum calcium, $X_{16'}$
6	Proteinuria	Fractures, X_8
7	Gender	WBC & total Serum protein, $X_7.X_{14'}$
8	Platelets	Gender, X_6
9		Platelets, X_3
10		Age, X_5

times of patients for identified risk factors. The rankings of risk factors of the Cox-PH model and the nonlinear statistical model are given in Table 9.10. We can see that there are a lot of similarities in risk factors (covariates) by the Cox-PH model and nonlinear statistical model. The major differences between the two models were the ranking order or position of the covariates. This supports the high quality and accuracy of our research findings about the models, which provides a fundamental justification for comparison. An extensive information about risk factors influencing the survival times of MMC can be obtained among other studies (American Cancer Society 2018; Bethesda 2020; World Health Organization 2014b).

In the current section of this chapter, we estimated survival function from the nonlinear statistical model and use it to predict the proportion of patient's survival of MMC, and compared it with that of the survival function of the Cox-PH model analysis of the survival time as a function of covariates (Mamudu and Tsokos 2021). We also proposed a novel algorithm or approach to survival analysis of cancer diseases based on statistical modeling.

9.9 Development of the Survival Function of the Nonlinear Statistical Model

We find the survival function of the survival times t^* predicted from the final proposed nonlinear statistical model with 300 survival times (Mamudu and Tsokos 2020a; b). The proposed nonlinear statistical model was given by

$$\begin{aligned} t_i^* = \exp\big(&-4.377 - 1.097X_1 + 0.332X_{3normal} + 0.949X_{4present} \\ &+ 0.016X_5 + 0.562X_{6femal} - 0586X_{8present} + 0.022X_{11} \\ &-1.268X_{13none} + 4.151X_{16'} - 0.252X_7X_{14'}\big), \end{aligned} \tag{9.5.1}$$

where $i = 1, 2, \ldots, 300$ and

$$X_j = \begin{cases} 1 - e^{-X_{j'}}, & \text{if } X < 0 \\ -1 + e^{X_{j'}}, & \text{otherwise, for } j = 14, 16. \end{cases} \quad (9.5.2)$$

From above model (9.5.1), we estimated $t_1^*, t_2^*, \ldots, t_{300}^*$ survival times from the risk factors that identified., i.e., t_1^* represent the survival time of patient 1 as a function of the risk factors, and t_n^* is the survival time of the nth patient influenced by the risk factors. To assess the distribution of the predicted survival timest^*, we displayed the descriptive statistics of the continuous survival times t^*. A detailed explanation of the implication of the value of the statistic (especially kurtosis and skewness) is given in Sect. 9.2 of this chapter (Mamudu and Tsokos 2020a; b). We can see that t^* is skewed, given by the higher value of skewness and kurtosis (Table 9.11).

The PDF distribution of t^* was also found to be three-parameter log-normal probability distribution, which was the same distribution for the original 48 patients survival times$t$. This was expected because the 300 bootstrap samples came from the 48 samples with the 3p-log-normal probability distribution, further justifying the high quality of the proposed nonlinear statistical model (Mamudu and Tsokos 2021). To further support the fact that t and t^* follow the same distribution, we conducted a nonparametric Kruskal–Wallis test (Corder and Foreman 2011; Kruskal and Wallis 1952) to compare the difference in two survival times t and t^*. As given in Table 9.13, the Kruskal–Wallis rank-sum test resulted in p-value $= 0.9066 \approx 1$, which is an indication of no difference between the two survival times. We estimated the parameters of the 3p-log-normal parameter distribution utilizing the maximum likelihood estimation (MLE) method as presented among studies (Cohen 1951; Hoare and Tsokos 2009; Mamudu and Tsokos 2020a, b). The results are given in Table 9.12.

Therefore, the 3p-log-normal pdf, $f(t^*)$, of the survival times of 300 MMC patients is given by

Table 9.11 Descriptive statistics of survival times t^* of multiple myeloma cancer

Survival time	Mean	Median	Std Err	Std Dev	Kurtosis	Skewness
t^*	22.07	15.68	1.14	19.59	1.46	1.46

Table 9.12 Parameter estimates for the 3p-log-normal pdf for t^*

Survival time	Location ($\hat{\gamma}$)	Scale ($\hat{\mu}$)	Shape ($\hat{\sigma}$)
t^*	4.4832	2.5603	1.0303

Table 9.13 Kruskal–Wallis rank-sum test of the difference between t and t^*

Type of test	Survival time	Data: list (t, t^*)
Kruskal–Wallis	Chi- squared $(\chi^{-2}) = 0.013776$	$p{-}value = 0.9066$

$$f\left(t^*|\gamma^*,\mu^*,\sigma^{*2}\right)=\begin{cases}0, & \text{if } t^*\le\gamma^*\\ \left(2\pi\sigma^{*2}\right)^{-\frac{1}{2}}\left(t_i^*-\gamma^*\right)^{-1}\exp\left(-\frac{1}{2}\left(\frac{\ln(t_i^*-\gamma^*)-\mu^*}{\sigma^*}\right)^2\right) & \text{if } t^*>\gamma^*\end{cases} \tag{9.5.3}$$

When we substitute the parameter estimates given in Table 9.12, we obtain

$$f\left(t^*|\gamma^*,\mu^*,\sigma^{*2}\right)=\begin{cases}0, & \text{if } t^*\le 4.4832\\ 0.38253\left(t_i^*-4.4832\right)^{-1}\exp\left(-\frac{1}{2}\left(\frac{\ln(t_i^*-4.4832)-2.5603}{1.0303}\right)^2\right), & \text{if } t^*>4.4832.\end{cases}$$

The pdf $f(t^*)$ plot of the survival times of the 300 MMC patients (t^*) is shown in Fig. 9.10. We can compute the probability that the survival time of an MMC patient will fall between a given time t_k^* and $t_{(k+1)}^*$. For instance, we can estimate the probability that a patient will survive between 20 and 40 months to be $P(20\le t^*\le 40)=0.025-0.008\approx 0.017$, as shown in Fig. 9.10. This can be interpreted as there was approximately a 1.7% probability that a patient survives between 20 and 40 months. Contrary, for survival times t, $P(20\le t\le 40)=0.019-0.007\approx 0.012$.

The 3p-log-normal cumulative distribution function, $F(t^*)$, of the survival times of t^* is given by

$$F_{T^*}\left(t^*|\gamma^*,\mu^*,\sigma^{*2}\right)=\frac{1}{\sqrt{2\pi}}\int_0^{t^*}\exp\left(-\frac{1}{2}z^2\right)dz=\Phi\left(\frac{ln\left(t_i^*-\gamma^*\right)-\mu^*}{\sigma^*}\right). \tag{9.5.4}$$

Replacing the parameter estimates from Table 9.12, we have

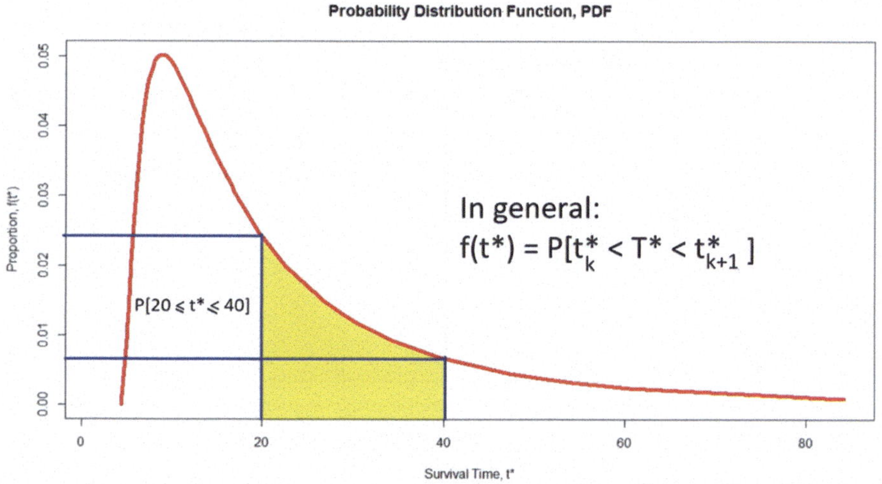

Fig. 9.10 Probability distribution function of survival time, t^* of multiple myeloma cancer

$$F_{T^*}\left(t^*|\gamma^*, \mu^*, \sigma^{*2}\right) = P\left[t^* \le T^*\right] = \Phi\left(\frac{ln\left(t_i^* - 4.4832\right) - 2.5603}{1.0303}\right),$$

where $\Phi(.)$ denotes standardized normal CDF. The CDF of the survival times t^* of MMC patients is shown in Fig. 9.11. Figure 9.11 depicts the probability that a patient with MMC survives up to a given time t^*. For instance, we can compute the probability that an MMC patient will survive up to time $t^* = 40$ months as; $F(t^* = 40) = P(t^* \le 40) \approx 0.82$. This implies about 82% chance that an MMC patient will survive up to 40 months. Further, we can compute the probability that a patient will survive after 40 months to be $P(t^* > 40) = 1 - F(t^* = 40) = 0.18$. From the original survival times t, $F(t = 40) = P(t \le 40) \approx 0.83$, and $P(t > 40) = 1 - F(t = 40) = 0.17$.

The survival function $\hat{S}(t^*)$ of t^* was obtained as

$$\hat{S}\left(t_i^*|\gamma^*, \mu^*, \sigma^{*2}\right) = 1 - F_{T^*}\left(t^*|\gamma^*, \mu^*, \sigma^{*2}\right) = 1 - \Phi\left(\frac{ln\left(t_i^* - \gamma^*\right) - \mu^*}{\sigma^*}\right). \tag{9.5.5}$$

We replace the estimates of the parameter given in Table 9.12, and we have

$$\begin{aligned}\hat{S}\left(t_i^*\middle|\gamma^*, \mu^*, \sigma^{*2}\right) &= 1 - F_{T^*}\left(t^*\middle|\gamma^*, \mu^*, \sigma^{*2}\right)\\ &= 1 - \Phi\left(\frac{ln\left(t_i^* - 4.4832\right) - 2.5603}{1.0303}\right),\end{aligned}$$

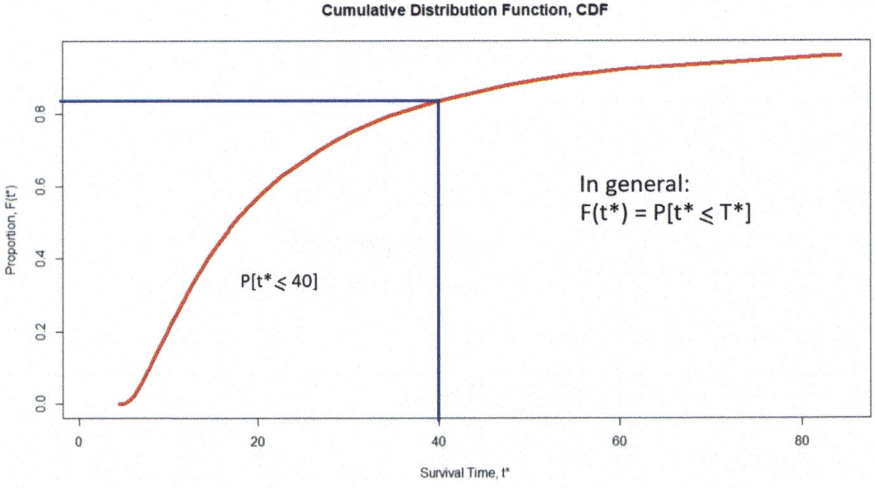

Fig. 9.11 Cumulative distribution function of survival time, t^* of multiple myeloma cancer

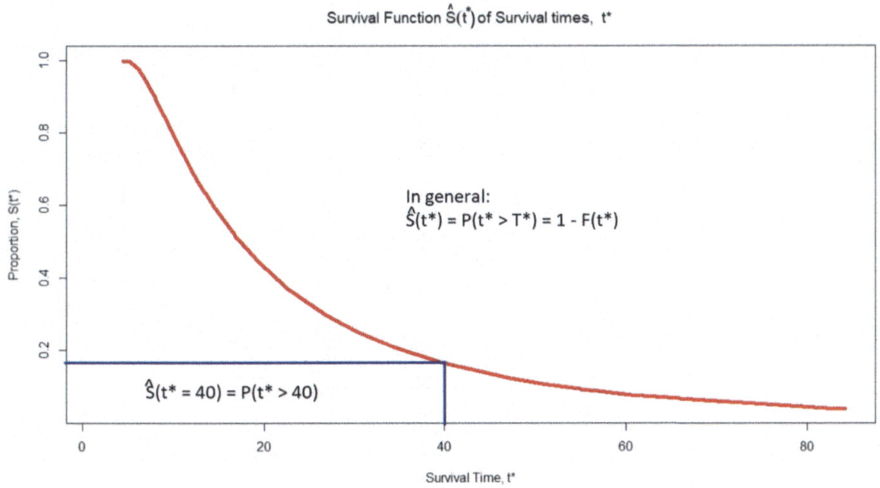

Fig. 9.12 Survival function of the survival times, t^* of multiple myeloma cancer patients

where $\Phi(.)$ denotes standardized normal CDF of t^* and $\hat{S}(t^*)$ represents the probability that a patient with MMC survives beyond a given time t^*. Using the survival function in Fig. 9.12, we can compute the probability that an MMC patient will survive beyond 40 months as, $\hat{S}(t^* = 40) = P(t^* > 40) \approx 0.18$. For t, $\hat{S}(t = 40) = P(t > 40) \approx 0.17$.

9.10 Algorithm for the New Nonlinear Statistical Modeling Approach to Survival Analysis

See Fig. 9.13.

9.11 Comparing the Survival Function of the Cox-PH Model with that of the Non-linear Statistical Model of Survival Times of Multiple Myeloma Cancer

As shown earlier, both the proposed Cox-PH model and the nonlinear statistical model were of high quality since they satisfy all the needed model assumptions and pass all the criteria for measuring the robustness and efficiency of a high-profile model. The Cox-PH model predicts the proportion or probability of survival at a given time as a function of identified significant risk factors or covariates, and the nonlinear statistical model predicts the survival time for a given risk factors. In

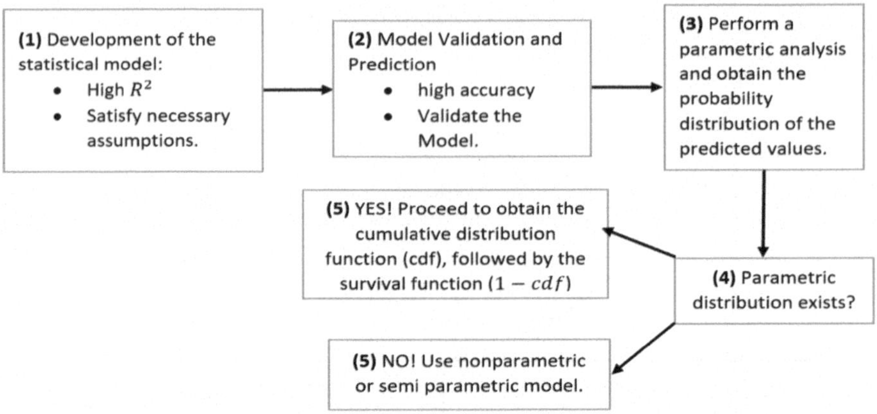

Fig. 9.13 Algorithmic flowchart for the development of the survival function of the nonlinear statistical model

Sect. 9.3 (Mamudu et al. 2020), we showed that Cox-PH model was of high importance given that it takes into account the influence of risk factors on the survival time, at the time patients were diagnosed with MMC. In this section, we compare the survival function of the Cox-PH model with that of the nonlinear statistical model of the MMC patients.

As shown in Fig. 9.14, we can see that the survival function of the Cox-PH model is below that of the nonlinear statistical model, that is, the nonlinear statistical model consistently estimates a higher survival probability of patients diagnosed with MMC than the Cox-PH model. The nonlinear statistical model survival function of the survival times t^* gave better estimates because it is based on a well-defined parametric probability distribution, which is more powerful than the survival function of the survival times t from the semi-parametric Cox-PH model.

9.12 Discussion

In this section of this contributed chapter, we estimated 300 survival times t^* from the proposed statistical model in Eq. (4.1) and assessed the probability distribution of survival. The PDF probability distribution of the t^* was found to be 3p-lognormal, which was the same as the probability distribution of the initial sample of 48 patients diagnosed with the MMC. The method of maximum-likelihood parameter estimation (Mamudu and Tsokos 2020a; b) was used to obtained the values of the parameters of the distribution $f_{T^*}(t^*)$ (see Table 9.12). We integrated PDF $f_{T^*}(t^*)$ with respect to t^* to obtain the CDF $F_{T^*}(t^*)$. The survival function $\hat{S}(t^*)$ was then obtained by subtracting the CDF from 1 (i.e., $1 - F_{T^*}(t^*)$). The CDF was used to estimate the survival probability up to a given time t^*, whereas the survival function estimated the probability of survival beyond a given time t^*. The $\hat{S}(t^*)$ of the statistical

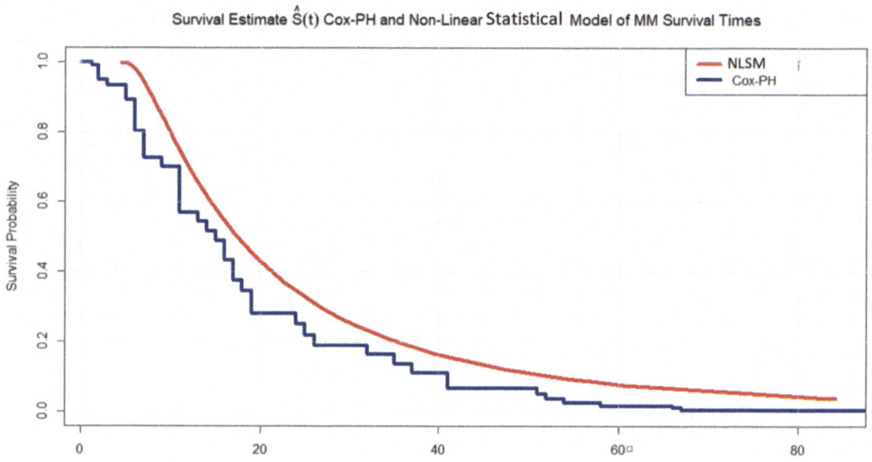

Fig. 9.14 Survival function of Cox-PH versus survival function nonlinear statistical model of MMC

nonlinear statistical model (Eq. 9.5.1) was compared with the $\hat{S}(t)$ of the Cox-PH model (Eq. 3.4), given by Fig. 9.14. The $\hat{S}(t^*)$ of the nonlinear statistical model gave a better estimate of the proportion of survival times of MMC patients than the $\hat{S}(t)$ of the Cox-PH model, with the $\hat{S}(t^*)$ of the nonlinear statistical model developed from a well-defined parametric distribution of the survival times of MMC patients.

We showed the ranks of the risk factors identified by the two models (see Table 9.10). The Cox-PH model identified and ranked the interaction between infections and serum calcium as fifth contributing factor to the survival of MMC. These two risk factors were identified individually by the nonlinear statistical model to be significantly contributing to the survival of MMC, ranked third and fourth, respectively. The ranking process of the Cox-PH model measured the prognostic effect of the risk factor on the survival of MMC using the hazard ratio. On the other hand, the nonlinear statistical model ranking of the risk factors measured the percentage of variation in the survival time explained by the identified risk factors, known as the coefficient of determination, R^2. Both the hazard function and the R^2 can play a similar role by measuring the impact of a given risk factor on the survival time. But R^2 along with the R^2_{adj} is much preferred because it measures the entire variability of the survival explained by the risk factors with a high degree of accuracy.

It is very important to recognize that both models are high profile considering the quality involved in the model building process. However, patients and medical personnel would be more interested and concerned about how long they can survive with MMC rather than the probability of surviving. Furthermore, knowing the percentage of contribution of a risk factor to the survival time can play a key role in the treatment process of MMC rather than whether it is a good or bad prognostic factor to the survival time. This makes the rank of the risk factors by the nonlinear statistical model more relevant. Moreover, it is more difficult to develop Cox-PH model by satisfying the assumptions and finding the interaction between covariates.

Altman and Stavola (1994) presented the practical problems in fitting a Cox-PH model. Ian Ford et al. (1995) also find inconsistencies in the development of Cox-PH model. With the criticisms about effectiveness of Cox-PH models, using the statistical model approach to survival analysis might be more appropriate which provides more flexibilities than the Cox-PH model. Our findings showed that we can obtain a better and more accurate estimation of the probability of a given survival time if we can find a well-defined parametric distribution that characterizes a given cancer survival data. Given that the objective would be to maximize the survival times, the nonlinear statistical model provides better strategy for enhancing the therapeutic/ treatment of MMC than the Cox-PH model.

9.13 Conclusion

In the present chapter, we have demonstrated that it could be misleading to adopt the nonparametric Kaplan–Meier for survival analysis of every cancer disease. As long as we can find a well-defined parametric probabilistic distribution for the survival times of a given cancer disease, we are deemed to obtain a robust estimate or probability of survival. We further developed a high-quality Cox-PH model that identified new risk factors associated with the length of survival (i.e., platelets, gender, white blood cells, and fractures; and the interaction term, infections and serum calcium) and accurately predict the proportion of survival time of patients with MMC. However, with the noticeable limitations and difficulty in developing and using the semi-parametric Cox-PH model, we proposed a high-quality and well-validated statistical model that predicts the real value of the survival times of MMC and has other vital usefulness beyond just the prediction of the survival time. We then developed a novel approach to survival analysis using the proposed statistical model.

This study provided a novel approach to survival analysis using a parametric approach of nonlinear statistical model, which offer a better estimation of the probability of survival times of patients diagnosed with MMC. The nonlinear statistical model does not only provide a better estimate of the survival probability, but it also has several usefulness, including: (1) It can predict the survival time of a patient by considering impact of the risk factors and the interaction term. (2) The risk factors and the interaction term can be ranked according to their percentage of contribution to the survival time. (3) We can perform for the optimization analysis of the survival time as a function of the risk factors and the interaction. (4) Confidence intervals for the survival time can be computed for inferential analysis. (5) We can also develop a survival function using parametric approach to obtain better survival estimates (i.e., the probability that a patient survives beyond a given time) than existing survival models. On the other hand, the Cox-PH model only useful for estimating the probability of survival. The contribution made in this chapter provides a novel approach and further improvement in cancer survival analysis to enhance research and policy intervention strategies for the treatment and prevention of cancer diseases, which can help to improve the length of survival of cancer patients.

Acknowledgements The similarity index is due to the fact that the content of the chapter is based primarily on the academic publications of the authors.

References

Akaike H (1974) A new look at the statistical model identification. IEEE Trans Autom Control 19(6):716–723
Alexanian R, Balcerzak S, Bonnet JD, Gehan EA, Haut A, Hewlett JS, Monto RW (1975) Prognostic factors in multiple myeloma. Cancer 36:1192–1201
Altman DG, De Stavola BL (1994) Practical problems in fitting a proportional hazards model to data with updated measurements of the covariates. Stat Med 13(4):301–341
American Cancer Society (2018) About multiple myeloma. The American cancer society medical and editorial content team. https://www.cancer.org/cancer/multiple-myeloma/about/what-is-multiple-myeloma.html
American Cancer Society (2020) About multiple myeloma. https://www.cancer.org/cancer/multiple-myeloma/about/what-is-multiple-myeloma.html
Bethesda M (2020) SEER cancer facts: myeloma. National Institute, Issue. https://seer.cancer.gov/statfacts/html/mulmy.html
Brody T (2016) Clinical trials: study design, endpoints and biomarkers, drug safety, and FDA and ICH guidelines. Academic press
Case LD, Kimmick G, Paskett ED, Lohman K, Tucker R (2002) Interpreting measures of treatment effect in cancer clinical trials. Oncologist 7(3):181–187
Cohen AC Jr (1951) Estimating parameters of logarithmic-normal distributions by maximum likelihood. J Am Stat Assoc 46(254):206–212
Corder GW, Foreman DI (2011) Nonparametric statistics for non-statisticians. Wiley
Costa G, Engle R, Taliente P (1969) Criteria for defining risk groups and response to chemotherapy in multiple myeloma. Proceedings of the american association for cancer research
Van de Donk NW, M. T., Poddighe PJ, Lokhorst HM, Zweegman S (2016b) Diag-nosis, risk stratification and management of monoclonal gammopathy of undetermined significance and smoldering multiple myeloma. Int J Lab Hematol. https://doi.org/38 Suppl 1: 11022. https://doi.org/10.1111/ijlh.12504. PMID 27161311
Durie B, Salmon SE, Moon TE (1980) Pretreatment tumor mass, cell kinetics, and prognosis in multiple myeloma
Durie BG, Salmon SE (1975) A clinical staging system for multiple myeloma correlation of measured myeloma cell mass with presenting clinical features, response to treatment, and survival. Cancer 36(3):842–854
Dutta AK, Hewett DR, Fink JL, Grady JP, Zannettino AC (2017) Cutting edge genomics reveal new insights into tumour development, disease progression and therapeutic impacts in multiple myeloma. Br J Haematol 178(2):196–208
Efron B, Tibshirani R (1993) An introduction to the bootstrap. 57 Boca Raton. CRC Press, FL
Feigl P, Zelen M (1965) Estimation of exponential survival probabilities with concomitant information. Biometrics 826–838
Ferri FF (2013) Ferri's clinical advisor 2014 e-book: 5 books in 1
Ford I, Norrie J, Ahmadi S (1995) Model inconsistency, illustrated by the Cox proportional hazards model. Stat Med 14(8):735–746
Harley J (1971) Ten years of experience in multiple myeloma at the West Virginia University Hospital. Morgantown, WV
Hoare A, Tsokos CP (2009) Statistical analysis and modeling of coral reef habitats. Nonlinear Anal Theory Methods Appl 71(12):e1360–e1369
Kalbfleisch JD, Prentice RL (2011) The statistical analysis of failure time data. Wiley

Kaplan EL, Meier P (1958) Nonparametric estimation from incomplete observations. J Am Stat Assoc 53(282):457–481
Kiang D, Goldman A, Fortuny I, Kheologi A, Kennedy B (1973) Prognostic factors in multiple-myeloma. Proceedings of the American association for cancer research
Kleinbaum DG, Klein M (2012) Survival analysis: a self-learning text, vol 3. Springer
Korde N, Kristinsson SY, Landgren O (2011) Monoclonal gammopathy of undetermined significance (MGUS) and smoldering multiple myeloma (SMM): novel biological insights and development of early treatment strategies. Blood, J Am Soc Hematol 117(21):5573–5581
Krall JM, Uthoff VA, Harley JB (1975) A step-up procedure for selecting variables associated with survival. Biometrics 49–57
Kruskal WH, Wallis WA (1952) Use of ranks in one-criterion variance analysis. J Am Stat Assoc 47(260):583–621
Kyle RA, Rajkumar SV (2008) ASH 50th anniversary review. Blood 111(6):2962–2972
Kyle RA, Bayrd ED, Bayrd EA (1976) The monoclonal gammopathies: multiple myeloma and related plasma-cell disorders. Thomas Publisher, Charles C
Mamudu L, Tsokos CP, Otunuga OE (2020) Survival analysis of multiple myeloma cancer (MMC) using the Cox-proportional hazard model. Med Clin Res 5(7).
Mamudu L, Tsokos C (2020a) Data-driven statistical modeling and analysis of the survival times of multiple myeloma cancer. Health Sci J 14(1):1–11
Mamudu L, Tsokos CP (2020b) Parametric and non-parametric analysis of the survival times of patients with multiple myeloma cancer. Open J Appl Sci 10(4):118–134
Mamudu L, Tsokos CP (2021) A new statistical modeling approach for survival analysis of cancer patients—multiple myeloma cancer. Open J Appl Sci 10(4):365–378
McCarthy PL, Holstein SA, Petrucci MT, Richardson PG, Hulin C, Tosi P, Bringhen S, Musto P, Anderson KC, Caillot D (2017) Lenalidomide maintenance after autologous stem-cell transplantation in newly diagnosed multiple myeloma: a meta-analysis. J Clin Oncol 35(29):3279
Merlini G, Waldenstrom JG, Jayakar SD (1980) A new improved clinical staging system for multiple myeloma based on analysis of 123 treated patients
National Cancer Institute (2017) Plasma cell neoplasms (including multiple myeloma) treatment. https://www.cancer.gov/types/myeloma/patient/myeloma-treatment-pdq#section/all
Rabb M, Podar K, Breitkreutz I, Richardson PG, Anderson KC (2009) Multiple myeloma. Lancet 374:324–339
Richard Peto J (1972) Asymptotically efficient rank invariant test procedures. J Royal Stat Soc, Ser A (135):185–207. https://doi.org/10.2307/2344317
Roberts DL, Dive C, Renehan AG (2010) Biological mechanisms linking obesity and cancer risk: new perspectives. Annu Rev Med 61:301–316
Sharma R, Bhandari R (2015) Skewness, kurtosis and Newton's inequality. Rocky Mt J Math 45(5):1639–1643
Siegel RL, Miller KD, Fuchs HE, Jemal A (2021) Cancer statistics, 2021. CA: A Cancer J Clin 71(1):7–33
Sonneveld P, Schmidt-Wolf IG, van der Holt B, El Jarari L, Bertsch U, Salwender H, Zweegman S, Vellenga E, Broyl A, Blau IW (2012) Bortezomib induction and maintenance treatment in patients with newly diagnosed multiple myeloma: results of the randomized phase III HOVON-65/GMMG-HD4 trial. J Clin Oncol 30(24):2946–2955
Therneau TM, Grambsch PM, Fleming TR (1990) Martingale-based residuals for survival models. Biometrika 77(1):147–160
Van de Donk N, Mutis T, Poddighe P, Lokhorst H, Zweegman S (2016a) Diagnosis, risk stratification and management of monoclonal gammopathy of undetermined significance and smoldering multiple myeloma. Int J Lab Hematol 38:110–122
World Health Organization (2014a) World cancer report 2014a pp Chapter 5.13
World Health Organization (2014b) World cancer report. pp Chapter 5.13. ISBN 978-9283204299

Chris P. Tsokos is Distinguished University Professor of Mathematics and Statistics at the University of South Florida. He received his B.S. in Engineering Science/Mathematics and his M.A. in Applied Mathematics from the University of Rhode Island. He was recognized with the Distinguished Alumni Excellence Award in Science and Technology. He received his Ph.D. in Statistics and Probability from the University of Connecticut. He was recognized with the Distinguished Alumni Award along with the recognition of the State of Connecticut General Assembly Award for outstanding public and philanthropic service to our country. He has also served on the faculties at Virginia Polytechnic Institute and State University and the University of Rhode Island. His research has extended into a variety of areas in the mathematical sciences, including stochastic systems, statistical models, reliability analysis, ecological systems, operations research, time series, Bayesian analysis, mathematical and statistical modeling of global warming, and both parametric and nonparametric survival analysis, cybersecurity, financial systems, among others. He is Author of more than 500 research publications in these areas. He is Author of more than 25 research monographs and books in mathematical and statistical sciences. He has been invited to lecture in several countries around the globe—Russia, People's Republic of China, India, Turkey, and most EU countries. He has mentored and directed the doctoral research of more than 75 postgraduate students and more than 100 Master of Science students. For the past five years, his research efforts have been focused on developing probabilistic analysis, parametric and nonparametric statistical models for breast, lung, brain, pancreatic, multiple myeloma, colon, and prostate cancer. His research aims are real data-driven and are oriented toward understanding these types of cancers and, most importantly, statistically identifying the attributable variables and interactions that cause such cancers. His research in cybersecurity analysis and modeling of computer systems' vulnerability is at the subject's frontiers. He is the recipient of several US patents in cybersecurity, health sciences, and geosciences, among others. He has served as Advisor, Consultant, and Lecturer for the US Army, the US Environmental Protection Agency, the US Air Force Office of Scientific Research, the US Navy, NASA, the Bureau of Land Management, and the American Cancer Society, among other US government agencies. He also has performed similar services for several public Fortune 500 companies. He was Co-founder and Director of the award-winning Urban Scholars Outreach Program (USOP) at USF for more than 20 years, whose objective was to offer free educational assistance to disadvantaged African-American and Hispanic students. USOP was started in 1998 and has helped hundreds of students to enter our universities. He is President of the International Federation of Nonlinear Analysts (IFNA), a not-for-profit global educational research organization with more than 125 participating countries. IFNA is an interdisciplinary professional organization globally promoting the understanding of related complex nonlinear problems and approaches to solutions from all disciplines. Our motto is "Understanding Through Global Diversity, Cooperation, and Collaboration". He is Member of several academic and professional societies. He is serving as Honorary Editor, Chief Editor, Editor, or Associate Editor of more than 20 international academic research journals. He is the recipient of many distinguished awards and honors, including Fellow of the American, Statistical Association, the International Statistical Institute, USF Distinguished Scholar Award, Sigma Xi Outstanding Research Award, USF Outstanding Undergraduate Teaching Award, USF Professional Excellence Award of the University Area Community Development Corporation, and the Time Warner Spirit of Humanity Award, among others.

Dr. Lohuwa Mamudu is Assistant Professor in the Department of Public Health at California State University Fullerton (CSUF), USA. He received a Ph.D. Statistics from the University of South Florida, USA. He is Statistician/Data Analyst/Data Scientist, Epidemiologist, and Research Consultant. He is Associate Editor of Preventing Chronic Disease Journal of the Centers for Disease Control and Prevention (CDC), USA. His research is data-driven and interdisciplinarily focused on statistical/analytical modeling and applications of qualitative and quantitative statistical and machine learning methods and algorithms. He collaborates with the National Institute of Health (NIH) and Tennessee Cancer Registry (TCR) as Freelance Data Analyst and

Researcher, working on cancer disease outcomes and disparities (breast, lung cancer, etc.), infectious disease, chronic disease, and immigration health disparities (depression, anxiety, and psychological distress) in the USA. He is Author/Co-author of several peer-reviewed research articles. He is Mentor/Advisor of several mentorship/advisory programs. He organizes, facilitates, and instructs in academic and research workshops. His long-term research goal and interest will focus on big data analytics and artificial intelligence (AI) in public health research.

Part IV
Digitalization

Chapter 10
Refined Machine Learning Approaches for Mask Policy Analysis

Lincy Y. Chen and John Tuhao Chen

Abstract Worldwide, the Covid-19 pandemic has created an imminent need for new public policies focused on infectious disease intervention, management, and prevention. Mask debates take center stage amongst these policies, but disagreements abound regarding the efficacy, fairness, and health effects of mandatory versus voluntary masking. Properly assessing these policies with available data necessitates statistical models. One of the main problems with statistical models is the misusage and improper interpretations. This is particularly true in the context of COVID-19 because of the high degree of random variation and dynamic evolution processes aimed at restraining the transmission of airborne viruses. For example, public health officials frequently employ logistic regression models to analyze odds ratios associated with social and behavioral consequences, which determine public health and management policies. However, the implementation of statistical analysis usually involves hidden model assumptions. Violating these model assumptions often invalidates data analytical results, and consequently misleads the direction of policy reforms. In this chapter, we put forth new and refined statistical instruments with relaxing model assumptions for public health policy analysis. After a discussion on updated machine learning techniques including range regression, weighted classification with KNN (k-nearest-neighbors) algorithm, and enhanced decision trees, we apply the refined hybrid statistical methods to analyze a set of social and behavioral data collected during the Covid-19 pandemic. The dataset contains survey information on social and behavioral characteristics in conjunction with mask policies (mandatory vs voluntary). Results of the refined data analysis cast new light on the effects of mandatory versus voluntary mask policies. Although the former (mandatory) is more effective and better controls the frequency of airborne virus transmission, the latter appears to be more friendly and associated with higher community warmth indices.

L. Y. Chen
School of Industrial and Labor Relations, Cornell University, Ithaca, NY, US

J. T. Chen (✉)
Department of Mathematics and Statistics, Bowling Green State University, Bowling Green, OH, US
e-mail: jchen@bgsu.edu

E. Çetin and H. Özen (eds.), *Healthcare Policy, Innovation and Digitalization*,
Accounting, Finance, Sustainability, Governance & Fraud: Theory and Application,
https://doi.org/10.1007/978-981-99-5964-8_10

Keywords Range regression · Weighted KNN · Decision tree · Covid-19 · Mask policy

10.1 Introduction

The Covid-19 pandemic has shown the world how an unexpected viral outbreak can rapidly create a global public health crisis. Within a few weeks of the first reported cases of Covid-19, the virus spread into cities, countries, ravaged over all corners of the globe, and turned into a catastrophe for humanity. As of May 2022, over 6.29 million people have died from this pandemic. Humanity was unprepared and helpless to face such an abrupt event. Amid the frustrating experiences of building lockdowns, travel restrictions, daily mortality counts, and hospital crises, ordinary individuals started discussing self-protection measures such as vaccinations and mask wearing. Mask wearing, which was originally reserved to the discourse of public health researchers, became a ubiquitous topic in daily news and routine conversations. In the United States, one of the most contentious issues was mandate versus voluntary mask policies. Proponents of mandatory mask policies argued that wearing a mask could effectively prevent airborne transmission, and mask mandates could significantly reduce the risk of inhaling infectious aerosols in enclosed areas. On the other hand, some people questioned the efficacy and social consequences of mask wearing. Mandatory mask policies quickly evolved from the public health sphere. Some viewed mask mandates as infringements on freedom and liberty. Others interpreted wearing a mask to signal that someone had Covid. Considering the digital era, it is critical to address the public perception with data information. Betscha et al. (2020) published a study regarding human being behaviors toward the mask policy in Germany, in which logistic regression model was applied to analyze the data.

Let $X_1, \ldots, X_k$ be k discrete predictors for a binary outcome y, where $y = 0, 1$ for the occurrence of the event (success versus failure), respectively. Denote $p_t = P(Y = 1|X = t)$ where t is one of the categories of the combinations of the domains of $X_1, \ldots, X_k$, denoted as $X_{(t)}$. The logistic regression model fits the data with the equation,

$$\log\left(\frac{p_t}{1 - p_t}\right) = \alpha + \beta X_{(t)},$$

where α is a constant in R and β is a real vector in R^k.

In data analytics, papers in application (such as Betsch et al. 2020) utilizing the logistic regression model, commonly focus on the binary outcome of the response variable y, without checking the plausibility of the model assumption. In fact, the underlying model assumption for the logistic regression reads.

$$y_t \sim \text{Bernoulli}(p_t),$$

where

$$p_t = \frac{e^{\alpha+\beta X_{(t)}}}{1+e^{\alpha+\beta X_{(t)}}}.$$

Such a model assumption is implausible if the regression model fails to include all the associated variables. To illustrate this point, consider a simple situation where $X_{(t)} = 0, 1$ for being exposed to smoking (a single risk factor) when the outcome variable y is mortality. In the logistic regression model, we have

$$\log\left(\frac{p_1}{1-p_1}\right) = \alpha + \beta X_{(t)} = \log\left(\frac{p_0}{1-p_0}\right) + \beta.$$

This implies that

$$\frac{\frac{p_1}{1-p_1}}{\frac{p_0}{1-p_0}} = e^{\beta}.$$

It means that the odds ratio between the smoking and non-smoking patients is a constant, which can only be plausible theoretically. In practice, the risk of the outcome (the death of a patient) is significantly associated with other factors such as the severeness of the disease, age, and morbidity. Thus, there is an inherent dilemma with the use of the logistic regression model to test for variable significance. On one hand, the validity of the model requires the assumption that *all* related factors are already specified in the model. On the other hand, had we known the significance of the risk factors, we would not need the logistic regression model for the purpose of testing significance.

Another common misusage of the underlying model assumption occurs in the application of the ordinary least squares (OLS) regression. When the response variable is discrete, the model assumption that $\varepsilon \sim N(0, \sigma^2)$ for

$$Y = \alpha + \beta X + \varepsilon$$

is implausible because the error term of the model is permissible only in a discrete domain, which violates the normality assumption.

Under this scenario, we need refined techniques to do proper data analysis. In the next section, we will discuss three refined analytic approaches that can be used to address some of the common problems in analyzing health-related policies. We then apply these techniques to analyze the German Covid-19 data.

10.2 Range Regression

Range regression is a data analytics tool that lets us use the ordinary data while relaxing the often-hidden assumption concerning the underlying distribution. It was first proposed (Comerota et al. 2012) in a medical data analysis when evaluating the status of venous disorders using CEAP (clinical, etiological, anatomical, and patho-physiological) scores for patients with thrombolysis and post-thrombotic syndrome. The method successfully detected a significant relationship between the average CEAP score and the percentage of clot lysis.

Consider any random vectors (X_i, Y_i) for $i = 0, 1, \ldots, n$, where X_i is the predictor and Y_i is the response. Partition the domain of the predictor into k categories, $T_1, \ldots, T_k$. Thus, the original dataset is partitioned into k sections, $(X_{j1}, Y_{j1}), \ldots, (X_{jm_j}, Y_{jm_j})$, for $j = 1, \ldots, k$. Denote z_j and t_j the average of the response and predictor, respectively, in category j. We have

$$z_j = \frac{1}{m_j} \sum_{p=1}^{m_j} Y_{\text{jp}} \quad \text{and} \quad t_j = \frac{1}{m_j} \sum_{p=1}^{m_j} X_{\text{jp}}.$$

In this way, the linear relationship between the predictor X and the response Y is represented by the corresponding average values $\left(t_j, z_j\right)$ with $j = 1, \ldots, k$.

The method of range regression was further discussed by Chen and Comerota (2012), as well as Chen (2014). Kerns and Chen (2016) derived the asymptotic normality property on the random variation term in range regression and theoretically proved that the method (after regrouping) produces stronger correlation coefficients between the index and the outcome variable. Chen and Hoppe (2017) discussed the intrinsic connection between simultaneous confidence region and range regression. Stoll (2019) further extended the method into nonparametric settings by using the sample median (instead of the sample mean) of the response variable after grouping.

The key feature of range regression focuses on the reduction of data variation by taking averages in each category of the predictor and the response. When the sample size in each category is large enough, as discussed by Kerns and Chen (2016), the central limit theorem guarantees the asymptotic normality of the sample means. Thus, the method does not require the distribution assumption on the response variable Y. Usually, the number of partitions and the categories is intrinsically determined by the characteristics of the predictor variable. When no additional information is available, cross-validation method can be used to select the optimal number of categories, as delineated in the following algorithm.

10.3 Range Regression Algorithm

1. Divide the data into a training set (75%) and a test set (25%).

2. For any k partitions on the domain of the predictor, compute the average values of the predictor (t_j) and the response variable (z_j) for each category.
3. Fit the linear regression with (t_j, z_j) for $j = 1, \ldots, k$.
4. Find the mean squared error (MSE) between the predicted and the observed.
5. Repeat Step-2 with different permissible values of k.
6. Select the k value that corresponds to the smallest MSE.

When the data of interest has intrinsic selection on the range, for instance, in the mask policy analysis with the German Covid-19 data, the survey participants' mask-wearing behavior is intrinsically classified in the six frequency categories: "not-applicable", "never", "rarely", "sometimes", "often", and "always". The ranges for the analysis are naturally selected as these six classes. In this case, there is no need to run the cross-sectional analysis for the ranges.

10.4 Weighted K-Nearest Neighborhood Classification

Another method of data analytics that does not require hidden assumptions on the underlying distribution of the data is the KNN (K-nearest neighborhood classification) approach. To enhance the conventional approach of KNN, we put forward a new approach called "weighted KNN" classification and discuss its theoretical property in this section.

Consider the partition of n subjects into k classes. Assume that each subject is characterized by a vector x_i for $i = 1, \ldots, n$. Let $C_1, \ldots, C_K$ be a partition of the index set $\{1, \ldots, n\}$. The conventional K-means clustering algorithm finds the partition for the following optimization problem.

$$\text{Minimize}_{C_1,\ldots,C_k}\left\{\sum_{k=1}^{K} W(C_k)\right\},$$

where $|C_k|$ denotes the number of elements in set C_k, and

$$W(C_k) = \frac{1}{|C_k|}\sum_{i\in C_k}\sum_{i'\in C_k}\sum_{j=1}^{p}(x_{\mathrm{ij}} - x_{i'j})^2.$$

Namely, the k clusters are constructed by minimizing the sum of the squared Euclidean distances featuring the n subjects. However, in real applications, some features are endowed with higher level of importance than other features. For instance, among the possible Covid-19 symptoms, fever, new loss of taste or smell, muscle or body aches, fatigue, trouble breathing, persistent pain or pressure in the chest, not all the symptoms carry the same weight toward the diagnosis of the disease. The new loss of taste or smell, trouble breathing, and chest pain are likely more of the warning signs for Covid-19, compared with fever or muscle aches (which may be caused by various reasons). Assume that each feature carries a weight ρ_i for the diagnosis of the disease, without any pre-information on the closeness of the features (such as what happened at the beginning of the pandemic), the "weighted KNN" approach is to

$$\text{Minimize}_{C_1,\ldots,C_K}\{\sum_{k=1}^{K} W(C_k)\},$$

where $W(C_k)$ is now a weighted quantity,

$$W(C_k) = \frac{1}{|C_k|}\sum_{i\in C_k}\sum_{i'\in C_k}\sum_{j=1}^{p}\rho_j(x_{\text{ij}} - x_{i'j})^2.$$

Notice that different from the conventional approach, the target function for optimization in the "weighted KNN" includes the weight ρ_j for the difference on the feature $x_{\text{ij}} - x_{i'j}$. To construct the optimization algorithm, we need to show the following result, which connects the problem of minimizing the sum of the weighted squared Euclidean distance for each cluster, to the problem of minimizing the weighted within-cluster variation.

Theorem 1 $\frac{1}{|C_k|}\sum_{i\in C_k}\sum_{i'\in C_k}\sum_{j=1}^{p} w_j(x_{\text{ij}} - x_{i'j})^2 = 2\sum_{i\in C_k}\sum_{j=1}^{p} w_j(x_{\text{ij}} - \overline{x}_{\text{kj}})^2$

Proof We start with the left-hand side of the equation by adding and subtracting the centroid of the initial cluster.

$$\begin{aligned}
&\frac{1}{|C_k|}\sum_{i\in C_k}\sum_{i'\in C_k}\sum_{j=1}^{p} w_j(x_{\text{ij}} - x_{i'j})^2\\
&= \frac{1}{|C_k|}\sum_{i\in C_k}\sum_{i'\in C_k}\sum_{j=1}^{p} w_j(x_{\text{ij}} - \overline{x}_{\text{kj}} + \overline{x}_{\text{kj}} - x_{i'j})^2\\
&= \frac{1}{|C_k|}\sum_{i\in C_k}\sum_{i'\in C_k}\sum_{j=1}^{p} w_j[(x_{\text{ij}} - \overline{x}_{\text{kj}})^2 + (\overline{x}_{\text{kj}} - x_{i'j})^2 - 2(x_{\text{ij}} - \overline{x}_{\text{kj}}((\overline{x}_{\text{kj}} - x_{i'j})]\\
&= 2\sum_{i\in C_k}\sum_{j=1}^{p} w_j(x_{\text{ij}} - \overline{x}_{\text{kj}})^2
\end{aligned}$$

With Theorem 1, we have the following algorithm for weighted features on classification of n subjects into K categories.

10.5 Weighted KNN Algorithm

1. Set up the initial clusters by randomly assigning the n subjects into K groups.

2. Computing: for each of the assigned group, $i, i = 1, \ldots, K$, compute the ith centroid by taking the mean vector of all the subjects in the ith group.
3. Reshuffling: with the K centroids from Step-2, for each subject, compute the weighted distance from the subject to each of the K centroids and reassign the subject to the centroid that has the closest weighted distance. In this way, all the subjects are reshuffled to the K clusters represented by the K-centroids.
4. Repeat Step-2 and Step-3, until no subject needs to be reassigned in Step-3.

To claim the validity of the weighted KNN algorithm discussed above, we need to show that the reshuffling process will terminate after a finite number of iterations.

Theorem 2 The weighted KNN algorithm converges after finite number of iterations.

Proof Consider any partition $\Theta_t = \{C_1, \ldots, C_k\}$ and the total within-cluster variation.

$$f(\Theta_t) = \sum_{k=1}^{K} \frac{1}{|C_k|} \sum_{i \in C_k} \sum_{i' \in C_k} \sum_{j=1}^{p} w_j (x_{\text{ij}} - x_{i'j})^2$$

If $f(\Theta_t)$ is the smallest, by Theorem 1, the smallest possible value is achieved, no point can be reassigned to a different set of the partition. The iteration process terminates.

If $f(\Theta_t)$ is not the smallest value, the right-hand side of Theorem-1 can be improved by rearranging the points around the centroids (the reshuffling step). Assigning each point to its closest centroid yields a new partition Θ_{t+1}, by Theorem-1, the within cluster variation for Θ_{t+1} is less than the one for Θ_t after reshuffling, thus,

$$f(\Theta_t) > f(\Theta_{t+1}).$$

Repeating the above process with the iterations $t = 1, 2, \ldots M$, we have a sequence

$$f(\Theta_1) > f(\Theta_2) > \ldots > f(\Theta_{n-1}) > f(\Theta_M).$$

Since the distance between the countable points are fixed, there are limited amounts of combinations for the partitions formed by the reshuffling step, as M increases, $f(\Theta_M)$ strictly decreases and approaches to its minimum value. This concludes the proof of Theorem 2.

With the weighted KNN algorithm and range regression algorithm, we can obtain the multiple range regression algorithm as follows. Consider (X_i, y_i) for $i = 1, \ldots, n$, where X_i is a vector representing features of subject i, and y_i is the corresponding clinical output of interest (such as body weight or systolic blood pressure). Different from the setting in the range regression section where the predictor is a value, the predictor in multiple range regression has a vector of features. Under this scenario, the weighted KNN can be applied to sort the n features X_i $i = 1, \ldots, n$ into K ranges before applying multiple range regression, as follows.

10.6 Multiple Range Regression Algorithm

1. Divide the data into a training set (75%) and a test set (25%).
2. For a permissible integer K, use the weighted KNN algorithm to group the n features into K categories.
3. Compute the average value of the response variable (z_j) for each category.
4. Arrange z_1, ..., z_K in an increasing order, denoted as $s_1, \ldots, s_K$.
5. Fit the linear regression with (j, s_j) for $j = 1$, ..., k.
6. Find the mean squared error (MSE) between the predicted and the observed.
7. Repeat Step-2 with different permissible values of k.
8. Select the k value that corresponds to the smallest MSE.

10.7 Enhanced Decision Trees

Another commonly applied method in data analytics without model assumption is decision tree. Like the setting in the preceding section, consider (X_i, y_i) for $i = 1, \ldots, n$ where X_i is a vector representing features of subject i, and y_i is the clinical output of interest (such as perceived warmth score for a mask policy or diagnosis outcome of Covid-19). Essentially, a decision tree finds splits of the predictor that minimize the overall residual sum of squares. Assume that the partition of the feature space is $C_1, \ldots, C_J$, the target function for optimization is

$$\text{RSS} = \sum_{j=1}^{J} \sum_{i \in C_j} (y_i - z_j)^2,$$

where z_j, as denoted in the section for range regression, is the average of the responses of subjects whose features fall in the set C_j. As pointed out in James et al (2017), the optimization process in the construction of a decision tree is usually the recursive binary splitting algorithm.

There is an intrinsic connection between the multiple range regression algorithm discussed in the preceding section and the decision tree method here. Notice that after the application of the weighted KNN algorithm, we order the mean responses corresponding to the K-categories and fit a linear regression line for multiple range regression. Instead of fitting a regression line, the tree approach fits a nonlinear function according to the split defined by the features of the subjects. For instance, when we have three categories C_1, C_2, C_3 with the corresponding mean responses z_1, z_2, z_3, the multiple range regression seeks the linear pattern between the ordered categories (1, 2, 3) and the responses $z_{(1)}, z_{(2)}, z_{(3)}$. However, the decision tree seeks the nonlinear pattern by identifying a split that separates the three categories into two (say $\{C_1, C_3\}$ versus $\{C_2\}$), then split the two categories into C_1 and C_3, as in the following Diagram 10.1.

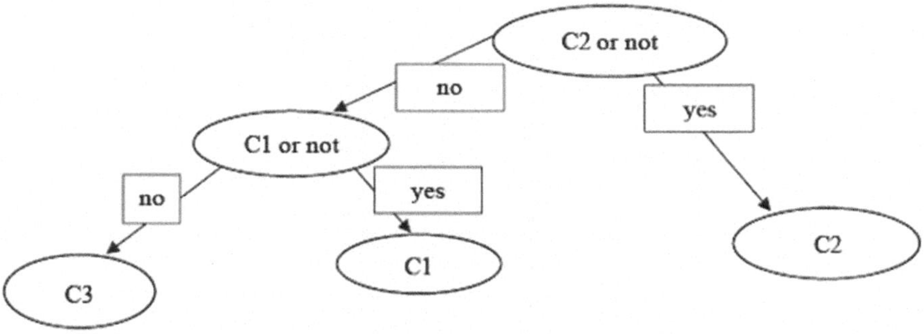

Diagram 10.1 A decision tree followed by weighted KNN algorithm

10.8 Analysis on Perceived Warmth Scores Associated with Mask Wearing

In this section, we shall apply the above-mentioned methods to analyze the impact of mask-wearing policy on the society. As described in Betsch et al. (2020), the dataset is from a weekly cross-sectional survey with 6973 German participants using online survey sampling from April 14, 2020, to May 26, 2020. Variables in the dataset includes age, gender, education, masking policy (voluntary or mandatory), perceived warmth score toward other people in a public gathering space such as supermarket, personal perception on mask policies, and the person's frequency on mask wearing. More specifically, the perceived warmth variable is a 21-point scale ranging from 1 (=very cold, zero-degree Fahrenheit) to 21 (=very warm, 100-degree Fahrenheit), the frequency of mask-wearing variable is a 6-point scale ranging from

"1" (=does not apply),
"2" (=never),
"3" (=rarely),
"4" (=sometimes),
"5" (=often), to
"6" (=always).

In the Betsch et al. (2020) paper, frequencies of participant's scale were converted into percentages, and OLS regression was used to analyze the relationship between perceived warmth score and the participant's frequency on wearing mask. One of the critical issues of their analysis is that OLS is valid only when the underlying distribution is normal. As in this scenario, for the countable 21-point scale on the warmth index and the 6-point scale on the frequency, the normality assumption for the error term of the OLS model is questionable.

Instead of using OLS regression, we plot the data to show that the normality assumption on the variation of the data is implausible. We apply the range regression method using the 6-point frequency scale as intrinsic range to detect the relationship that, as the frequency of mask wearing increases, the warmth score decreases toward

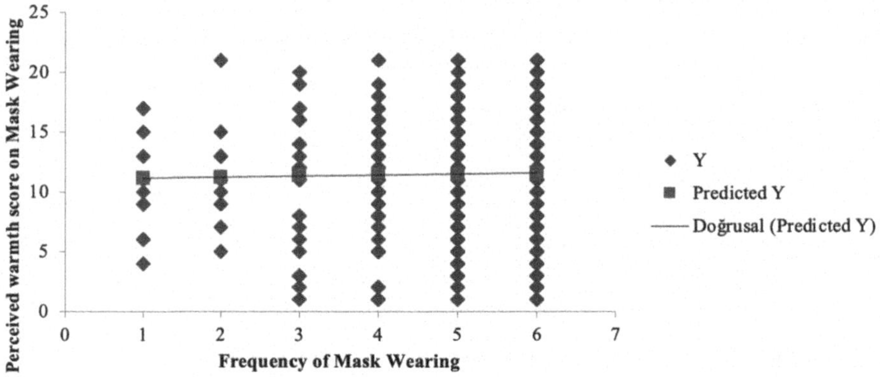

Fig. 10.1 OLS regression on warmth to mask wearing (R-square = 0.000359)

people who do not wear mask; however, as the frequency of mask wearing increases, the corresponding warmth score increases toward people who wear mask.

As shown in Fig. 10.1, when the frequency of mask wearing is treated as the predictor for the warmth index toward other people wearing mask, the data variation is large, and there is essentially no observable pattern between the frequency of the participant's mask wearing versus his/her perceived warmth score toward other people wearing mask during the pandemic. The sample R-square is 0.0003591, indicating extremely low variation on the warmth score explained by the variation in the participant's mask-wearing behavior.

Similar pattern appears to the plot of participant's perceived warmth score toward others who did not wear mask in the public area. As shown in Fig. 10.2, the perceived warmth score toward shoppers not wearing mask varies dramatically across different frequencies of the mask-wearing behaviors of the survey participants at the categories of "not applicable" and "never". However, in the three levels of frequency categories: "sometimes", "often", and "always", the perceived warmth scores toward no mask wearing takes every value from "1", the coldest, to "21", the warmest. The plot does not indicate any observable pattern between the perceived warmth score and the participants' habits on mask wearing. The R-square is 0.129, which implies that relatively little of the variation on the warmth score toward no mask wearing is explained by the variation in the mask-wearing habits of participants.

It should be noted that the insignificance in the OLS regression models only shows that OLS regression is unable to detect data evidence that reveals existing relationships due to data variation. The failure of OLS regression (partially due to the implausible model assumptions on normality) in detecting the relationship *does not mean there is no relationship* between the perceived warmth score and the mask-wearing habit of the participants. In fact, when we consider the mean perceived warmth score associated with each level of frequency on mask wearing using range regression, the relationships emerge.

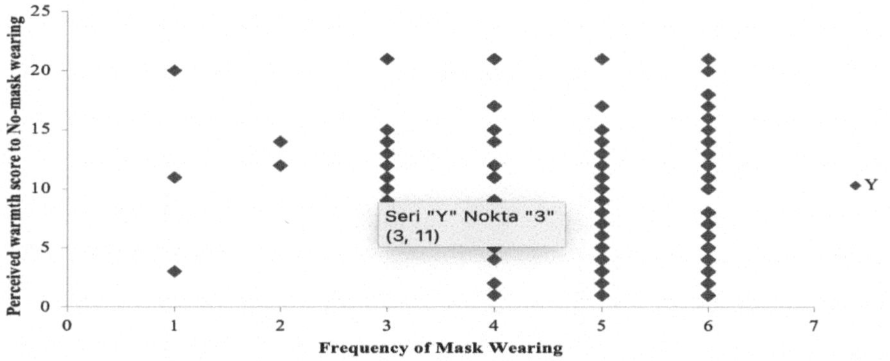

Fig. 10.2 Plot on perceived warmth to non-mask (R-square = 0.129)

As shown in Fig. 10.3, with the method of range regression, it is visually clear that participants who frequently wear masks tend to appreciate and have higher perceived warmth scores toward shoppers who also wear masks. The linear pattern is observable, and the R-squared is at the level of 0.530, which is much higher than the R-square value of 0.000359 in the corresponding OLS regression.

Figure 10.4, on the other hand, shows that the perceived warmth score decreases (gets colder) toward shoppers who did not wear masks as the survey participant's mask-wearing frequency increases. As given in Table 10.2, the p-value is 0.026 (< 0.05) supporting the hypothesis on common human behavior appreciating the same social behavior. The R-squared in this situation is 0.749, indicating a relatively strong association between the mean perceived warmth score and the mask-wearing frequency (Table 10.1).

Comparing the two range regression results in Figs. 10.3 and 10.4, the data supports the hypothesis that people wearing masks tend to appreciate others wearing masks (measured by the increase on perceived warmth index). At the same time,

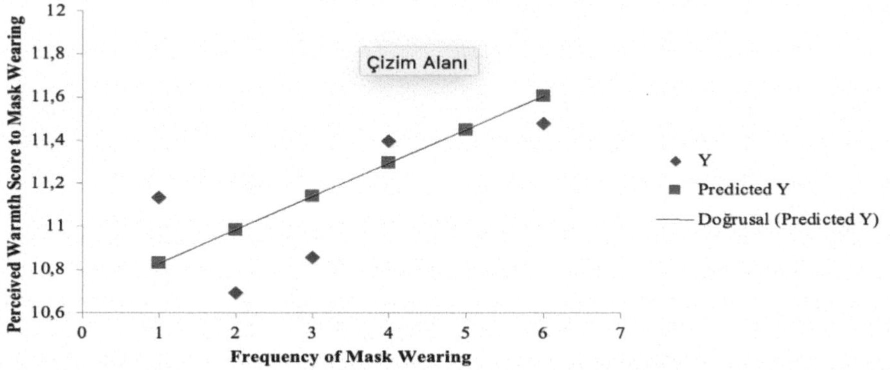

Fig. 10.3 Range regression warmth to mask wearing (R-square = 0.530)

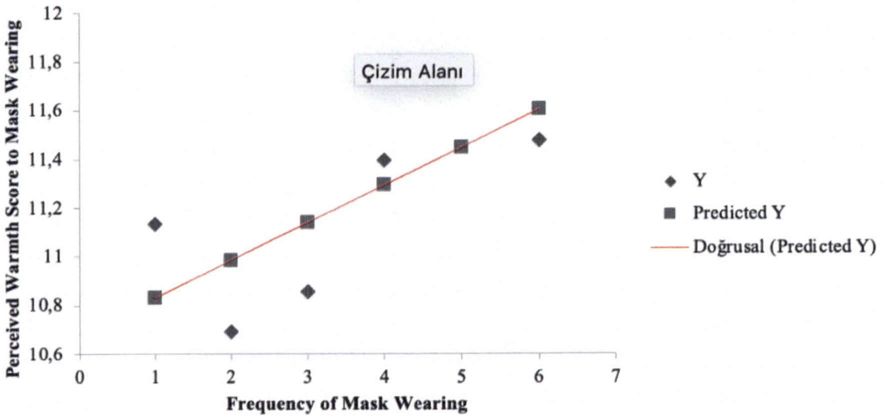

Fig. 10.4 Range regression on perceived warmth to no-mask (R-square = 0.749)

Table 10.1 Output of range regression on perceived warmth toward no mask wearing

Regression statistics						
Multiple R	0.86525502					
R square	0.74866624					
Adjusted R square	0.6858328					
Standard error	1.41570623					
Observations	6					
ANOVA						
	df	*SS*	*MS*	*F*	*Significance F*	
Regression	1	23.880516	23.880516	11.9150925	0.02601109	
Residual	4	8.01689653	2.00422413			
Total	5	31.8974125				
	Coefficients	*Standard error*	*t Stat*	*P-value*	*Lower 95%*	*Upper 95%*
Intercept	14.2574328	1.31795078	10.8178796	0.00041423	10.5982148	17.9166508
Mask freq	−1.1681613	0.33841852	−3.4518245	0.02601109	−2.1077618	-0.2285609

people who rarely wear a mask may appreciate people who similarly do not wear masks. Comparing the two estimated intercepts on the fitted range regression lines, participants who have never worn a mask tend to have similar mean perceived warmth scores on mask wearing (10.7 for mask wearing and 13 for not wearing). However, survey participants who always wear a mask are much warmer toward mask wearing (11.5 for mask wearing and 6.27 for not -wearing) than not wearing. The mean perceived warmth index is almost doubled for mask-wearing shoppers. In other

Table 10.2 Output of range regression on perceived warmth toward mask wearing

Regression statistics						
Multiple R	0.72776106					
R square	0.52963615					
Adjusted R square	0.41204519					
Standard error	0.30627583					
Observations	6					
ANOVA						
	df	*SS*	*MS*	*F*	*Significance F*	
Regression	1	0.42250235	0.42250235	4.50405498	0.1010827	
Residual	4	0.37521953	0.09380488			
Total	5	0.79772189				
	Coefficients	*Standard error*	*t Stat*	*P-value*	*Lower 95%*	*Upper 95%*
Intercept	10.6743647	0.28512728	37.4371919	3.04E-06	9.88272446	11.4660049
Mask freq	0.15538015	0.07321393	2.1222759	0.1010827	−0.0478943	0.3586546

words, there are much colder perceptions toward shoppers who did not wear masks during the pandemic.

The above analysis coincides with the implications of the decision tree in Fig. 10.5. Notice that the first split of the decision tree on the perceived warmth score occurs at whether other shoppers wear masks.

The second split is whether the survey participants wear masks, which is followed by the third split on whether the environment was under voluntary mask policies or mandatory mask policies. While the decision tree provides a big picture on the perceived warmth index, the range regression produces an explicit relationship on the change of the responses corresponding to the change of mask-wearing habits of the survey participants. It effectively reduces the impact of data variation across ranges (Fig. 10.6).

In summary, our paper has provided three new insights. First, it has demonstrated the shortfalls in applying regular OLS techniques to Covid survey data. The primary concern is the failure of the normality assumption with discrete variables. Second, our paper has proposed a new method of regression techniques, multiple range regression, that can be used even without the normality assumption. Finally, our paper has applied this technique to existing Covid datasets. The new analysis has shown that people are warmer toward mask-wearing behavior that is more akin to their own behavior (i.e., people who wear masks perceive those who wear masks more warmly, and people who do not wear masks perceive those who do not wear masks more warmly). Covid-19 has disrupted all our lives. To move forward, we need the right policy tools. The

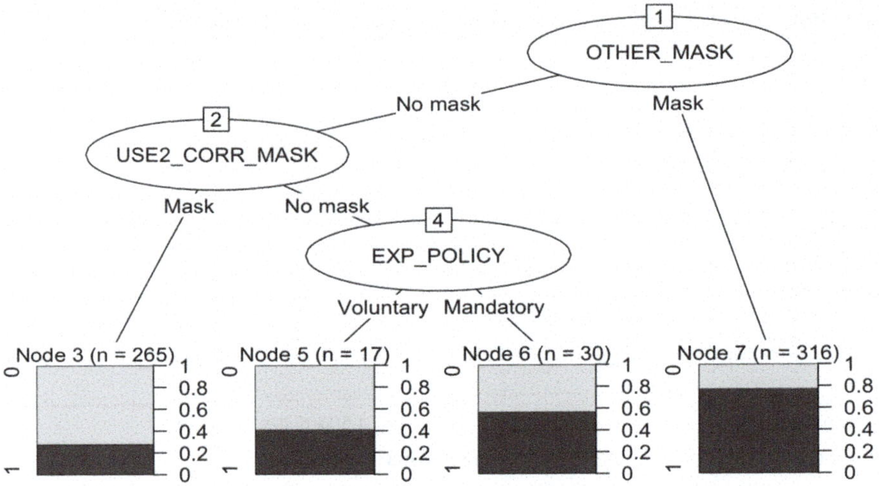

Fig. 10.5 Decision tree on the perceived warmth scores related to mask policy

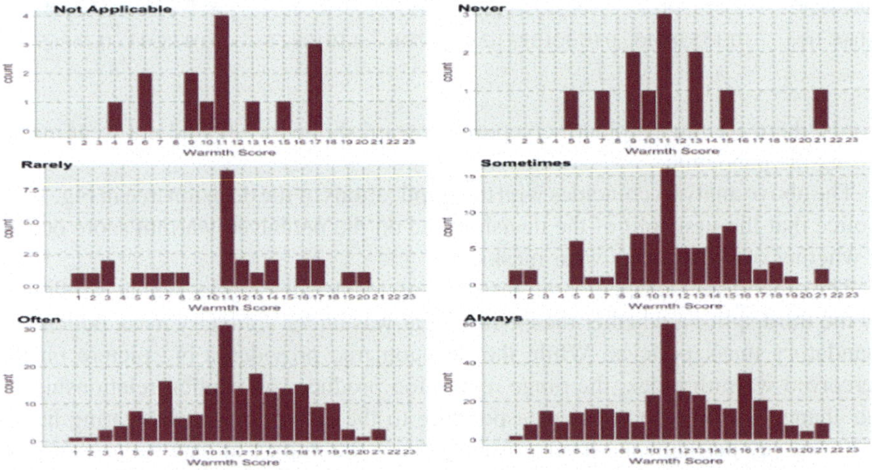

Fig. 10.6 Histogram of perceived warmth scores across categories for range regression

new method of multiple range regression may better enhance Covid data analysis and steer policy makers in the right direction.

References

Betsch C, Korna L, Sprengholza P, Felgendreffa L, Eitzea S, Schmida P, Böhm R (2020) Social and behavioral consequences of mask policies during the COVID-19 pandemic. PNAS (proc Natl Acad Sci) 117(36):21851–21853

Chen T, Hoppe F (2017) Simultaneous confidence intervals and regions. Wiley StatsRef: statistics reference online (ISBN 9781118445112), published online, at: http://onlinelibrary.wiley.com/doi/https://doi.org/10.1002/9781118445112.stat05961.pub2/full

Chen J, Comerota A (2012) Detecting the association between residual thrombus and post-thrombotic. Classification of chronic venous disease with range regression. Rev Recent Clin Trials 7:329–334

Chen JT (2014) Multivariate Bonferroni-type inequalities, theory and applications, Chapman and Hall, CRC, Boca Raton

Comerota A, Grewal N, Martinez J, Chen J et al (2012) Postthrombotic morbidity correlates with residual thrombus following catheter-directed thrombolysis for ilifemoral deep vein thrombosis. J Vascular Surg 55:768–773

James G, Witten D, Hastie T, Tibshirani (2017) An introduction to statistical learning with application in R. Springer, New York

Kerns L, Chen J (2016) A note on range regression analysis. J Appl Probabil Stat 11(2):19–27

Stoll KE (2019) Methodologies for missing data with range regressions. Mathematics Ph.D. dissertations, 37. https://scholarworks.bgsu.edu/math_diss/37

Lincy Y. Chen is skillful in programming and data analytics. She has coding expertise in R, Python, and other computing packages related to big data analysis. Lincy participated in various research projects, including the analysis of international food market data with time series. She published papers with MidStory on data analytics. Lincy She also served as a key personal in a project cleaning and analyzing a big dataset on Asian American quality of life in Texas, USA, as well as in a Cornell project analyzing education, culture, and environment factors associated with African American population in the city of Buffalo, USA.

Dr. John Tuhao Chen completed postdoctoral training at McMaster University, Canada, after graduating from the University of Sydney, Australia. He joined Bowling Green State University after working as a vVisiting assistant Assistant professor Professor at the University of Pittsburg. During his tenure at BGSU, Dr. Chen he has published two books and 68 peer-reviewed research papers in statistics, probability, biostatistics, and medical research. He was a Visiting Professor in the Department of Statistics at the University of California, Berkeley (2018), and in the department of Biostatistics at the University of Michigan, Ann Arbor (2010). Dr. Chen served in NSF funding panelists and NSF Committee of Visitors.

Chapter 11
Allocating Capacity for Office and Virtual Visits in Chronic Care Settings

Xiao Yu, Armağan Bayram, Yuchi Guo, and Gökçe Kahvecioğlu

Abstract Access to care is an important measure, especially for chronic care patients who require regular doctor appointments to control their chronic condition. Due to limitations on the number of available office appointments and due to additional precautions during the pandemic, patients may have limited access to chronic care; consequently, they may not receive the preferred treatment. In an effort to improve access to care for all patients during the pandemic, most healthcare providers have introduced virtual appointments as an alternative to traditional office appointments. In this study, to investigate the utilization of virtual appointments in the context of chronic care, we take into account a cohort of patients receiving chronic care and develop a capacity allocation model. We use an open migration network model to analyze the patients' flow between different types of appointments in chronic care and use the newsvendor model to investigate how the capacity is allocated between the virtual and office appointments. By considering patients' disease progressions, we propose policies based on our findings to make more systematic capacity allocation decisions.

Existing similarity (38%) with the literature comes from authors' previous journal and conference papers. (According to Turnitin report, 16% similarity is about the article and internet source of (Yu, X., Bayram, A. Managing capacity for virtual and office appointments in chronic care. Health Care Manag Sci 24, 742–767 (2021). https://doi.org/10.1007/s10729-021-09546-4).

X. Yu
Industrial and Manufacturing Systems Engineering, University of Michigan at Dearborn, Dearborn, USA
e-mail: xiaoyuu@umich.edu

A. Bayram (✉) · Y. Guo
Industrial and Manufacturing Systems Engineering, University of Michigan at Dearborn, Dearborn, USA
e-mail: armagan@umich.edu

Y. Guo
e-mail: yuchig@umich.edu

G. Kahvecioğlu
Supply Chain Optimization Technologies, Amazon, Seattle, WA, USA

E. Çetin and H. Özen (eds.), *Healthcare Policy, Innovation and Digitalization*, Accounting, Finance, Sustainability, Governance & Fraud: Theory and Application,
https://doi.org/10.1007/978-981-99-5964-8_11

Keywords Virtual appointments · Chronic care · Migration network · Newsvendor model · Capacity allocation

11.1 Introduction

Approximately 50% of people in the U.S. have at least one chronic condition, and around 75% of the entire healthcare spending constitutes chronic care management (Wu and Green 2000). Through effective management of chronic diseases, their negative effects on the total economic costs and overall patients' health can be mitigated. (Centers for Disease Control and Prevention 2020). The effective management of chronic care involves the diagnosis and the treatment of persistent diseases like diabetes and asthma and is essential to keep a chronic disease under control (Better Health Channel 2020). Hence, to improve the patient's life quality and health state, it is required for chronic care patients to have a regular and sufficient number of visits with physicians. Given the limitations on the physicians' capacity and the limitations of resources, virtual appointments are introduced as a recent healthcare delivery approach for providing timely and efficient service which can be integrated with office appointments (Siwicki 2020).

Virtual appointments are an alternative, cost-effective, and convenient way of providing continuous care remotely (i.e., using phone, video, and email consultations) (Greenhalgh et al. 2016). Similar to office appointments, virtual appointments can be used to diagnose patients' health states, and then the proper treatment can be provided accordingly (Bayram et al. 2020). There can be several and varying uses of virtual appointments. For example, they can be used to substitute for a scheduled office appointment, they can be used to complement office appointments by enabling follow-up patients' status after office appointments, and they can be used to diagnose/triage patients to refer them to the appropriate care. Due to this flexibility and convenience of virtual appointments, the need for their use has increased significantly such that they improve access to care for patients who cannot have in-person visits frequently (i.e., who are having poor physical conditions and are living in rural areas) (McGrail et al. 2017). More specifically, the adaptation of virtual appointments by physicians has increased by 340% since 2015 (Roth 2019) and its use has been widespread in the US, especially after the COVID-19 outbreak. Telemedicine makes an effective effort to examine, diagnose, and treatment of patients' health during the COVID-19 pandemic (Li et al. 2020; Triantafillou and Rajasekaran 2020). According to reports, the use of virtual appointment platforms has quickly expanded by around 110% since January 2020 (TMF 2020), and the stored health information on these platforms can benefit future examinations (Patel et al. 2020). It has been supported by clinic staff and patients to continue to integrate virtual appointments with office appointments in the future (Smithson et al. 2021).

Despite its benefits and its widespread use, it is challenging to integrate virtual and office appointments and make operational decisions due to the differences between virtual and office appointments in terms of the efficiency of their treatments, their costs, and the duration of their appointments (Bashshur et al. 2014). More specifically, allocating available capacity effectively leads to the cost-effective use of the clinics' resources, and it can ensure the regular and consistent care of patients to keep their chronic diseases under control. On the other hand, if there is not an efficient policy for capacity allocation, this can cause capacity overbookings or idle available resources, which will have a negative impact on both patients' health status and clinics' long-run profits. As a result, it is critical to establish effective policies to identify capacity allocation policies for both virtual and office appointments.

For this purpose, in this study, we integrate virtual appointments with office appointments and answer the following research questions:

- Under steady-state conditions, what is the expected number of patients in virtual and office appointments?
- What should be the optimal capacity allocation policy between different type of appointments that maximizes long-term average revenue of a health clinic?

To answer the above questions, we develop the migration network model and the newsvendor model as our methodology. Our objective is first to define the expected number of patients at different appointments at the steady-state conditions and then, to characterize the optimal capacity allocation policy among these appointments. We first construct a migration network model to study the patient flow and determine the number of patients who need virtual and office appointments in the steady state. The output of the migration network model is used in the newsvendor model by identifying the best capacity allocation policy that maximizes long-term average profits. One of the models is a newsvendor model without capacity constraints, while the second one considers capacity constraints. We analytically identify the ideal capacity of both appointments.

The remainder of the paper is structured as follows: In Sect. 11.2, we present the relevant literature. Section 11.3 describes the migration network model and two newsvendor models without and with capacity constraints. Finally, our conclusions are outlined in Sect. 11.4.

11.2 Literature Review

Our research builds on existing research related to operations management in chronic care delivery where the studies in this area are limited (Kucukyazici et al. 2011; Kucukyazici and Verter 2013; Deo et al. 2013; Leung et al. 2018; Balcik and Smilowitz 2020; Cheng et al. 2020; Yu and Bayram 2021). Among these studies, Kucukyazici and Verter (2013) provide a summary of the most relevant papers in the area of community-based chronic care delivery.

Our research is also related to issues with capacity planning, scheduling, and resource allocation in healthcare, which is related to allocating scarce resources to meet patient demand. There are several studies in this area, and an extensive review of the relevant study is summarized (Hulshof et al. 2012; Gür and Eren 2018; Zhu et al. 2019; Marynissen and Demeulemeester 2019; Leeftink et al. 2020). Among this literature, Stolletz and Brunner (2012) address the assignment of limited office appointment capacity to minimize the paid-out hours by building the scheduling model. In a different study, Choi and Wilhelm (2014) propose a newsvendor-based model to construct a block scheduling policy in order to minimize the total projected surgical lateness costs. In another study in this area, the ideal number of beds for a dialysis clinic is determined in order to optimize the clinic's overall profits (Lee and Zenios 2009). In another relevant study, Li et al. (2016) build a migration network and newsvendor model to investigate the optimal capacity for community-based services and nursing homes. In contrast to the research mentioned above, our analysis focuses on two different appointments (i.e., virtual and office appointments) with various levels of treatment effectiveness. We also take patient flow and disease progression into consideration. More related to our work, Bayram et al. (2020) investigate the scheduling problem of patients for virtual and office appointments by building a discrete-time Markov model. However, their study lacks investigation of the cost of virtual and office appointments which is essential to analyze the overall earnings of the clinics. Moreover, different from these studies methodologically, we also use the migration network model to investigate the patient flow.

Relevant to our methodology, queuing network models are widely applied in the analysis of healthcare systems to simulate the healthcare processes investigate the flow of patients, and analyze the system efficiency measures. Among the relevant studies, Green and Savin (2008) model a physician appointment scheduling system as a single-server queuing system in which backlog size and the probability of getting the same-day appointment are used to identify the acceptable range of patient panel size. In another single-server queuing model, Liu and Ziya (2014) model the appointments scheduled for a provider and investigate the optimal panel size and the service capability decision that maximizes the net reward function. As an illustration of a more comprehensive queuing network model, Creemers and Lambrecht (2011) present a general G/G/s queuing network model to study the whole hospital network. Using the decomposition approach, they divide the whole network into a set of single queuing systems that can be analyzed independently, and the results of every single queuing system can be aggregated and utilized to evaluate the effectiveness of the healthcare system as a whole. Different from the listed studies, we consider an incapacitated model where each node is an./G/∞ to analyze the patient scheduling decisions where we integrate our queuing network model with the optimization model to find optimal follow-up rates for virtual and office appointments. Moreover, our study differs from the listed literature as we also integrate the disease progression and modeling in the queuing network structure. We summarize the contributions of our paper in Table 11.1.

Table 11.1 Summary of the contributions

Contributions
Consideration of different types of appointments
Inclusion of the disease progression
Integration of the disease progression in the queueing network structure
Integration of the queueing network model with the optimization models

11.3 The Mathematical Model

In this section, we build mathematical models for a health clinic that provides virtual and office appointments. To replicate patient patterns within the network, we first introduce the migration network model. Then, we describe the capacity allocation models. Figure 11.1 illustrates the methods outlined.

11.3.1 Migration Network Model

Let $\mathcal{I}$ represent the set appointments where $i \in \mathcal{I} = \{\text{office appointments } (o), \text{virtual appointments } (v)\}$. Following the Poisson process, it is considered that new patients arrive with an arrival rate of $\lambda_i, \quad i \in \mathcal{I}$ for virtual and office appointments, respectively. The duration of appointments is considered as service times, and for both appointments, service times follow the exponential distribution.

We define $\mu_i, i \in \mathcal{I}$ as the service rate of virtual and office appointments. More specifically, $\frac{1}{\mu_i}$ represents the time required for appointment type $i \in \mathcal{I}$. During each appointment, the physician suggests a time for the next appointment $i \in \mathcal{I}$ depending on the patients' health status. By following this recommendation, patients are scheduled for the next appointment. Hence, we consider that the time between two

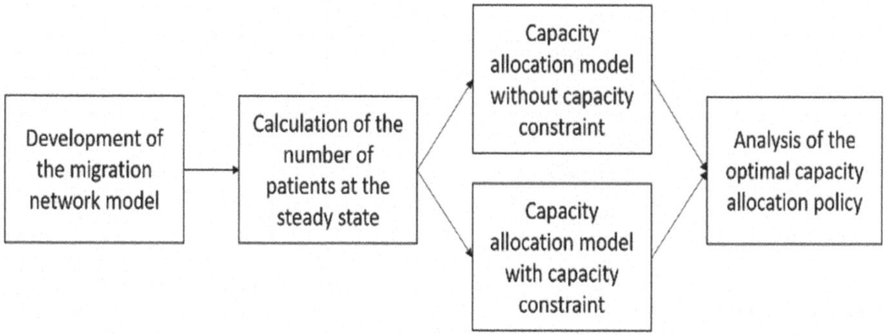

Fig. 11.1 Illustration of the methodology

appointments also follows the exponential distribution with an average of $\frac{1}{\sigma_i}, i \in \mathcal{I}$ for virtual and office appointments. Then, $\sigma_i, \forall i \in \mathcal{I}$ represents the visit frequency (follow-up rate) for each appointment. In chronic care, some patients may decide to change their programs, clinics, or physicians. Thus, patients may exit the system with a rate of δ which is independent of the other processes. To efficiently model the differences between virtual and office appointments, we consider disease progression over time. More specifically, we consider that patients' health status can deteriorate over time. Hence, let $\mathcal{K}$ be the set of states of health where $k \in \mathcal{K} = \{\text{healthy } (0), \text{sick } (1)\}$. To build the migration network model, $\mathcal{L}$ denotes the set of nodes and we consider 4 nodes $l \in \mathcal{L}$, namely: (i) patients scheduled for virtual appointments (V), (ii) patients scheduled for office appointments (O), (iii) patients who are in the health status of "0" and not scheduled for any appointment (NS_0), (iv) patients who are in the health status of "1" and not scheduled for any appointment (NS_1). By using the above description, we depict the migration network in Fig. 11.2 as follows.

We depict the patient flow between each node in Fig. 11.2. We believe that during the patient visit, both virtual and office appointments provide diagnosis and treatment. Regardless of patients' health state before the appointment, we assume that patients will be in "0" (i.e., healthy) health state after the office appointment. On the other hand, since virtual appointments are conducted remotely, we assume that occasionally the patients' diagnosis and treatment may not be adequate. Hence, we consider that with probability P patients will be in state "0" (i.e., healthy), and with probability $1 - P$ patients will be in state "1" (i.e., sick) after the virtual appointment. As mentioned before, we further consider disease progression. Hence, we consider patients' health states will change from state "0" to state "1" at a rate of γ over time. We illustrate this change in the migration network from node NS_0 to node NS_1. Let

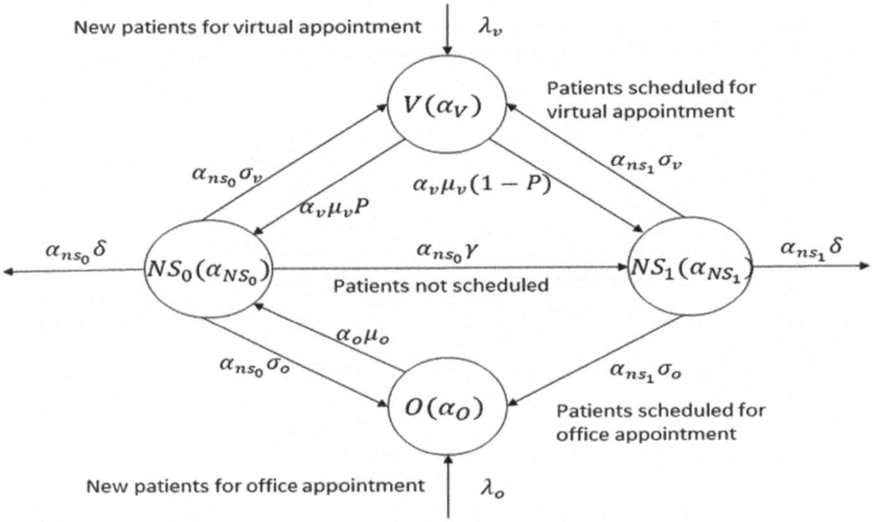

Fig. 11.2 Migration network model

α_l represent the expected number of patients at each node $l \in \mathcal{L}$ in the steady state. Then, according to Fig. 11.2, we write the following traffic equations for each node $l \in \mathcal{L}$:

$$\mu_v \alpha_v - \sigma_v \alpha_{ns_0} - \sigma_v \alpha_{ns_1} = \lambda_v \tag{11.1}$$

$$-\mu_v P \alpha_v + (\sigma_v + \sigma_o + \delta + \gamma)\alpha_{ns_0} - \mu_o \alpha_o = 0 \tag{11.2}$$

$$-\mu_v (1 - P)\alpha_v - \gamma \alpha_{ns_0} = 0 \tag{11.3}$$

$$-\sigma_o \alpha_{ns_0} - \sigma_o \alpha_{ns_1} + \mu_o \alpha_o = \lambda_o \tag{11.4}$$

These equations show that the inflow to node $l \in \mathcal{L}$ and the outflow from node $l \in \mathcal{L}$ are equal. Equations (11.1)–(11.4) consist of four equations with four unknowns, and we solve the traffic equations to determine each α_l value as follows:

$$\alpha_v = \frac{\sigma_v(\lambda_v + \lambda_o)}{\delta \mu_v} + \frac{\lambda_v}{\mu_v} \tag{11.5}$$

$$\alpha_{\text{ns0}} = \frac{(P\delta + \sigma_o + P\sigma_v)\lambda_v + (\delta + \sigma_o + P\sigma_v)\lambda_o}{\delta(\delta + \gamma + \sigma_o + \sigma_v)} \tag{11.6}$$

$$\alpha_{\text{ns0}} = \frac{\left[(1 - P)\delta + \gamma + (1 - P)\sigma_v)\lambda_v\right] + \left[\gamma + (1 - P)\sigma_v\right]\lambda_o}{\delta(\delta + \gamma + \sigma_o + \sigma_v)} \tag{11.7}$$

$$\alpha_o = \frac{\sigma_o(\lambda_v + \lambda_o)}{\delta \mu_o} + \frac{\lambda_o}{\mu_o} \tag{11.8}$$

Equations (11.5)–(11.8) specify the number of patients at each node in steady-state in which the nodes' capacity is infinite. In Theorem 1,[1] we characterize some of the structural properties using these equations.

Theorem 1 *Number of patients at each particular node in the migration network model exhibits the following properties*:

(a) *Number of patients at node V is a linear increasing function of the virtual follow-up rate (σ_v), while it is independent of the office follow-up rate (σ_o).*
(b) *Number of patients at node O is a linearly increasing function of the office follow-up rate (σ_o), while it is independent of the virtual follow-up rate (σ_v).*
(c) *Number of patients at node NS_0 (i.e., patients who are not scheduled for an appointment and in the health status of "0") is an increasing concave function of office follow-up rate (σ_o), while it is a decreasing convex function of the virtual follow-up rate (σ_v) if the following condition holds: $P\gamma\lambda_o + P\gamma\lambda_v <$*

[1] We include all proofs in the Appendix.

$(1-P)(\lambda_o\sigma_o + \lambda_v\sigma_o + \delta\lambda_o)$, and an increasing concave function of the virtual follow-up rate (σ_v) otherwise.

(d) *Number of patients at node* NS$_1$ *is a decreasing convex function of office follow-up rate (σ_o), while it is an increasing concave function of the virtual follow-up rate (σ_v) if the following condition holds: $P\gamma\lambda_o + P\gamma\lambda_v < (1-P)(\lambda_o\sigma_o + \lambda_v\sigma_o + \delta\lambda_o)$, and a decreasing convex function of the virtual follow-up rate (σ_v) otherwise.*

In Theorem 1, we demonstrate how the number of patients at each node is affected by the virtual and office follow-up rates (i.e., σ_v and σ_o). As expected, as the virtual follow-up rate σ_v increases, the number of patients scheduled for virtual appointments increases at the same rate. The same relation holds between office follow-up rate σ_o and the number of patients scheduled for office appointments. We further analyze how the number of patients who are not scheduled but are waiting at home in controlled and uncontrolled health states are impacted by virtual and office follow-up rates. The number of patients in the controlled health state increases with a concave structure, whereas the number of patients in the uncontrolled health state declines with a convex structure when the office follow-up rate σ_o increases. On the other hand, other parameters also affect the impact of virtual follow-up rate on the number of non-scheduled patients. If condition $P\gamma\lambda_o + P\gamma\lambda_v < (1-P)(\lambda_o\sigma_o + \lambda_v\sigma_o + \delta\lambda_o)$ is met, the number of patients in the controlled health state decreases (convex structure) as the virtual follow-up rate increases, while the number of patients in the uncontrolled health state increases (concave structure).

In order to evaluate how to optimally assign the capacity, we also take the clinics' revenue into account in the following section. To this end, let x_l denote the number of patients at node *l*, and π represents the steady-state Poisson probability distribution for each node *l* and is defined as follows (Kelly 2011):

$$\pi_l(x_l = x) = \frac{\alpha_l^x}{x!} \Big/ \sum_{n=0}^{\infty} \frac{\alpha_l^n}{n!} = e^{-\alpha_l}\frac{\alpha_l^x}{x!}, \quad l \in \mathcal{L} \tag{11.9}$$

11.3.2 Unconstrained Capacity Allocation Model

In this section, we construct a newsvendor model to identify the optimal policy to allocate available capacity for virtual and office appointments. Our objective is to maximize a health clinic's long-run average revenues. Let $M = (M_o, M_v)$ be the decision variable, where M_o and M_v represent the capacity of the virtual and office appointments, respectively. We use $r_i, i \in \mathcal{I}$ to represent the marginal profit for each patient receiving any of the appointments. To define the fixed cost that is associated with each unit of capacity for both virtual and office appointments, we use $c_i, i \in \mathcal{I}$ to represent the unit fixed costs. The total fixed cost of capacity is defined as $c_oM_o + c_vM_v$, which is independent of the patient flow. Although there is a capacity limit for each appointment, in some cases, patients can be overbooked for appointments (i.e., the scheduled number of patients can exceed the capacity) with higher costs. To define the cost of overbooking, we use $f_i, i \in \mathcal{I}$ as the penalty cost of overbooked appointments, which is higher than the standard cost for non-overbooked ones, and the revenue earned. Hence, when the capacity is overbooked, the marginal profit will be $r_i - f_i < 0, i \in \mathcal{I}$. We also consider in our model that $r_i > c_i$, where the marginal profit of an appointment is higher than its fixed cost. Then our model can be defined as follows:

$$\max A(M) \tag{11.10}$$

Subject to

$$M \in N_+ \tag{11.11}$$

where

$$\begin{aligned} A(M) &= \lim_{T\to\infty} \frac{1}{T}\{ \sum_{i\in\mathcal{I}} \int_0^T r_i \min[x_i(t), M_i]\mathrm{dt} \\ &\quad - \sum_{i\in\mathcal{I}} \int_0^T c_i M_i \mathrm{dt} - \sum_{i\in\mathcal{I}} \int_0^T f_i (x_i(t) - M_i)^+ \mathrm{dt} \qquad (11.12) \\ &= \sum_{i\in\mathcal{I}} A_i(M_i) \qquad (11.13) \end{aligned}$$

$$= \sum_{i\in\mathcal{I}} (r_i - c_i)\alpha_i - c_i \mathbb{E}_{\pi_i}(M_i - x_i)^+ - (f_i + r_i - c_i)\mathbb{E}_{\pi_i}(x_i - M_i)^+ \tag{11.14}$$

According to the steady-state distribution π_i, $\mathbb{E}_{\pi_i}(x_i)$ represents the expected number of patients at node *i*. $A_i(M_i)$ denotes the individual objective function for node *i*. The first term in Eq. (11.14) is the marginal profit, the second term is the opportunity cost of unused capacity, and the last term is the patient overflow cost Eq. (11.11) ensures that the optimal capacity is a positive integer.

Because of the independence between the virtual and office processes, maximizing the objective function *A(M)* is equivalent to maximizing the sub-objective $A_i(M_i)$ separately. The differential and the second-order differential of $A_i(M_i)$ are:

$$\begin{aligned} \Delta A_i(M_i) &= A_i(M_i + 1) - A_i(M_i) \\ &= f_i + r_i - c_i - (f_i + r_i)\pi_i(x_i \le M_i) \end{aligned} \tag{11.15}$$

$$\begin{aligned} \Delta^2 A_i(M_i) &= \Delta A_i(M_i + 1) - \Delta A_i(M_i) \\ &= -(f_i + r_i)[\pi_i(x_i \le M_i + 1) - \pi_i(x_i \le M_i)] < 0 \end{aligned} \tag{11.16}$$

The smallest positive integer $M_i = M_i^{\min}$ is the optimal capacity at node i, and it makes $\Delta A_i(M_i) \leq 0$, which leads to the maximum of $A_i(M_i)$. Hence, we define $M^* = \left(M_o^{\min}, M_v^{\min}\right)$ as the optimal solution to the capacity allocation problem without capacity constraints.

$$M_i^{\min} = \min\left\{M_i \geq 0 : \pi_i(x_i \leq M_i) \geq \frac{f_i + r_i - c_i}{f_i + r_i}\right\} \tag{11.17}$$

11.3.3 Capacity Allocation Model with Capacity Constraint

Our initial model assumes that the capacity is unlimited. However, in practice, there is a capacity limit for healthcare clinics. Hence, in this section, we extend the unconstrained capacity allocation model (shown in Sect. 11.3.2) by including the capacity constraint. Adding a capacity constraint will not affect the flow equations of the migration network, which means the objective function remains the same. *TC* stands for the total capacity available to allocate for virtual and office appointments. Then, our capacity allocation model, which maximizes the long-run average profit, with capacity constraint can be defined as follows:

$$\max A(M) \tag{11.18}$$

Subject to

$$M_o + M_v \leq \text{TC} \tag{11.19}$$

According to Eq. 11.9, the total number of virtual and office appointments must not exceed the total limited capacity. M_o and M_v are no longer independent because of the capacity constraint. Hence, we propose an algorithm to characterize a capacity allocation policy, which is given in Algorithm 1. M_o^{tep} and M_v^{tep} are the intermediate variables. They denote the number of virtual and office appointments at each iteration.

The updated partial differentials of the objective functions are shown below:

$$\begin{aligned} A'(M_o) &= \frac{\partial A(M)}{\partial M_o} \\ &= A(M_o + 1, M_v) - A(M_o, M_v) \\ &= f_o + r_o - c_o - (f_o + r_o)\pi_o(x_o \leq M_o) \end{aligned} \tag{11.20}$$

$$\begin{aligned} A'(M_v) &= \frac{\partial A(M)}{\partial M_v} \\ &= A(M_o, M_v + 1) - A(M_o, M_v) \end{aligned}$$

$$= f_v + r_v - c_v - (f_v + r_v)\pi_v(x_v \leq M_v) \quad (11.21)$$

Algorithm 1 Optimal Capacity Allocation Based on the Partial Differential

1. Set $M_o^{\text{tep}} = 0$ and $M_v^{\text{tep}} = 0$;
2. Calculate $A'\left(M_o^{\text{tep}}\right)$ and $A'\left(M_v^{\text{tep}}\right)$; if $A'\left(M_o^{\text{tep}}\right) \leq 0$ and $A'\left(M_v^{\text{tep}}\right) \leq 0$, then go to step 5;
3. If $A'\left(M_o^{\text{tep}}\right) \leq A'\left(M_v^{\text{tep}}\right)$, then $M_v^{\text{tep}} \leftarrow M_v^{\text{tep}} + 1$; otherwise, $M_o^{\text{tep}} \leftarrow M_o^{\text{tep}} + 1$;
4. If $M_o^{\text{tep}} + M_v^{\text{tep}} < \text{TC}$, then go to step 2; otherwise, go to step 5;
5. $M_o^* \leftarrow M_o^{\text{tep}}$, $M_v^* \leftarrow M_v^{\text{tep}}$

The M_o^* and M_v^*, variables in Algorithm 1 represent the final allocated capacity for virtual and office appointments, respectively. $M_{\text{TC}}^* = \left(M_o^*, M_v^*\right)$ is defined as the solution to the capacity allocation problem with limited capacity, *TC*.

Algorithm 1 calculates the marginal profit by adding one additional unit to the threshold capacity. The capacity is allocated for the type of appointment having the higher marginal gain. When the assigned capacity reaches the total capacity for all appointments Algorithm 1 terminates.

11.4 Conclusion

Virtual appointments are gaining ground rapidly and they are used to complement and substitute for office appointments. Many health clinics and hospitals are integrating virtual appointments with office appointments and migrating to virtual health services. Especially after the COVID-19 outbreak both the need and the use of virtual appointments have increased rapidly and dramatically. The identification of the optimal capacity allocation policy for virtual and office appointments, however, poses a significant but challenging operational problem.

In this study, the disease progression is taken into account while modeling how chronic patients move between virtual and office appointments using a migration network. We provide various structural characteristics of the migration network model and demonstrate the relationship between the number of patients and the follow-up rates. The optimal capacity allocation policy for scheduling both virtual and office appointments that maximizes the clinic's long-term average revenue is then determined by building newsvendor models both without and with the consideration of capacity restrictions.

Appendix 11.1. Proofs

Proof of Theorem 1 Item A In the migration network model, the number of patients at node V (i.e., patients scheduled for virtual appointments) is an increasing function of the virtual follow-up rate (σ_v), while it is not dependent on the office follow-up rate (σ_o).

To show the above statement we take the derivative of α_v with respect to σ_v and σ_o.

$$\frac{\partial \alpha_v}{\partial \sigma_v} = \frac{(\lambda_v + \lambda_o)}{\delta \mu_v} \tag{11.28}$$

$$\frac{\partial \alpha_v}{\partial \sigma_o} = 0 \tag{11.29}$$

As shown in Eq. (11.28), the first-order derivative of α_v with respect to σ_v is a positive value which states that α_v is an increasing function of σ_v. On the other hand, the first-order derivative of α_v with respect to σ_o is equal to zero which states that α_v does not depend on σ_o.

Proof of Theorem 1 Item B In the migration network model, the number of patients at node O (i.e., patients scheduled for office appointments) is an increasing function of office follow-up rate (σ_o), while it is not dependent on the virtual follow-up rate (σ_v).

To show the above statement we take the derivative of α_o with respect to σ_o and σ_v.

$$\frac{\partial \alpha_o}{\partial \sigma_o} = \frac{(\lambda_v + \lambda_o)}{\delta \mu_o} \tag{11.30}$$

$$\frac{\partial \alpha_o}{\partial \sigma_v} = 0 \tag{11.31}$$

As shown in Eq. (11.30), the first-order derivative of α_o with respect to σ_o is a positive value which states that α_o is an increasing function of σ_o. On the other hand, the first-order derivative of α_o with respect to σ_v is equal to zero which states that α_o does not depend on σ_v.

Proof of Theorem 1 Item C In the migration network model, the number of patients at node NS_0 (i.e., patients who are not scheduled for an appointment and in the health status of "0") is an increasing concave function of office follow-up rate (σ_o), while it is a decreasing convex function of the virtual follow-up rate (σ_v) if the following condition holds: $P\gamma\lambda_o + P\gamma\lambda_v < (1 - P)(\lambda_o\sigma_o + \lambda_v\sigma_o + \delta\lambda_o)$, and an increasing concave function of the virtual follow-up rate (σ_v) otherwise.

To show the above statement we first take the derivative of α_{ns_0} with respect to σ_o.

$$\frac{\partial \alpha_{ns_0}}{\partial \sigma_o} = \frac{\delta\lambda_v + \gamma\lambda_o + \gamma\lambda_v + \lambda_o\sigma_v + \lambda_v\sigma_v - P\lambda_o\sigma_v - P\lambda_v\sigma_v - P\delta\lambda_v}{\delta(\delta + \gamma + \sigma_o + \sigma_v)^2} \tag{11.32}$$

$$= \frac{\gamma\lambda_o + \gamma\lambda_v + (1 - P)\lambda_o\sigma_v + (1 - P)\lambda_v\sigma_v + (1 - P)\delta\lambda_v}{\delta(\delta + \gamma + \sigma_o + \sigma_v)^2} \tag{11.33}$$

$$= \frac{\gamma\lambda_o + \gamma\lambda_v + (1-P)(\lambda_o\sigma_v + \lambda_v\sigma_v + \delta\lambda_v)}{\delta(\delta+\gamma+\sigma_o+\sigma_v)^2} \tag{11.34}$$

As shown with Eq. (11.34), the first-order derivative of α_{ns_0} with respect to σ_o is a positive value which states that α_{ns_0} is an increasing function of σ_o. To show that α_{ns_0} is concave with respect to σ_o we take its second derivative as follows:

$$\frac{\partial^2\alpha_{ns_0}}{\partial^2\sigma_o} = \frac{-2(\delta\lambda_v + \gamma\lambda_o + \gamma\lambda_v + \lambda_o\sigma_v + \lambda_v\sigma_v - P\lambda_o\sigma_v - P\lambda_v\sigma_v - P\delta\lambda_v)}{\delta(\delta+\gamma+\sigma_o+\sigma_v)^3} \tag{11.35}$$

$$= \frac{-2(\gamma\lambda_o + \gamma\lambda_v + (1-P)(\lambda_o\sigma_v + \lambda_v\sigma_v + \delta\lambda_v))}{\delta(\delta+\gamma+\sigma_o+\sigma_v)^3} \tag{11.36}$$

As shown with Eq. (11.36), the second-order derivative of α_{ns_0} with respect to σ_o is a negative value which states that α_{ns_0} is a concave function with respect to σ_o.

Next, we take the derivative of α_{ns_0} with respect to σ_v.

$$\frac{\partial\alpha_{\mathrm{ns}_0}}{\partial\sigma_v} = \frac{P\lambda_o\sigma_o - \lambda_o\sigma_o - \lambda_v\sigma_o - \delta\lambda_o + P\lambda_v\sigma_o + P\delta\lambda_o + P\gamma\lambda_o + P\gamma\lambda_v}{\delta(\delta+\gamma+\sigma_o+\sigma_v)^2} \tag{11.37}$$

$$= \frac{P\gamma\lambda_o + P\gamma\lambda_v - (1-P)\lambda_o\sigma_o - (1-P)\lambda_v\sigma_o - (1-P)\delta\lambda_o}{\delta(\delta+\gamma+\sigma_o+\sigma_v)^2} \tag{11.38}$$

$$= \frac{P\gamma\lambda_o + P\gamma\lambda_v - (1-P)(\lambda_o\sigma_o + \lambda_v\sigma_o + \delta\lambda_o)}{\delta(\delta+\gamma+\sigma_o+\sigma_v)^2} \tag{11.39}$$

As shown with Eq. (11.39), the first-order derivative of α_{ns_0} with respect to σ_v is positive if $P\gamma\lambda_o + P\gamma\lambda_v > (1-P)(\lambda_o\sigma_o + \lambda_v\sigma_o + \delta\lambda_o)$, and negative otherwise. Next, we take its second derivative as follows:

$$\frac{\partial^2\alpha_{ns_0}}{\partial^2\sigma_v} = \frac{-2(P\lambda_o\sigma_o - \lambda_o\sigma_o - \lambda_v\sigma_o - \delta\lambda_o + P\lambda_v\sigma_o + P\delta\lambda_o + P\gamma\lambda_o + P\gamma\lambda_v)}{\delta(\delta+\gamma+\sigma_o+\sigma_v)^3} \tag{11.40}$$

$$= \frac{-2(P\gamma\lambda_o + P\gamma\lambda_v - (1-P)(\lambda_o\sigma_o + \lambda_v\sigma_o + \delta\lambda_o))}{\delta(\delta+\gamma+\sigma_o+\sigma_v)^3} \tag{11.41}$$

As shown with Eq. (11.41), the second-order derivative of α_{ns_0} with respect to σ_v is a negative value if $P\gamma\lambda_o + P\gamma\lambda_v > (1-P)(\lambda_o\sigma_o + \lambda_v\sigma_o + \delta\lambda_o)$ which states that α_{ns_0} is a concave function with respect to σ_v. Otherwise, it will be a positive value and so will be a convex function.

Proof of Theorem 1 Item D In the migration network model, the number of patients at node NS_1 (i.e., patients who are not scheduled for an appointment and in the health status of "1") is a decreasing convex function of office follow-up rate (σ_o), while it is an increasing concave function of the virtual follow-up rate (σ_v) if the following condition holds: $P\gamma\lambda_o + P\gamma\lambda_v < (1-P)(\lambda_o\sigma_o + \lambda_v\sigma_o + \delta\lambda_o)$, and a decreasing convex function of the virtual follow-up rate (σ_v) otherwise.

To show the above statement we first take the derivative of α_{ns_0} with respect to σ_o.

$$\frac{\partial \alpha_{\mathrm{ns}_1}}{\partial \sigma_o} = \frac{-(\delta\lambda_v + \gamma\lambda_o + \gamma\lambda_v + \lambda_o\sigma_v + \lambda_v\sigma_v - P\lambda_o\sigma_v - P\lambda_v\sigma_v - P\delta\lambda_v)}{\delta(\delta+\gamma+\sigma_o+\sigma_v)^2} \tag{11.42}$$

$$= \frac{-(\gamma\lambda_o + \gamma\lambda_v + (1-P)\lambda_o\sigma_v + (1-P)\lambda_v\sigma_v + (1-P)\delta\lambda_v)}{\delta(\delta+\gamma+\sigma_o+\sigma_v)^2} \tag{11.43}$$

$$= \frac{-(\gamma\lambda_o + \gamma\lambda_v + (1-P)(\lambda_o\sigma_v + \lambda_v\sigma_v + \delta\lambda_v))}{\delta(\delta+\gamma+\sigma_o+\sigma_v)^2} \tag{11.44}$$

As shown with Eq. (11.44), the first-order derivative of α_{ns_1} with respect to σ_o is a negative value which states that α_{ns_1} is a decreasing function of σ_o. To show that α_{ns_1} is convex with respect to σ_o we take its second derivative as follows:

$$\frac{\partial^2 \alpha_{\mathrm{ns}_1}}{\partial^2 \sigma_o} = \frac{2(\delta\lambda_v + \gamma\lambda_o + \gamma\lambda_v + \lambda_o\sigma_v + \lambda_v\sigma_v - P\lambda_o\sigma_v - P\lambda_v\sigma_v - P\delta\lambda_v)}{\delta(\delta+\gamma+\sigma_o+\sigma_v)^3} \tag{11.45}$$

$$= \frac{2(\gamma\lambda_o + \gamma\lambda_v + (1-P)(\lambda_o\sigma_v + \lambda_v\sigma_v + \delta\lambda_v))}{\delta(\delta+\gamma+\sigma_o+\sigma_v)^3} \tag{11.46}$$

As shown with Eq. (11.46), the second-order derivative of α_{ns_1} with respect to σ_o is a positive value which states that α_{ns_1} is a convex function with respect to σ_o.

Next, we take the derivative of α_{ns_1} with respect to σ_v.

$$\frac{\partial \alpha_{\mathrm{ns}_1}}{\partial \sigma_v} = \frac{-(P\lambda_o\sigma_o - \lambda_o\sigma_o - \lambda_v\sigma_o - \delta\lambda_o + P\lambda_v\sigma_o + P\delta\lambda_o + P\gamma\lambda_o + P\gamma\lambda_v)}{\delta(\delta+\gamma+\sigma_o+\sigma_v)^2} \tag{11.47}$$

$$= \frac{-(P\gamma\lambda_o + P\gamma\lambda_v - (1-P)\lambda_o\sigma_o - (1-P)\lambda_v\sigma_o - (1-P)\delta\lambda_o)}{\delta(\delta+\gamma+\sigma_o+\sigma_v)^2} \tag{11.48}$$

$$= \frac{-(P\gamma\lambda_o + P\gamma\lambda_v - (1-P)(\lambda_o\sigma_o + \lambda_v\sigma_o + \delta\lambda_o))}{\delta(\delta+\gamma+\sigma_o+\sigma_v)^2} \tag{11.49}$$

As shown with Eq. (11.49), the first-order derivative of α_{ns_1} with respect to σ_v is negative if $P\gamma\lambda_o + P\gamma\lambda_v > (1-P)(\lambda_o\sigma_o + \lambda_v\sigma_o + \delta\lambda_o)$ and positive otherwise. Next, we take its second derivative as follows:

$$\frac{\partial^2 \alpha_{\text{ns}_1}}{\partial^2 \sigma_v} = \frac{2(P\lambda_o\sigma_o - \lambda_o\sigma_o - \lambda_v\sigma_o - \delta\lambda_o + P\lambda_v\sigma_o + P\delta\lambda_o + P\gamma\lambda_o + P\gamma\lambda_v)}{\delta(\delta + \gamma + \sigma_o + \sigma_v)^3} \tag{11.50}$$

$$= \frac{2(P\gamma\lambda_o + P\gamma\lambda_v - (1 - P)(\lambda_o\sigma_o + \lambda_v\sigma_o + \delta\lambda_o))}{\delta(\delta + \gamma + \sigma_o + \sigma_v)^3} \tag{11.51}$$

As shown with Eq. (11.51), the second-order derivative of α_{ns_1} with respect to σ_v is a positive value if $P\gamma\lambda_o + P\gamma\lambda_v > (1 - P)(\lambda_o\sigma_o + \lambda_v\sigma_o + \delta\lambda_o)$ which states that α_{ns_1} is a convex function with respect to σ_v. Otherwise, it will be a negative value and so will be a concave function.

References

Balcik B, Smilowitz K (2020) Contributions to humanitarian and non-profit operations: equity impacts on modeling and solution approaches. In: Women in industrial and systems engineering. Springer, Cham, pp 371–390

Bashshur RL, Shannon GW, Smith BR, Alverson DC, Antoniotti N, Barsan WG, Bashshur N, Brown EM, Coye MJ, Doarn CR, Ferguson S, Yellowlees P (2014) The empirical foundations of telemedicine interventions for chronic disease management. Telemedicine and e-Health 20(9):769–800

Bayram A, Deo S, Iravani S, Smilowitz K (2020) Managing virtual appointments in chronic care. IISE Trans Healthcare Syst Eng 10(1):1–17

Better Health Channel (2020) Managing long-term illness and chronic conditions. Available at: https://www.betterhealth.vic.gov.au/health/ServicesAndSupport/managing-long-term-illness-and-chronic-conditions

Centers for Disease Control and Prevention (2020). Health and economic costs of chronic diseases. https://www.cdc.gov/chronicdisease/about/costs/index.htm.

Cheng CY, Lee YC, Huang CH, Wu HH (2020) Assessing nurses' overall satisfaction of patient safety culture from a regional teaching hospital in Taiwan. Int J Ind Syst Eng 36(4):537–548

Choi S, Wilhelm WE (2014) An approach to optimize block surgical schedules. Eur J Oper Res 235(1):138–148

Creemers S, Lambrecht M (2011) Modeling a hospital queueing network. In: Queueing networks. Springer, Boston, MA, pp 767–798

Deo S, Iravani S, Jiang T, Smilowitz K, Samuelson S (2013) Improving health outcomes through better capacity allocation in a community-based chronic care model. Oper Res 61(6):1277–1294

Green LV, Savin S (2008) Reducing delays for medical appointments: a queueing approach. Oper Res 56(6):1526–1538

Greenhalgh T, Vijayaraghavan S, Wherton J, Shaw S, Byrne E, Campbell-Richards D, Bhattacharya S, Hanson P, Ramoutar S, Gutteridge C, Hodkinson I, Morris J (2016) Virtual online consultations: advantages and limitations (VOCAL) study. BMJ Open 6(1):e009388

Gür Ş, Eren T (2018) Scheduling and planning in service systems with goal programming: literature review. Mathematics 6(11):265

Hulshof PJ, Kortbeek N, Boucherie RJ, Hans EW, Bakker PJ (2012) Taxonomic classification of planning decisions in health care: a structured review of the state of the art in OR/MS. Health Syst 1(2):129–175

Kelly F, Koizumi P (2011) Reversibility and stochastic networks. Cambridge University Press

Kucukyazici B, Verter V (2013) Managing community-based care for chronic diseases: the quantitative approach. In: Operations research and health care policy. Springer, New York, NY, pp 71–90

Kucukyazici B, Verter V, Mayo NE (2011) An analytical framework for designing community-based care for chronic diseases. Prod Oper Manag 20(3):474–488

Lee DK, Zenios SA (2009) Optimal capacity overbooking for the regular treatment of chronic conditions. Oper Res 57(4):852–865

Leeftink AG, Bikker IA, Vliegen IMH, Boucherie RJ (2020) Multi-disciplinary planning in health care: a review. Health Syst 9(2):95–118

Leung E, Chau CW, Lee A, Chen YF, Lee DT (2018) Integrated care as a strategic solution for active aging in the community: tools and models. Sustain Health Long-Term Care Solutions Aging Popul 145–160

Li P, Liu X, Mason E, Hu G, Zhou Y, Li W, Jalali MS (2020) How telemedicine integrated into China's anti-COVID-19 strategies: case from a national referral center. BMJ Health Care Inf 27(3).

Li Y, Zhang Y, Kong N, Lawley M (2016) Capacity planning for long-term care networks. IIE Trans 48(12):1098–1111

Liu N, Ziya S (2014) Panel size and overbooking decisions for appointment-based services under patient no-shows. Prod Oper Manag 23(12):2209–2223

Marynissen J, Demeulemeester E (2019) Literature review on multi-appointment scheduling problems in hospitals. Eur J Oper Res 272(2):407–419

McGrail KM, Ahuja MA, Leaver CA (2017) Virtual visits and patient-centered care: results of a patient survey and observational study. J Med Internet Res 19(5):e177

Patel PD, Cobb J, Wright D, Turer RW, Jordan T, Humphrey A, Kepner AL, Smith G, Rosenbloom ST (2020) Rapid development of telehealth capabilities within pediatric patient portal infrastructure for COVID-19 care: barriers, solutions, results. J Am Med Inf Assoc 27(7):1116–1120

Roth M (2019) Physician telehealth usage wallops early ehr adoption rates. Available at: https://www.healthleadersmedia.com/innovation/physician-telehealth-usage-wallops-early-ehr-adoption-rates

Siwicki B (2020) Telemedicine during COVID-19: benefits, limitations, burdens, adaptation. Health Care News

Smithson R, Roche E, Wicker C (2021) Virtual models of chronic disease management: lessons from the experiences of virtual care during the COVID-19 response. Aust Health Rev 45(3):311–316

Stolletz R, Brunner JO (2012) Fair optimization of fortnightly physician schedules with flexible shifts. Eur J Oper Res 219(3):622–629

TMF (2020) COVID-19 and the rise of telemedicine. Available at: https://medicalfuturist.com/covid-19-was-needed-for-telemedicine-to-finally-go-mainstream/

Triantafillou V, Rajasekaran K (2020) A commentary on the challenges of telemedicine for head and neck oncologic patients during COVID-19. Otolaryngol-Head Neck Surg 163(1):81–82

Wu SY, Green A (2000) Projection of chronic illness prevalence and cost inflation. Santa Monica, CA: RAND Health, 18

Yu X, Bayram A (2021) Managing capacity for virtual and office appointments in chronic care. Health Care Manag Sci 24:742–767. https://doi.org/10.1007/s10729-021-09546-4

Zhu S, Fan W, Yang S, Pei J, Pardalos PM (2019) Operating room planning and surgical case scheduling: a review of literature. J Comb Optim 37(3):757–805

Xiao Yu is a research engineer in the healthcare field. He has received his M.S. degree in the Industrial Engineering Department at UM-Dearborn. His research interests include operations research, healthcare applications, production management, and data analysis.

Armağan Bayram is an assistant professor in the Department of Industrial and Manufacturing Systems Engineering at the University of Michigan—Dearborn. She worked as a postdoctoral fellow in the Department of Industrial Engineering and Management Sciences at Northwestern University. She received her Ph.D. in Management Science from the University of Massachusetts Amherst and M.S. and B.S. degrees in Industrial Engineering from Istanbul Technical University. Dr. Bayram's research interests include the development of stochastic models and solution methods for capacity and resource allocation problems.

Yuchi Guo is a Ph.D. student in the Industrial Engineering Department at UM-Dearborn. Her research interests include operations research, healthcare applications, and retail operations.

Gökçe Kahvecioğlu is Research Scientist at Supply Chain Optimization Technologies at Amazon. Before joining Amazon, she worked as a Senior Analyst in the Revenue Management division at United Airlines. She completed her B.S. and M.S. in Industrial Engineering at Sabancı University and earned her Ph.D. in Industrial Engineering and Management Sciences at Northwestern University. Her research spans several areas of stochastic and combinatorial optimization for data science and supply-chain management applications.

Chapter 12
Collaborative Systems Analytics to Advance Clinical Care: Application to Congenital Cardiac Patients

Eva K. Lee

Abstract This chapter reports the operations research advances in the Edelman finalist work on "Collaborative Systems Analytics: Establishing Effective Clinical Practice Guidelines for Advancing Congenital Cardiac Care." The clinical advances and results of this project have been reported elsewhere. This paper highlights the OR-analytic advances and briefly summarizes the clinical implementations, results, and impacts. Specifically, we devised a customizable model and decision support framework that combines systems modeling, simulation-optimization decision analytics, clustering, and machine learning within a collaborative learning paradigm to help hospitals pinpoint key factors on practice variation, and design clinical practice guidelines (CPGs) for rapid implementation to improve the outcomes of congenital heart defects surgeries. The OR-analytic collaborative learning framework described herein is generalizable and is applicable for numerous domains. Within healthcare, it enables systems redesign, quality improvement, resource allocation, and clinical support and decision advances. The computational engine facilitates systems and process optimization. The results improve efficiency of healthcare delivery, reduce costs and wastes while improving quality of life of patients. A critical contribution is that the system offers an effective, flexible way to study adequate numbers of patients across multiple sites with uncommon diseases through a common infrastructure for recruiting, monitoring, and following patients whose conditions will be characterized in a standard fashion. The modeling-computational framework facilitates the design of a common CPG, and its successful implementation with documented and measurable clinical outcomes. Such a framework permits a flexible clinical transformative environment that can accommodate practice variance while enabling care teams to identify critical system pathways for multiple-site clinical care and process improvement. The hypothesis testing and dissemination of findings allows for rapid

E. K. Lee (✉)
Center for Operations Research in Medicine and HealthCare, The Data and Analytics Innovation Institute, Atlanta, Georgia
e-mail: evalee-gatech@pm.me

USA NSF I/UCRC Center for Health Organization Transformation, Washington, USA

Industrial and Systems Engineering, Georgia Institute of Technology, Atlanta, Georgia

E. Çetin and H. Özen (eds.), *Healthcare Policy, Innovation and Digitalization*, Accounting, Finance, Sustainability, Governance & Fraud: Theory and Application,
https://doi.org/10.1007/978-981-99-5964-8_12

learning and adoption at multiple sites with a shortened duration and only a fraction of the budget as opposed to the conventional randomized clinical trials. Hence, the OR-analytic collaborative learning framework can serve as a blueprint for other clinical and process-improvement initiatives.

Keywords Congenital heart defects · System process modeling · Machine-learning · Clustering · Simulation–optimization decision analytics · Resource allocation · Mixed-integer programming · Collaborative learning · Early extubation · Quality of life · Length of stay · Improved delivery efficiency and effectiveness · Healthcare savings

12.1 Introduction

Congenital heart defects (CHDs) are the most common birth defects, occurring in around 1% of births. They are the most common cause of infant deaths due to birth defects, and survivors often face health issues into adulthood, including issues with growth and eating, developmental delays, difficulty with exercise, heart rhythm problems, heart failure, and sudden cardiac arrest or stroke (American Academy of Pediatrics). An estimated 1 million U.S. children and about 1.4 million U.S. adults are living with CHDs (Centers for Disease Control and Prevention). Therefore, any improvement in treating these defects is of paramount importance to the quality of life of these patients.

Most procedures that remedy the impact of complex CHDs require open-heart surgeries on the newborn infants. These procedures are complex and require a large and diverse team of expert providers. As recently as 10–15 years ago, most of these newborns did not survive (Gilboa et al. 2010). The ability to care for these patients improved tremendously with the discovery of prostaglandin and technical innovations with echocardiography, interventional cardiac catheterization, and neonatal surgery (Anand et al. 2016). The expansive growth in therapies and technology available for treating CHDs led to a concomitant need for collaborative care teams that include a combination of surgeons, primary care providers, pediatric intensivists, cardiologists, anesthesiologists, and neonatologists. However, variances in surgical practices to treat patients with CHDs among different healthcare centers have led to inconsistent surgical outcomes, some with negative consequences for patients (Anand et al. 2016).

Understanding the effect of treatment approaches on outcomes is best learned from clinical trials. Yet, in the past 25 years, fewer than 40 randomized clinical trials have been carried out on patients with congenital or acquired heart disease. One major barrier is the small number of individuals with a particular CHD at any one center. This has been compounded by the lack of resources to provide national coordination to study outcomes of different treatment approaches used at various centers (Pediatric Heart Network). Unlike the situation for children with cancer, for whom the National Cancer Institute funded clinical trial networks, prior to year

2000, no sustained infrastructure was available for conducting clinical trials and other studies for children with congenital heart disease, a condition that is more common than pediatric cancer.

In 2001, the National Heart, Lung, and Blood Institute established the Pediatric Heart Network (PHN) with clinical research as its primary objective. The PHN has nine core U.S. centers (made up of large academic pediatric heart centers) and sites in Canada and South Korea. The PHN is the anchor of a translational continuum, from discoveries in the lab to clinical applications on patients. Through a multiple-center approach, it works to advance the care of individuals with congenital heart defects and pediatric acquired heart disease by conducting large, definitive studies that inform clinical practice and meet regulatory requirements. The PHN seeks to employ an operations research-analytic collaborative learning model to lessen practice variance as a potential means to improve patient outcomes.

This project showcases the transformation that can happen when data-driven operations research (OR) is applied to improve the outcomes of CHD surgeries in a coordinated effort involving multiple pediatric heart hospitals. Working with the PHN, our operations researchers devised a customizable model and decision support framework that couple systems modeling, simulation–optimization decision analytics, clustering, and machine-learning within a collaborative learning paradigm to help hospitals pinpoint key factors on practice variance, and to design clinical practice guidelines (CPGs) to improve the outcomes of CHD surgeries.

Specifically, a new CPG regarding early extubation (i.e., early removal of breathing apparatus) after surgery was established and implemented successfully in five PHN core centers, with five other centers serving as controls. The day of extubation is a critical time during an intensive care unit (ICU) stay. Extubation is usually undertaken once the patient has demonstrated sufficient respiratory drive. Extubation failure occurs in 5–10% of patients and is associated with poor outcomes. There is evidence that extubation failure can directly worsen patient outcomes independently of the severity of the underlying illness (Thille et al. 2013). Hospitals establish protocols to guide extubation. However, there is no guarantee that a patient will extubate successfully even when these guidelines are followed. Numerous studies have been conducted to understand factors influencing early extubation (Davis et al. 2004; Harris et al. 2014; Mahle et al. 2016a).

The post-CPG implementation results in all sites were positive; early extubation rates increased from 12 to 67%, the median duration of postoperative intubation decreased from 21.2 h to 4.5 h, and the length of stay (LOS) for patients in ICUs decreased from 68.5 h to 51.0 h. Overall, the five hospital sites experienced LOS reductions ranging from 12 to 35%, decreased time to oral feeds from 30.7 h to 19.2 h, and an earlier discontinuation of IV analgesics (37–55% depending on drug type). The implementation resulted in cost savings of approximately 27%, or $13,500 per surgical procedure for children with tetralogy of Fallot on average. It reduced clinical care costs by 65%, pharmacy costs by 46%, laboratory costs by 44%, and imaging costs by 32%.

The novelty of our OR work has four main aspects that, to the best of our knowledge, have not been incorporated in previous methods or studies. This work was the first to (1) intertwine systems modeling, simulation, and optimization into one system to simultaneously analyze resources, decision-making, and patient care among institutions to reveal exact points of practice variance, (2) advance unsupervised learning and supervised learning to uncover clinical outcome variance patterns, and then pinpoint and rank discriminatory factors in influencing and predicting outcomes, (3) engage operation researchers to observe, analytically determine, and prioritize potential areas in which clinicians should focus their collaborative learning efforts to advance surgical outcomes, and (4) use operations research as the central analytic tool to drive the design of a CPG for a clinical trial.

12.2 Background

Numerous studies have shown that surgical outcomes differ among centers that treat congenital heart disease. One possible reason for these differences is that the current healthcare environment is characterized by substantial practice variations, some of which have negative consequences in the delivery of recommended care (Chassin et al. 1987; Kahn et al. 1988; Casparie, 1996; Damoiseaux et al. 1999; Shah et al. 2010; Song et al. 2010). Because of the innate variation in patient symptoms, patient diseases, provider training, provider experience, health system design, and resource allocation, variations in healthcare cannot be eliminated completely (Peterson et al. 1997; Grytten et al. 2003; McGlynn et al. 2003; Mangione-Smith et al. 2007; Berrington et al. 2009). The diverse care teams of providers for CHD magnifies such variances. However, understanding the causes and effects of such variations can help providers and healthcare organizations avoid practices that negatively impact outcomes.

Practice variance is the difference in care provided to two similarly diagnosed patients. In many cases, it causes consistent differences in hospital LOS and clinical outcomes. Practice variance leads to variations in care quality, inefficient use of resources, fewer patients being treated, increased costs for both patients and hospitals, poor hospital utilization, and increased patient susceptibility (e.g., to diseases and infections). Hence, practice variance is an important issue to analyze as a means to optimize healthcare outcomes and delivery.

Numerous studies have been made to identify the nature of practice variation. Some studies have found associations between physicians (Wood and Angus 2001; Veloski et al. 2005), geographic locations, and institutions (Casparie 1996; Lee et al. 2012b). Others suggest how to optimize appropriate levels of practice variation with resource utilization (Wood and Angus 2001; Willson et al. 2001) and decision-making processes (Veloski et al. 2005). Additional research has suggested that implementing practice guidelines at multiple clinical institutes can improve standardized care and reduce highly variable results (Tucker et al. 2007; O'Connor et al. 1996; Ogrinc et al. 2008; Sekimoto et al. 2004). Comparing resources, decision-making, and patient care among institutions might reveal the exact points of practice variance and help identify critical factors to reduce it; to date, this has not yet been examined.

12.2.1 Challenges and Objectives

The incidence of congenital heart disease is at least triple that of childhood cancers and is substantially greater than pediatric AIDS. Yet conducting clinical and translational research on pediatric heart disease remains challenging. The major barriers include the small numbers of individuals with a particular CHD at any one center, differences in treatment approaches, and lack of resources to provide national coordination of collaborative research. The PHN approach attempts to address these barriers by offering an effective, flexible way to study adequate numbers of patients with uncommon diseases through a common infrastructure for recruiting, monitoring, and following patients.

Collaborative learning is the process by which two or more people attempt to learn something together (Dillenbourg 1999). As opposed to individual learning, collaborative learning promotes sharing experiences, resources, skills, and techniques to enhance performance (Bruffee 1993; Dillenbourg 1999). Collaborative learning is based on a model whereby knowledge can be created within a group in which members interact and share experiences about a specific process (Chiu 2000). Its application in medicine is more closely aligned with benchmarking, a method of comparing services or outcomes at an individual center when compared with other leading centers (Camp 1989; Bradner 1991; Povey 1997; Vorhies and Morgan 2005). Experience garnered from organizations that specialize in adult cardiac care, such as the New England Cardiovascular Disease Study Group and the Michigan Society of Thoracic and Cardiovascular Surgeons Quality Collaborative, suggest that many beneficial institutional practices can be disseminated and adopted across participating centers (O'Connor et al. 1996; Prager et al. 2009). Within the field of pediatric cardiology, the National Pediatric Cardiology Quality Improvement Collaborative has established a national network that employs both providers and parents to promote clinical research with the objective of improving outcomes (Anderson 2011; Baker-Smith et al. 2011; Brown et al. 2011; Jacobs et al. 2011; Schidlow et al. 2011; Menon et al. 2013).

However, the success of the collaborative learning process requires careful coordination and commitment among the various stakeholders. As such, the potential of this process is yet to be fully realized in pediatric medicine and the broader medical community. Congenital cardiac care can benefit particularly from collaborative learning because of the unique patient population. Specifically, the complexity and heterogeneous nature of the disease processes and the geographically dispersed sites at which care is provided create challenges for more traditional research efforts. In addition, collaboration is a characteristic of the multidisciplinary nature of the clinical teams that care for these patients. Collaborative learning requires the quick evaluation and dissemination of practices that are deemed to be beneficial to these patients. This is particularly true for complex processes that require the input of experts in many disciplines.

PHN has completed multiple clinical trials, and the implemented results show dramatic variations in clinical practice and outcomes at participating centers (Stark et al. 2000; Ohye et al. 2011; Pasquali et al. 2012; Newburger et al. 2014). Thus, the PHN's interest in reducing practice variance as a means to improve patient outcomes provided motivation to explore the use of our OR-driven collaborative learning model.

This project focused on systems modeling, simulation–optimization decision analytics, clustering and machine-learning, and collaborative learning for modeling practice variance, optimizing clinical care processes, and uncovering key factors that influence outcomes. The team aimed to improve the outcomes of CHD surgeries by establishing best clinical practice guidelines (CPGs) that can be adopted rapidly across multiple hospitals.

During the period prior to the implementation (2012–2013), the average intubation time following surgery ranged from 17.75 h to 35.42 h across the four sites. *The goal was to achieve extubation within six hours of the patient admission to the intensive care unit following infant cardiac surgery* (Mahle et al. 2016a, b). Clinical outcome measures included: total duration of mechanical ventilation, reintubation within 48 h, duration of sedation, cumulative doses of sedation, time to the first introduction of enteral feeds, and postoperative hospital LOS.

Children with CHD face a lifelong risk of health problems, such as issues with growth and eating, developmental delays, difficulty with exercise, heart rhythm problems, heart failure, and sudden cardiac arrest or stroke. Shorter mechanical ventilation times allow babies to resume normal feeding earlier, which lessens their risks of medical complications. Early extubation and shorter hospital stays for patients reduce exposure to critical care therapies and indwelling devices, which subsequently reduce the risk of hospital-acquired infections (HAIs). HAIs are one of the 10 leading causes of death in the United States. Furthermore, a shorter hospital stay, and thus less medication, X-rays, and daily tests, improves overall quality of life and cumulatively saves billions of dollars in subsequent CHD management.

12.3 Methods and Design of Study

With our novel OR-collaborative framework design in mind, the team tackled patient-care process variation, particularly from the provider standpoint in postoperative care of congenital heart surgery. The team found that in each hospital, physicians, nurses, and other hospital staff interacted in different ways with each other and with patients. They also used resources differently. Specifically, the study involved six major components:

1. Observe and document process and workflow and collect hospital data, identify postoperative care processes, standardized protocols, and highlight practice variations in clinic and care-management workflow. We focused on establishing postoperative care processes in cardiac intensive care units (ICUs), cardiac step-down units (CSUs), patient handoff from the operating room to the ICU, and patient handoff from the ICU to the CSU in each hospital.

Some hospitals followed strict institutional protocols in managing patient care, while others employed flexible processes based on experience. We found that as the care process became more complex and involved different types of providers, communication and role definition became more important. We studied the existence of institutional standards and the compounded effects of protocols.

2. Design a system model to establish a system process map and understand practice variations among multiple hospitals and their contribution to different outcomes based on different resources, decision-making processes, and care-management styles. Each hospital possessed different care resources. For example, a care team might consist of physicians, respiratory therapists, and registered nurses, and would make different care decisions based on its resource availability. These variations often resulted in different care plans for similar conditions.
3. Perform unsupervised learning to characterize LOS and outcome factors across multiple sites and design a simulation–optimization decision model to identify and categorize key factors that affect LOS and outcomes. Combining on-site observations with hospital data, we analyzed factors that could affect the total LOS and clinical outcomes in each hospital and categorized them. Using the categorization, we demonstrated how the LOS relates to the clinical outcomes and may be improved through changes in clinical practice.
4. Perform supervised learning to uncover and rank discriminatory factors that can predict and influence the outcome. We designed a novel machine-learning model and algorithm to ascertain the Gini Importance among all features and determine their significance in predicting and influencing LOS and outcome.
5. Conduct collaborative learning across sites with multiple stakeholders to reach a consensus on an effective CPG. We applied collaborative learning using multi-disciplinary team site visits and information sharing, along with the predictive results, to determine decision options, impact on clinical care, and subsequent outcome expectations. This permitted rapid and structured fact finding, the dissemination of expertise, and transformative changes among the sites. We established a CPG with prescriptive instructions to carry out the best practices that we identified.
6. Implement and evaluate performance. We conducted a multicenter clinical trial to determine whether the CPG can reduce the duration of endotracheal intubation following infant heart surgery.

Our study team, comprised of pediatric cardiologists, cardiac intensivists, cardiac surgeons, and a group of operations researchers and informaticians, completed week-long site visits at five pediatric heart centers, Children's Healthcare of Atlanta (Lee 2012a), C.S. Mott Children's Hospital (Lee 2012b), Children's Healthcare of Philadelphia (Lee 2012c), Texas Children's Hospital (Lee 2012d), and Primary Children's Hospital in Utah (Lee 2013a) in 2012 and 2013.

12.3.1 Process and Workflow Observation and Data Collection

Using multifactorial studies, we observed and recorded data and information in the operating room process, handoff from the operating room to the ICU, care-team resources and staffing in the ICU, ICU rounds, ICU care process, chest closure in the ICU, chest tube removal in the ICU, extubation in the ICU, handoff from the ICU to the CSU, and discharge coordination and family education in each hospital. The observations focused on key factors including (1) process and outcomes, (2) personnel and resource availability, and (3) decision-making processes. We captured activities from various personnel across the entire care continuum. Resource availability included team composition, number and skill set of care members, staff shift and hours of care, and resource coordination and support for patient care. Decision-making was noted throughout the postoperative care process, including decisions pertaining to daily rounds, medication, sedation, laboratory tests, x-rays, chest closures, tube removal, extubation, handoff from ICU to CSU, and discharge process.

We obtained clinical protocols at each site and interviewed key personnel in four main areas: ICU patient-care teams, surgical teams, CSU and discharge teams, and administrative leaders. The ICU personnel included the attending and fellow physicians on the care teams, nurse practitioners, respiratory therapists, pharmacists, nutritionists, charge nurses, and registered nurses. The surgical teams included the attending and fellow surgeons, attending and fellow anesthesiologists, nurse practitioners, charge nurses, registered nurses, and physician assistants. The CSU and discharge teams included the attending physicians, charge nurses, registered nurses, social workers, and child-life specialists, if available. The administration leaders consisted of the chiefs of cardiology, directors of the ICU and CSU, chiefs of surgical teams, head nurses, directors of research, research physicians, and research nurses. The interviews focused on the same key factors (process and outcomes, resource availability, and decision-making process) pertaining to the direct observation and data-collection processes. By design, the observations and interviews allowed us to correlate the actual processes and tasks that were undertaken versus the perceived steps that the personnel followed.

12.3.2 Systems Process: Interdependencies and Practice Variance

We next created a patient-care system process map to help us in understanding and analyzing the interplay of care and coordination. Specifically, for each site, we mapped out the pathway for patient care, following a patient from the handoff from the operating room to the ICU until that patient was discharged. After identifying the patient-care steps, we overlaid key personnel involved with each patient-care step and the available resources that supported the task onto the system process map. Next, we highlighted decision-making points and tasks on the system process map. Finally, we created an overall postoperative care process common to all five sites from a systems viewpoint. Through the common system process map, we were able to capture and compare variation points among the five sites.

Combining observations and data from each hospital, we established a computer system simulation model for further analysis. First, we identified seven major system factors that influenced LOS and outcomes in each hospital:

- *Pre-Op*: Before surgery, patients usually stay in the ICU for a day to prepare for surgery. We set this stage as pre-op.
- *Surgery*: The length of surgery is typically between 2.5 h and 6 h.
- *Extubation*: Intubation is the process of inserting an endotracheal tube through the mouth and then into the airway (Cleveland Clinic). This is done so that a patient can be placed on a ventilator to assist with breathing during anesthesia, sedation, or severe illness. Extubation is the removal of the breathing apparatus. Extubation may be performed in the operating room after surgery. However, some patients may spend recovery time in the ICU and extubation is then performed in the ICU.
- *Tube removal*: This factor links to surgical care in the ICU. The patient has tubes on both hands for fluids, drips, and medication, the arterial line for pressure monitoring, and fluids infusing; the central venous catheter used to receive vasopressors (e.g., epinephrine, dopamine), total parenteral nutrition (TPN), hemodynamic monitoring, and high-risk drugs (e.g., promethazine, antibiotics); the Foley catheter for urine draining; the mediastinal chest tube to assist the clearance of blood from the pericardial space, and the gastrointestinal tube for feeding. The surgical team may be involved in deciding to remove the tubes and close the chest at various times during the patient's ICU stay. This step is complex and varied across the sites.
- *ICU care*: Patients recover in the ICU after surgery. The ICU care team takes care of the patients and updates their care plans each day. Based on our observations, each hospital had a different approach to designing the patient-care plans. The ICU care factor is complex because it involves the personnel who make the decision, the types of decisions made, the processes used to carry them out, and the types of providers who carry them out.
- *Discharge planning*: Each hospital has its own family education plan and checklist. We focused on when and where each hospital begins discharge and education planning.
- *Step-down care*: Patients are moved to the step-down area for recovery. Step-down care is less intensive; nonetheless, patients may spend extensive time in the step-down area.

12.3.3 *Unsupervised Learning and Machine-Learning: Predicting LOS, Patient Outcome Characteristics, and Ranking Discriminatory Factors*

Below, we discuss our use of clustering and discriminant analysis via a mixed-integer programming (DAMIP) classification model.

Clustering: To understand the LOS correlation across the five sites, we first performed an expectation–maximization (EM) clustering algorithm on the surgical cases to categorize the LOS into related groups. This allowed us to explore its interdependencies with other clinical processes and outcomes.

Machine-learning via an optimization-based support vector machine: DAMIP is a general-purpose optimization-based classification model that is suitable for establishing reliable predictive rules for a broad variety of biological and medical applications (Feltus et al. 2003; Lee and Wu 2007; Querec et al. 2009; Nakaya et al. 2011, 2015; Lee et al. 2012b, c, 2015, 2016a, b; Lee, 2016, 2017; Koczor et al. 2013; Kazmin et al. 2017). The predictive framework simultaneously incorporates (1) the ability to classify any number of distinct groups; (2) the ability to incorporate heterogeneous and temporal types of attributes as input; (3) a high-dimensional data transformation that reduces dimension, and minimizes noise and errors; (4) the ability to incorporate constraints to limit the rate of misclassification, and a reserved-judgment region that provides a safeguard against overtraining (which tends to lead to high misclassification rates from the resulting predictive rule), and (5) a successive multistage classification capability to handle data points placed in the reserved-judgment region (Lee 2016; Lee and Ega 2022). Thus far, among all the support vector machines designed by researchers and commercial vendors, only DAMIP combines all these properties within a single mathematical model.

We have discussed DAMIP in great detail in our previous papers (Lee 2016, 2017; Lee et al. 2015, 2016a; Lee and Egan 2022). Below, we include a brief excerpt description of its overall schema designed for this study.

Using the comprehensive data and clustering results, we applied DAMIP to uncover patient characteristics, including processes, resource needs, treatment outcomes, and decision activities to establish the predictive rules. A significant contribution of our work is that it is the first study in which demographics, socio-economic status, clinical information, processes and operations, disease behavioral patterns, and decision-making processes are employed simultaneously as attributes within a machine-learning framework. It is also the first time such an approach has been applied to multiple pediatric sites to uncover variances and outcome characteristics.

The computational design of our machine-learning framework utilizes a wrapper approach. Specifically, we applied pattern recognition based on our recent advances on text mining for unstructured clinical notes to the input features (Hagen et al. 2013), and designed a combinatorial feature-selection algorithm to identify discriminatory features. Next, we integrated the DAMIP supervised learning classification module with the feature-selection algorithm. The feature-selection, classification, and cross-validation procedures are wrapped so that the feature-selection algorithm searches through the space of feature subsets using the cross-validation accuracy from the DAMIP classification module as a measure of goodness. The small subset of features returned from the machine-learning analysis can be viewed as critical patient and clinical factors that drive service, outcome, and LOS characteristics. We sought to identify the rule that offers the best predictive capability. This provided feedback to the PHN stakeholders for prioritization of factors and consensus design of the CPGs.

In this supervised classification approach, the group status of each patient in the training set is obtained from the clustering results. The training set consisted of data, which we extracted from the hospital database, for a set of patients whose LOS and group status (via clustering) were known. We then entered the training data into the machine-learning framework. Through the feature-selection algorithm, we selected a subset of features to form a classification rule; we then used this rule to perform tenfold cross-validation on the training set to obtain an unbiased estimate.

To gauge the predictive power of the rule, we performed blind prediction on an independent set of patients who had never been used in feature selection. We ran each patient through the rule, which then returned a status. We compared this status to the actual status of the patient (actual LOS and clustering-group status). Hence, we always compared our prediction with the actual outcome to measure predictive accuracy.

DAMIP employs a 0–1 variable to denote if an entity is classified correctly. The model seeks to maximize the number of correct classifications while constraining the percentage of misclassifications in each group. It employs a nonlinear function to transform the feature space vectors into the group space vectors, which govern the placement of an entity into each of the a priori groups or the reserved-judgment region. Constraints are included to ensure that an entity is assigned to exactly one group. The misclassifying constraint allows the users to preset the desirable misclassification levels as overall errors for each group, pairwise errors, or overall errors for all groups together. With the reserved judgment in place, the mathematical system ensures that a solution that satisfies the preset errors always exists. Lee et al. (2003); Lee (2007), Lee and Wu (2007), Brooks and Lee (2010, 2014), and Lee and Egan (2022) detail some variations of the DAMIP modeling and its theoretical and computational contributions.

DAMIP's special characteristics include the following: (1) the resulting classification rule is strongly universally consistent, given that the Bayes optimal rule for classification is known (Brooks and Lee 2010); (2) the misclassification rates using the DAMIP method are consistently lower than with other classification approaches when tested on both simulated and real-world data (Gallagher et al. 1997; Lee et al. 2003, 2012b, c, 2016a, b; Lee 2016, 2017; Lee and Egan 2022); (3) the DAMIP classification rules appear to be insensitive to the specification of prior probabilities, yet capable of reducing misclassification rates when the number of training entities from each group is different; and (4) the DAMIP model generates stable and robust classification rules regardless of the proportions of training entities from each group (Lee et al. 2003, 2007, 2010, 2012b, 2012c, 2016a, 2016b; Lee and Wu 2007; Querec et al. 2009; Brooks and Lee 2010, 2014; Koczor et al. 2013; Nakaya et al. 2011, 2015;

Kazmin et al. 2017; Lee 2017; Lee and Egan 2022). Thus, we determined it to be suitable for this PHN multicenter analysis.

Mathematically, DAMIP has been proven to be NP-complete (Brooks and Lee 2010, 2014). We tailored the model for our PHN study and solved the multigroup instances for this study using advances in hypergraph theory (Lee et al. 2022) and new combinatorial heuristics algorithms to rapidly obtain solutions.

12.3.4 System Simulation and Overall Performance Optimization

Armed with the results from machine-learning, we designed a system simulation–optimization model to gauge the significance of the identified features along the patient-care continuum. In addition to simulating the systems performance (LOS and outcomes of the surgical cases), we also optimized the entire process to determine which key features provided the maximum return on investment (ROI). Although there is significant financial return associated with outcome improvement, our ROI focused mostly on the outcomes of these infants, their well-being, LOS, usage of drugs, and cognitive impact because of the surgical procedures.

Mathematically, we formulated this as a general multiple-objective, multiple-decision process, and resource-allocation problem. The multiple-objective function minimizes the LOS, while optimizing the performances of each process within each unit (e.g., operating room, ICU, step-down), the decision choices, and the adaptability and return of overall performance under multiple resource-allocation scenarios (e.g., availability of various personnel). The system optimizes the overall outcome performance based on the investment made (in the changes of various processes based on our machine-learning findings) and the effect on different processes and the global return. Such a modeling platform remains computationally challenging and is not yet commercially available. We leveraged our previous successes using this framework and expanded the design to accommodate this CHD study.

Simulation–Optimization Environment with Nonlinear Mixed-Integer Program for Optimizing Outcome Performance

We expand the simulation–optimization computation framework from our previous work (Lee et al. 2009, 2012a, 2013, 2015), which has been designed and used successfully in public health emergency response logistics and mass dispensing, optimizing guest-flow at theme parks, and hospital emergency department patient care advances. While the mathematical formula is similar to our previous work, we caution that the simulation–optimization in this PHN environment is more complex because it simultaneously addresses the practice variance involving processes, patient care, decision-making, and resources.

Let $\mathbf{R}$ be the set of resource groups, $\mathbf{T}_r$ be the set of resource type in resource group $r \in \mathbf{R}$, and $\mathbf{S}$ be the set of services and decisions in the clinical process. We use the decision variables $x_{ijr} \in \mathbf{Z}_+$ to denote the number of resource type i in resource group r assigned to service/decision j, $\forall r \in \mathbf{R}, i \in \mathbf{T}_r, j \in \mathbf{S}_{ir}$ with associated unit cost of k_{ijr}.

The functions *w(x), p(x), u(x)* are used to represent various services and decisions within the postoperative care process. Examples include the timing of extubation, the step-down preparation, the decision to do certain tests, the discharge and feeding education timing, the time of sedation, and the choices and amount of medication used. These are key components in our study because we aim to pinpoint practice variance across multiple sites and their effect on outcomes.

Let $\overline{m_{ijr}}$ and $\underline{m_{ijr}}$: be the maximum and minimum number of resources of type i in resource group r that may be assigned to service j. $r \in \mathbf{R}, i \in \mathbf{T}_r, j \in \mathbf{S}_{ir}$, and n_{ir} the number of available resources of type i in resource group r. $r \in \mathbf{R}, i \in \mathbf{T}_r$. Mathematically, we formulate a general representation of the multiple-objective, multiple-decision, and resource-allocation problem as follows:

Min	$z = f\left(\sum_{j \in \mathbf{S}} g_j, c, \theta\right)$		(N0)
s.t	$\underline{m_{ijr}} \leq x_{ijr} \leq \overline{m_{ijr}},$	$\forall r \in \mathbf{R}, i \in \mathbf{T}_r, j \in \mathbf{S}_{ir}$	(N1)
	$\sum_{j \in \mathbf{S}_{ir}} x_{ijr} \leq n_{ir},$	$\forall r \in \mathbf{R}, i \in \mathbf{T}_r$	(N2)
	$w(x)_j \leq w_{max}$ $p(x)_j \leq p_{max}$ $u_{min} \leq u(x)_j \leq u_{max}$	$\forall j \in \mathbf{S}$	(N3)
	$\theta(x) \geq \theta_{min}$ $c(x) \leq c_{max}$		(N4)
	$x_{ijr} \in \mathbf{Z}_+$	$\forall r \in \mathbf{R}, i \in \mathbf{T}_r, j \in \mathbf{S}_{ir}$	(N5)

The objective function minimizes the length of stay *c*, the performance of each process within each unit (services and decisions *w, p, u*, in operating room, ICU, step-down, etc.), and the adaptability and return of overall performance (θ) under multiple decisions and resource allocations (x) (e.g., availability of personnel, equipment).

Constraint sets (N1) and (N2) describe the resource availability for each service or decision. (N3) ensures that services satisfy clinical guidelines. For example, the timing for wound cleaning must satisfy clinical protocol guidelines, the dosage of medication should not exceed the prescribed limit, etc. The set of constraints for (N3) is large and diverse, reflecting the complex clinical care team and the services needed to provide for the CHD population. (N4) constrains two of the clinical outcome metrics: the desirable outcome with quality of adjusted life year and risks of infection, and the length of hospital stay.

The cost of each service or decision j is represented by $g_j\left(\sum_{i,r} k_{\text{ijr}}x_{\text{ijr}},\ w_j,\ p_j,\ u_j\right)$.

And the resulting objective function $f\left(\sum_{j\in \mathbf{S}} g_j, c, \theta\right)$ are not expressible in closed form. The system optimizes the overall outcome performance using length of hospital stay as a surrogate measurement based on the investment made (in the changes of various processes based on our machine-learning findings) and the effect on different processes and the global return.

Given a clinical process configuration with various service distributions for each process and decision point and associated clinical outcome metrics and risk complications, the system simulates the entire postoperative care process to acquire the length of stay and the system performance under this configuration. The output simulation statistics are then input into the stochastic NMIP optimization model where performance and resources are optimized. The resulting optimization output is entered back into the simulation to obtain the next system performance metrics. The simulation–optimization iterates until no improvement is achieved.

12.3.5 Collaborative Learning and the Design of Clinical Practice Guidelines

Although approaches that include collaborative learning and site visits have been applied on a local level in other surgical fields (O'Connor et al. 1996; Prager et al. 2009), they have not yet been applied to pediatric cardiac care at the national level. Furthermore, engaging system operation researchers to observe, analytically determine, and prioritize potential areas in which providers should focus their collaborative learning efforts had not yet been done. Therefore, we designed this study to test the feasibility of collaborative learning; our ultimate goal was to develop a CPG that would minimize variation among the participating sites. This process included data sharing and meta-analytics, benchmarking, machine-learning to prioritize potential areas that should be changed, site visits, and the dissemination of best practices at five pediatric heart centers. Thus, our objectives were to change practice, test hypotheses, and improve patient outcomes. Figure 12.1 shows the collaborative learning process. The multi-site effort was initiated after successful completion of an in-depth study at one site.

Based on site reports (Lee 2012a, b, c, 2012d, 2013a) and OR-analytic objective analysis, we stratified practice variations among centers for relevant clinical practices into three tiers (high, moderate, and low). We then correlated clinical outcomes to relevant clinical practice using standard tests of significance, including linear and logistic regression, contingency tables, rank sum tests (Wolf et al. 2016), and the machine-learning results.

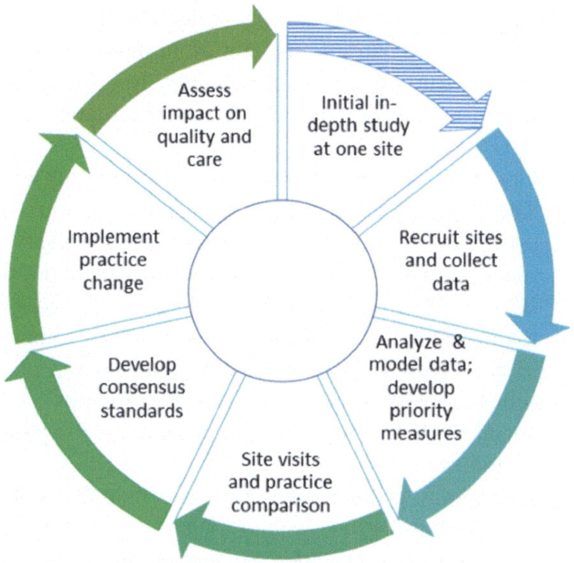

Fig. 12.1 The flowchart illustrates the seven steps that comprise the collaborative learning process

Using the comprehensive data collected at the site visits, the group conducted extensive discussions and worked to establish a consensus on the areas on which it should focus. Its objective was to facilitate practice change and implementation and then observe the potential impact on quality and patient care. The approach the group used for its analysis comprised two stages: (1) an objective OR-analytic approach to mine the data, process interdependencies, and analyze outcomes (as we describe above), and (2) a clinical approach employing subjective practice experience and deep learning to establish a hypothesis for potential implementation.

12.4 Computational Results, Implementation, and Achieved Outcome

We carried out an intensive data-collection process and spent one year collecting data for the initial site (Children's Healthcare of Atlanta). Each week, the team consisted of 10 to 14 operations research and industrial engineering students. Over the course of the year, 34 students were involved. Observation protocols were developed regarding patients, process, decisions, resources, outcome metrics, and other critical factors that had to be recorded. Comment blocks were used to include anything unusual not listed in our design file.

Data collected include the following seven categories of information for which we designed an interactive web portal to manage.

1. Operative (surgical) notes generated by the clinical team of surgeons, anesthesiologists, and cardiologists for more than 100 patients. These notes documented precise surgical procedures, medication, unexpected events, prognosis expectation, and care coordination and requirements as a patient was moved to the ICU.
2. Operative observation notes generated by our team for more than 200 patients. These notes documented all aspects of the surgical procedure, including processes, timeline, events, personnel involved, compliances, skills, environment, individual and team interactions, behavior, and operating room traffic.
3. ICU notes of daily morning rounds for 50 weeks. The team covered about 15 to18 patients daily. Morning rounds involve visits, evaluations, and discussions of each patient by a cardiology intensivist, a physician assistant, a cardiology fellow, a bedside nurse, a respiratory therapist, a pharmacist, a dietician, and a social worker. Selectively, we observed some patients over their entire ICU stay period.
4. Day-to-day process notes for more than 200 patients. This involved processes such as extubation, wound cleaning, and tube removal. We documented the procedures that providers performed on the patients.
5. Interviews of 30 providers. We interviewed all the surgeons, cardiovascular intensivists, respiratory therapists, a pharmacist, most of the bedside nurses, and two social workers.
6. Videos of 20 surgeries.
7. Electronic medical records (EMRs) for these patients.

For the four other participating sites, we streamlined the observation process to achieve the same purpose. These visits were week-long events in which the team observed and documented ICU activities for 18–24 h daily. The observation focused on team composition, communication, use and adherence to protocols, resources, and practice variations. At each site, we obtained operative notes for 50 patients, operative observation notes for 15–20 patients, notes of daily morning rounds for five days for about 20–25 patients each day, medication documents (from EMRs) for 50 patients, videos for five surgeries, and interviews of 10 providers.

12.4.1 Practice Variance: Processes, Care Coordination, Decision Making, and Resource Usage

We identified numerous points of process, resources, and decision variations, and ample opportunities for improvement in postoperative care. Detailed findings for each site and the comparison are summarized and reported in Lee, 2012a, b, c, 2012d, and 2013a.

Each site managed between 20 and 30 ICU beds. Table 12.1 shows an abbreviated sample version of the practice variance in care coordination, team composition, team collaboration, decision-making processes, protocols, and procedural and operational processes. Most morning rounds spanned under 10 min per patient; however, rounds took 30 min per patient at one site.

Figure 12.2 highlights the practice variance at each site on the common postoperative care process map.

We observed the following variances: communications in the handoff from the operating room to the ICU (team-to-team versus individual-to-individual); surgical decision-making in the ICU (surgical team versus intensivist); surgical personnel performing chest closure and tube removal versus the ICU team performing these procedures; presence or absence of early extubation policy; ICU care plan (conservative versus dynamic aggressive); different personnel in charge of extubation in the ICU; family education, and timing of discharge (ICU versus CSU). Figure 12.3 shows differences in the process order at each site.

Observing such a diverse spectrum of practice variance can be overwhelming. The goal of practice variance reduction is to allow for the greatest flexibility while

Table 12.1 Summarizes ICU and step-down unit similarities and differences based on protocols and observations

	ICU care	Step-down unit care
Similarities	Rounding team composition Weekend coverage Team-to-team handoff Blood transfusion parameters Postoperative: • Fluid management • Inotropic support	High-flow nasal cannula allowed Chest tubes allowed Select intravenous inotropes allowed Discharge coordinator involvement Pharmacist educator
Differences	Overnight coverage Neonatal feeding advancement protocol Ventilator weaning protocol Shift-change composition/collaboration and process Extubation decision • Where, when, and who Extubation action • When, who Medication ordering decision timing Collaboration policies between ICU team and surgical team • Surgical team involvement in decision process • Chest closure / chest tube removal personnel • Handoff personnel (how and who) • Surgical decision in ICU Timing on update patient-care plans X-ray (when/how to order) Lab test (when to order) Sedation decision (who and how) Feeding (when and how)	Arterial line allowed Care-team composition Discharge education and timeline Sedative protocol Feeding protocol Patient care

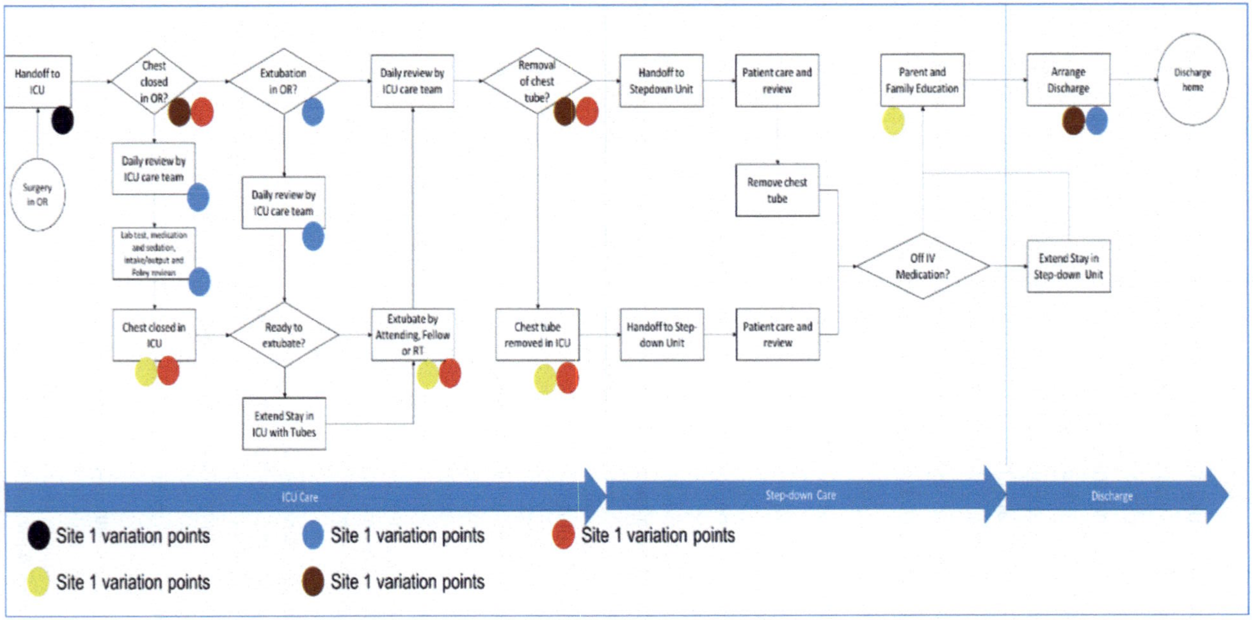

Fig. 12.2 We captured this figure from an interactive visualized process map that we built. The tool allows users to drill down (using mouse clicks) at each level as the figure expands to reveal details

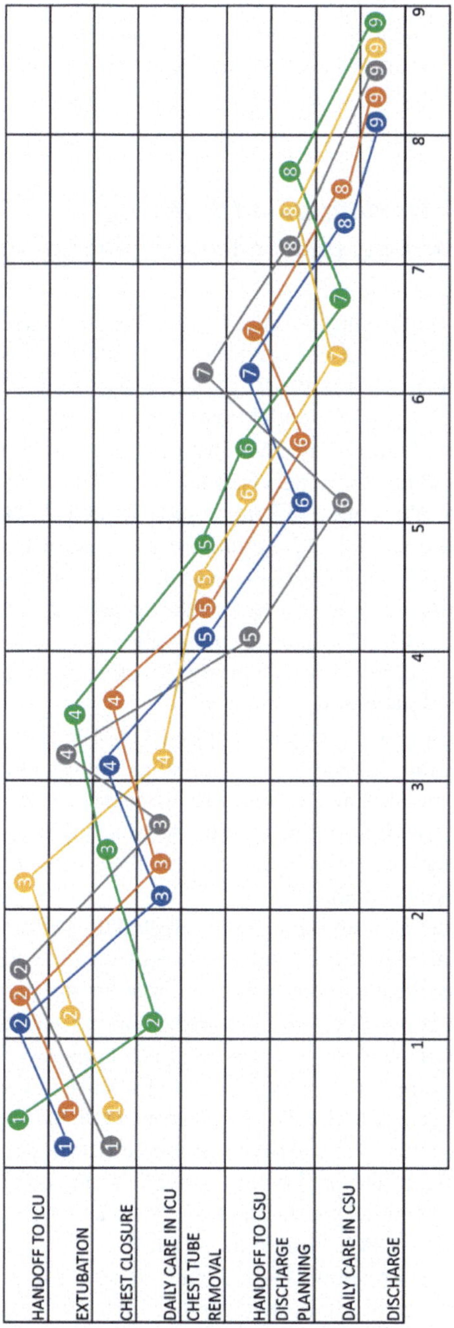

Fig. 12.3 This figure shows nine important steps for postoperative processes and their sequences at each site

still ensuring the best outcome and LOS by *identifying the minimum set of critical factors that will have the greatest global impact (and ROI) on outcome and LOS*. It is also essential to identify a critical set of compliance standards that can minimize and safeguard against medical errors. Quality assurance across sites and ease of training must also be ensured.

12.4.2 Expectation–Maximization Clustering, Machine-Learning, and Systems Optimization

EM clustering performed on the surgical cases of the five sites yielded three distinct clusters (Fig. 12.4) based on the LOS for tetralogy of Fallot and coarctation repair (in days): Short: LOS < 6.7, Medium: $6.7 \leq$ LOS < 8.8, Long: LOS ≥ 8.8. The distribution is roughly 31.3%, 26.7%, and 42.0%, respectively.

Using the three groups identified by EM clustering, we randomly selected 66% of these surgical cases as input for supervised learning. The remaining 34% were used for blind prediction. We compared the results from DAMIP to those from 24 commonly used classifiers. Table 12.2 summarizes some of these comparisons. DAMIP performed best, with 95% tenfold cross-validation unbiased estimate, and over 90% on blind-prediction accuracy for each of the three groups. Bayesian network also returned good results with 80.0% tenfold cross-validation unbiased estimate, and over 75% blind-prediction accuracy for each group. All other classifiers suffered from imbalanced data and scored unevenly.

In addition to classifying LOS, we also applied DAMIP to uncover groups of patients with various outcome characteristics, including length of intubation, amount of sedative used, time to feed, time to hand off to critical care units, etc. These patient-process-resource-decision characteristics became input to the simulation–optimization system where we optimized the clinical processes and decisions to achieve the best outcome performance.

We ran feature selection on 24 classifiers and selected the top 50% of the features among all models. We calculated the Gini coefficient on the selected features to determine the significance of each in predicting and influencing the LOS (Wolf et al. 2016). The Gini coefficient measures the inequality among values of a frequency distribution. A Gini coefficient of zero expresses perfect equality, where all values are the same (https://en.wikipedia.org/wiki/Gini_coefficient). Table 12.3 shows that step-down unit care, ICU care, and early extubation were the top three features. ICU and step-down care (Table 12.1) involve varying care protocols, multiple levels of resource utilization and staffing, and complex medication and care coordination. Their similarities and complexities among the five sites made them less appealing as targets for establishing a new CPG (Wolf et al. 2016).

Time to extubation, however, showed significant variations in practice approach, staffing models, and timing across all the institutions (Table 12.1). Data regarding the duration of postoperative ventilation, which we collected from institutional databases

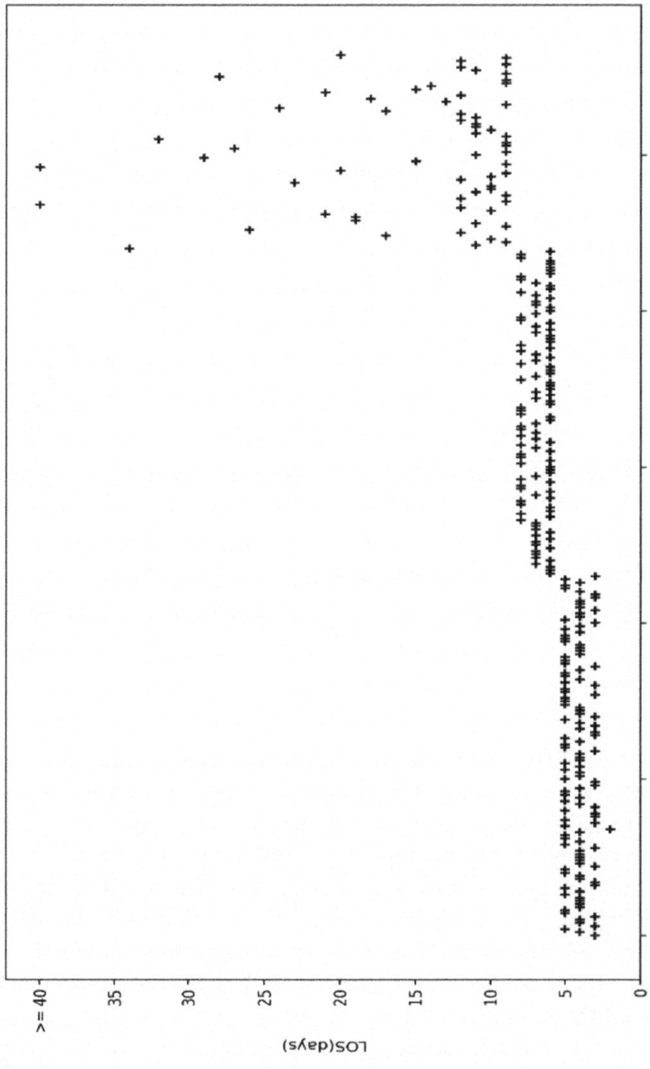

Fig. 12.4 LOS clusters from pre-CPG are used to identify pattern characteristics of patients. We list each surgery along the x-axis. The y-axis corresponds to the length of stay in days

Table 12.2 Compares machine-learning results to identify the smallest set of discriminatory factors that predict patient LOS. We performed feature selections on the training set. Once we established a rule, we used it to blind predict a new set of patients to test its accuracy

	Training Set: 66% surgical cases			Blind-prediction set: 34% surgical cases		
	Tenfold cross validation accuracy			Blind-prediction accuracy		
Classification method	Short LOS (%)	Medium (%)	Long (%)	Short (%)	Medium (%)	Long (%)
Linear discriminant analysis	86.3	15.5	76.3	89.1	11.3	59.3
Naïve Bayesian	51.6	57.0	61.6	51.7	59.2	89.2
Bayesian network	81.4	80.9	81.1	75.9	77.2	77.2
Support vector machine	86.5	9.0	76.5	76.2	8.1	70.0
Logistic Regression	76.5	5.9	80.5	66.3	8.3	77.3
Classification tree	66.6	4.4	59.2	66.3	3.0	53.0
Random forest	76.9	1.5	67.5	76.3	1.9	68.9
Nearest Shrunken centroid	62.7	50.0	64.5	48.7	64.7	57.0
DAMIP	**95.9**	**95.1**	**96.3**	**92.2**	**90.5**	**93.5**

Table 12.3 Shows the Gini coefficient for each factor in predicting the LOS

Factor	Extubation	Tube removal	ICU care	Step-down care	Discharge plan
Total Gini coefficient	82	72	89	93	70

for a sample period of 2010–2011, showed an institutional median duration of mechanical ventilation post tetralogy of Fallot (TOF) repair of 19.8 h with ranges between 0 and 26.1 h, and an institutional median duration of mechanical ventilation following coarctation repair of 20.6 h with ranges between 0 and 27.8 h.

Time to extubation is a process that can be adopted relatively easily for a new protocol. Moreover, it affects the duration of mechanical ventilation, LOS, and other clinical metrics; these impacts are measurable. Extubation can also simultaneously affect the time it takes for removal of other tubes and ICU care management and will have a downstream effect on step-down care.

Based on our analysis of machine-learning results, we developed a clinical strategy to reduce LOS and improve outcomes by shortening the duration of mechanical ventilation. Because most of these patients now survive to hospital discharge, total LOS has become a common surrogate endpoint for measuring performance, as well as time in the ICU, time of mechanical ventilation, and amount of vasoactive support required (Table 12.4).

Table 12.4 Summarizes the distributions (based on observations, historical data, and interviews) used for each process at each site

	Pre-OP	Surgery	Extubation	Tube removal	ICU Care	Discharge planning	Step-down care	Median LOS
Site 1	1 day	2.5–6 h	0 ~ 3 days	0 ~ 2 days	2 days	~ 1 day	1.5 days	8 days
	Fixed		Tri(0,1.2, 2.5)	Tri(0, 1.2, 2)	N(2, 0.25)	Tri(0.5, 0.8, 1)	Fixed	
Site 2	1	2.5–6 h	0 ~ 1	0 ~ 1.5	1.5	~ 1	1	5
	Fixed		Tri(0, 0.3, 1)	Tri(0,0.8,1.5)	N(1.5, 0.25)	Tri(0.5, 0.8, 1)	Fixed	
Site 3	1	2.5–6 h	1 ~ 4	0 ~ 5	3 ~ 6	~ 1	1.5	11
	Fixed		Tri(1, 2, 4)	Tri(0, 2.5, 5)	Tri(3,3.5,6)	Tri(0.2, 0.8, 1)	Fixed	
Site 4	1	2.5–6 h	0 ~ 2	0 ~ 3	2 ~ 4	~ 2	2 ~ 4	14
	Fixed		Tri(0, 1, 2)	Tri(0, 1.5, 3)	Tri(2,3.5,4)	Tri(0.5, 1.5, 2)	Tri(2, 3, 4)	
Site 5	1	2.5–6 h	0 ~ 1.5	0 ~ 2	1 ~ 2	~ 1	1 ~ 2	6
	Fixed		Tri(0, 0.8, 1.5)	Tri(0,1.25,2)	Tri(1,1.25,2)	Tri(0.5, 0.8, 1)	Tri(1, 1.2, 2)	

We analyzed our recommended strategies using the simulation–optimization computational model to determine the potential improvement over the current practice. Table 12.4 summarizes the distributions (based on observations, historical data, and interviews) used for each process at each site. Pre-op takes approximately one day for most sites. The model was designed to quantify how various factors influence LOS. Within the system simulation model, each patient is characterized by the service distribution, process characteristics, interplay of providers and processes, and patient disease symptoms.

To simulate early extubation, we made changes to all sites according to the earliest extubation distribution—Site 2—and gauged the potential gains in various clinical metrics. Figure 12.5a shows a reduction in LOS, which ranges from 10.0% to 29.4% across the five active sites and could have a significant impact on the long-term health and cognitive development of the patients. Furthermore, Fig. 12.5b shows that early extubation also has a positive impact on other clinical outcomes.

12.4.3 Collaborative Analytics to Establish the Clinical Practice Guidelines

In clinical practice, the results of processes are often utilized as clinical outcomes. For example, intubation is a process, whereas the length of intubation provides a clinical outcome measure on the condition of the patients. However, they can be subject to underlying variance at the site. This is the crux of CPG—trying to improve clinical

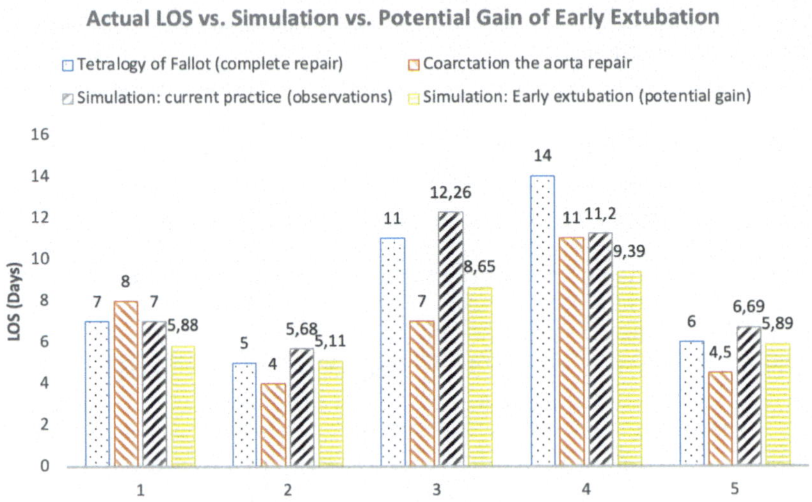

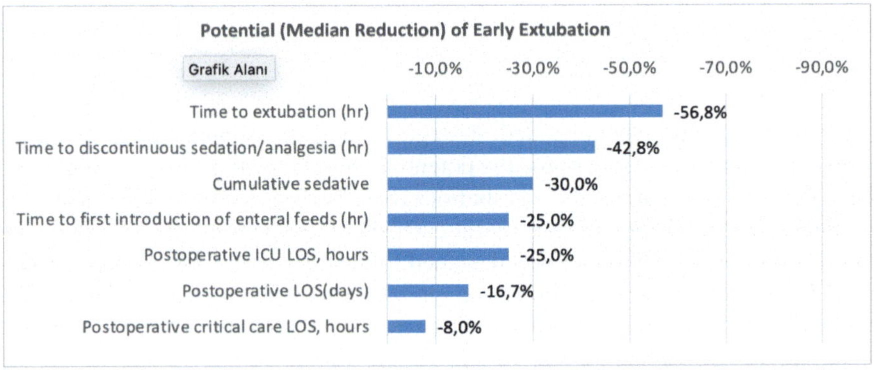

Fig. 12.5 **a** The graph illustrates potential gains of early extubation, and its systems influence on the overall LOS. **b** The graph illustrates potential gains of early extubation and its systems effects on various clinical outcome metrics

outcomes by identifying the differences in the outcomes attained at each site and determining a target goal of clinical outcomes (for all sites) to achieve (Mahle et al. 2016b). Data from questionnaires, subjective practice experience of participating providers, deep collaborative learning, and OR-analytic objective analysis formed the basis for developing and implementing a CPG to promote early extubation following infant cardiac surgery. Characteristics of clinical practices that are appropriate for collaborative learning include practices that (1) vary among centers; (2) correlate with significant clinical outcomes; (3) are systems based; (4) are difficult to disseminate solely by publications; and (5) may result in reduced resource usage.

Using these criteria, the PHN collaborative team identified eight clinical practices for potential collaborative initiatives: discharge process, timing of delayed sternal closure, sedation and analgesia practices, postoperative feeding, postoperative ventilation and extubation, postoperative monitoring, anesthetic management, and family teaching. Combining the objective machine-learning results and deep learning analysis, the team chose early extubation for consensus CPG design and multiple-site implementation.

12.4.4 Implementation via Multicenter Clinical Trial and the Outcomes Achieved

The goal of the CPG is to extubate patients within six hours of their admission to the ICU following two infant cardiac surgical procedures: complete repair of TOF and repair of isolated coarctation of the aorta (CoA) via thoracotomy. Secondary outcomes include total duration of mechanical ventilation, rate of reintubation within 48 h, duration of sedation, cumulative doses of sedation, time to first introduction of enteral feeds, and postoperative hospital LOS (Mahle et al. 2016b).

Patients with TOF and CoA tend to have an uncomplicated early postoperative course and typically have physiologies that would both permit and benefit from early extubation. The CPG was designed to promote an approach to anesthesia, operating room, and ICU practices that permit early extubation (Mahle et al. 2016b; Wolf et al. 2016). Figure 12.6 shows the clinical study design. The CPG is shown in the Appendix. Four centers participated in the new CPG (active sites), one center served as the model site, and five other centers served as controls.

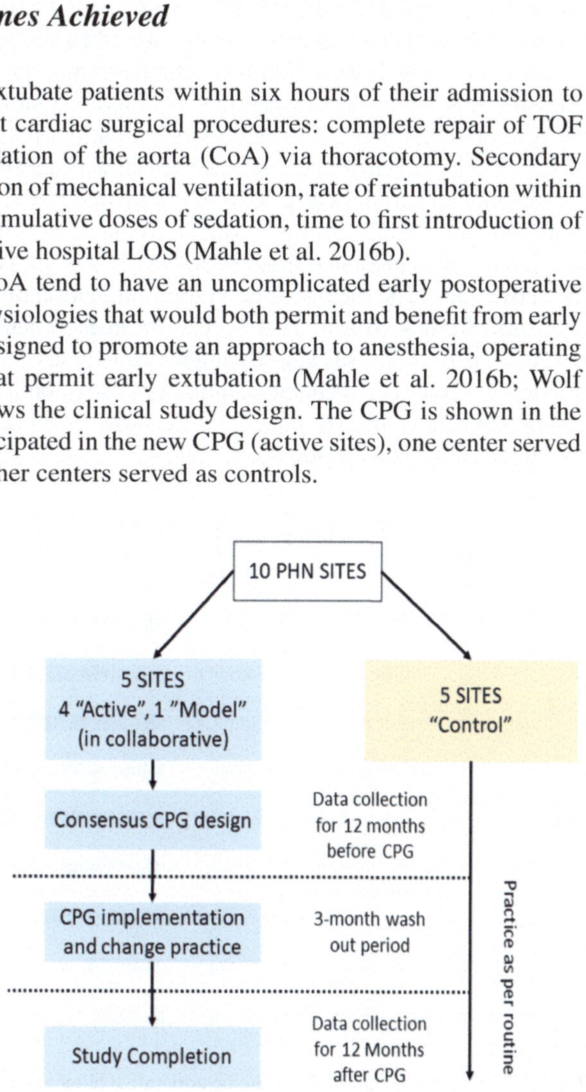

Fig. 12.6 The graph shows the flow of our clinical study design

12.4.5 *Outcome Improvement and Individual Site Performance*

We compared outcomes during the 12 months prior to and following the CPG implementation. After the implementation, we observed an increase in early extubation rates from 11.8% to 66.9%, and a decrease in the median duration of postoperative intubation from 21.2 h to 4.5 h. There was also a tendency toward shorter ICU LOS (68.5 h prior to the implementation versus 51.0 h following it). Overall, the four active hospital sites experienced LOS reductions ranging from 12 to 35%, decreased time to oral feeds, and an earlier discontinuation of IV analgesics (Mahle et al. 2016b).

In our machine-learning analysis, the Gini importance score suggested that ICU care and step-down care could affect LOS more significantly than early extubation. We also predicted that early extubation can affect ICU care management and may have a positive downstream effect on step-down care (Fig. 12.5b). Indeed, we found several clinical variables related to ICU care that were impacted by the early extubation CPG. Specifically, the time to first introduction of enteral feeds following surgery decreased from a median (range) of 30.7 h (21.9 h–49.3 h) to 19.2 h (10.9 h–27.8 h) at the active sites. The median time to discontinuation of all continuous IV analgesics decreased from 43.6 h to 19.3 h. The mean face, legs, activity, cry, consolability (FLACC) pain score was slightly higher (1.3 versus 2.1). We also found a reduction (among the TOF cohort) in the cumulative dose of both opioids and benzodiazepines among the active sites. In Fig. 12.7, we show nine postoperative clinical outcomes for all the subjects we studied.

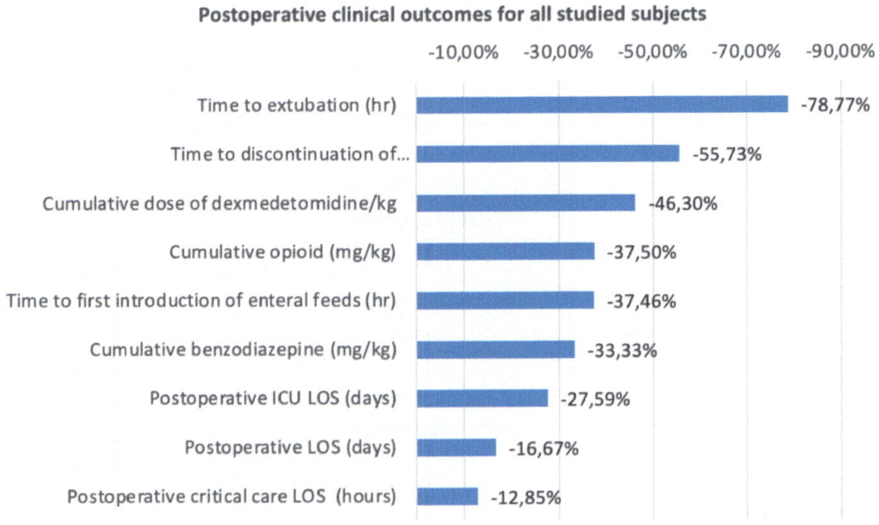

Fig. 12.7 This figure summarizes the postoperative clinical outcomes for all patients studied. We compare the median before and after CPG

Understanding the details of how centers implemented the CPG and their outcomes is important in supporting the dissemination and implementation of this early extubation CPG to other interested sites.

We analyzed patient characteristics and outcomes (e.g., intubation time, LOS) across these sites and assessed both absolute and relative pre-CPG to post-CPG changes to account for different starting points across sites (Bates et al. 2019). Figure 12.8a displays the median duration of intubation at each site (x-axis) in the pre-CPG period (solid) versus post-CPG period (diagonal strips). In the pre-CPG period, the median intubation time varied across the four active sites, ranging from 17.75 h to 35.42 h; thus, the sites had different starting points. In the post-CPG period, all sites had a statistically significant decrease after implementing the CPG ($p < 0.001$ for all). The post-CPG intubation times ranged from 0.3 h to 5.13 h across the sites. There was significant variation in the magnitude of change considering the starting point, as Fig. 12.8a shows, ranging from a relative decrease from 73 to 99%. Of note, Site 3 started with the highest median duration of intubation and ended up with the lowest post-CPG.

Figure 12.8b shows the percentage of babies extubated early (defined as within six hours of ICU admission) post-CPG. The rate increased significantly at all sites. Differences across sites were significant in the pre-CPG period but were no longer significant in the post-CPG period.'

Our results show a drastic increase in early extubation, driving intubation below six hours for all sites. The sites uniformly achieve high compliance of early extubation, regardless of their initial starting point.

12.4.6 Implementation Strategies and Compliance

We conducted and analyzed semi-structured interviews with team members from each site to understand their implementation strategies. Table 12.5 briefly summarizes the interview data with the various implementation strategies highlighted. Bates et al., presented detailed breakdowns of each of the implemented steps (2019).

Table 12.5 shows that all sites shared some common implementation strategies (1–5) and the sites used some unique strategies. In particular, Site 3, which had the greatest reduction in intubation duration, included ancillary staff and fellows on the core team, conducted monthly in-person data reviews, collected supplemental data to address nursing concerns, and created additional care protocols to support the CPG. These findings are important for dissemination of the CPG across the nation (and internationally).

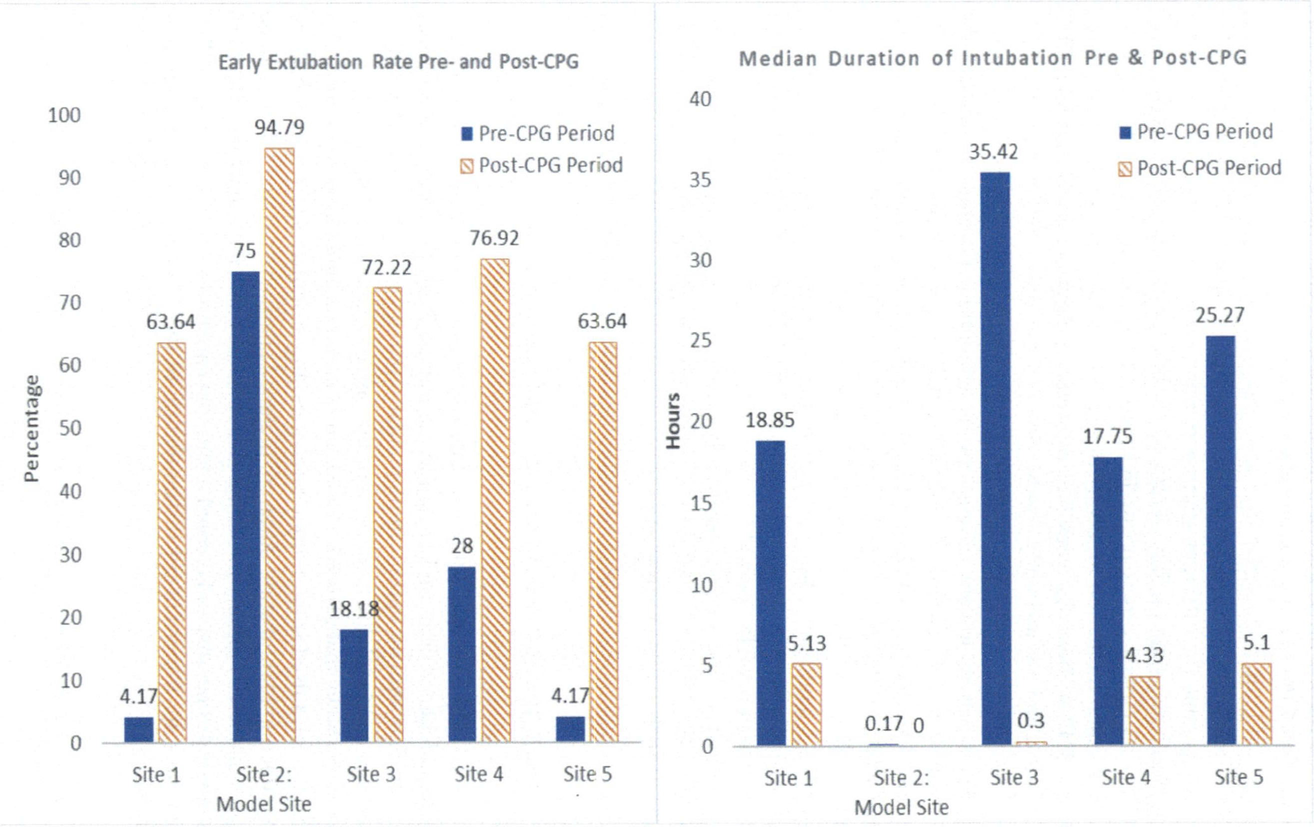

Fig. 12.8 **a** The figures compare the pre-CPG versus post-CPG early extubation rates (left). **b** The median duration of intubation 240 patients (right)

Table 12.5 Lists early extubation CPG implementation strategies used at four sites

Strategy	Site			
	3	5	1	4
1. Valued collaborative learning	■	■	■	■
2. Learned from round-robin site visits	■	■	■	■
3. At least one team member with formal QI training	■	■	■	■
4. Core team: ICU MD, ICU RN, anesthesiologist, RT	■	■	■	■
5. CPG annotated to fit local practice	■	■	■	■
6. Core team: ancillary staff, fellows	■			
7. Monthly in-person data reviews	■			
8. Collection of supplemental data to address nursing concerns	■			
9. Creation of additional care protocols to support CPG	■			
10. Monthly email updates on data to core team			■	■
11. One-on-one coaching for resistant faculty			■	
12. Structured debrief after cases where early extubation was not done		■		

12.4.7 Impact on Healthcare Cost

The field of medicine has seen a recent emphasis on both improving the quality of care and reducing costs (or optimizing healthcare value). The collaborative learning study demonstrated that sites that implemented the early extubation CPG in patients undergoing TOF and CoA repair had increased early extubation rates, decreased time to oral feeds, and earlier discontinuation of IV pain medications, and tended to have shorter ICU LOS across all the participating subjects and shorter LOS for TOF patients. We further evaluated the impact of this early extubation CPG on hospital costs and assessed which cost components were most affected (McHugh et al. 2019).

Our data source consisted of linked clinical information from this collaborative learning study and the Children's Hospital Association Inpatient Essentials Database, which captures resource utilization data. Patient-level data were linked using the method of probabilistic matching of indirect identifiers, which has been previously tested and verified in this population.

Costs were evaluated across four active sites (not counting the model site) and five control sites in the year prior to and following the CPG implementation (with a three-month washout in both periods). A washout period is defined as the time between treatment periods. Our analysis utilized generalized linear mixed-effects models, and accounted for within-center clustering, the skewed distribution of costs, and important patient characteristics that differed between the active and control groups. An econometric technique known as a difference-in-difference approach was used to assess whether any changes observed in active sites were above and beyond secular trends in the control sites.

This analysis included 428 eligible patients (McHugh et al. 2019). Figure 12.9a displays the results for the TOF and CoA cohorts. We observe a significant decrease in costs between the pre-CPG and the post-CPG periods, with 24% reduction for TOF and 12% for CoA.

Patients with TOF are often discharged shortly after birth and readmitted electively for repair at around four to six months of age. Many patients with CoA who require surgery in infancy will undergo such surgery in the first few weeks of life. While early extubation may be an important step in postoperative recovery, other factors such as advancement of feeding may be more important determinants of the discharge pathway. Also, the majority of the costs for both groups are related to postoperative care. We hypothesize that costs associated with noncardiac newborn care of patients with critical CoA may mask changes from a CPG focused on postoperative care. On average, CoA patients underwent surgery on their third day in the hospital with 62% of total hospital costs occurring on or after the day of surgery. On average, TOF patients underwent surgery on their first day in the hospital with 66% of total hospital costs occurring on or after the day of surgery.

Figure 12.9b shows the specific categories of cost reduction in the TOF cohort. The percentage change from the pre-CPG to post-CPG period is shown on the y-axis. We observe a significant reduction in clinical, pharmacy, laboratory, and imaging costs; room and board and supply costs are less significant. This analysis confirms the findings of reductions in sedative and IV analgesics usage (Amula et al. 2019; Mahle et al. 2016b; McHugh et al. 2019).

12.5 Benefits and Impacts

CHDs are the most common birth defects, occurring in approximately 1% of births. Yet conducting research in this area remains challenging. The major barriers include a small sample size of specific heart defects at any one hospital, treatment variance, and insufficient resources for national coordination of collaborative research. This data-driven OR-analytic collaborative work with the Pediatric Heart Network and the subsequent implementation and successes address these major challenges. Weaving OR-analytic tools within a collaborative learning environment, the multidisciplinary team succeeded in establishing an effective early extubation CPG for rapid implementation among multiple large pediatric heart centers and achieving improved postoperative outcomes.

This unique collaboration offers exciting scientific and medical advances that are far-reaching and immensely impactful to the quality of life of pediatric heart patients and their families. The generalizability of our analytic collaborative framework and

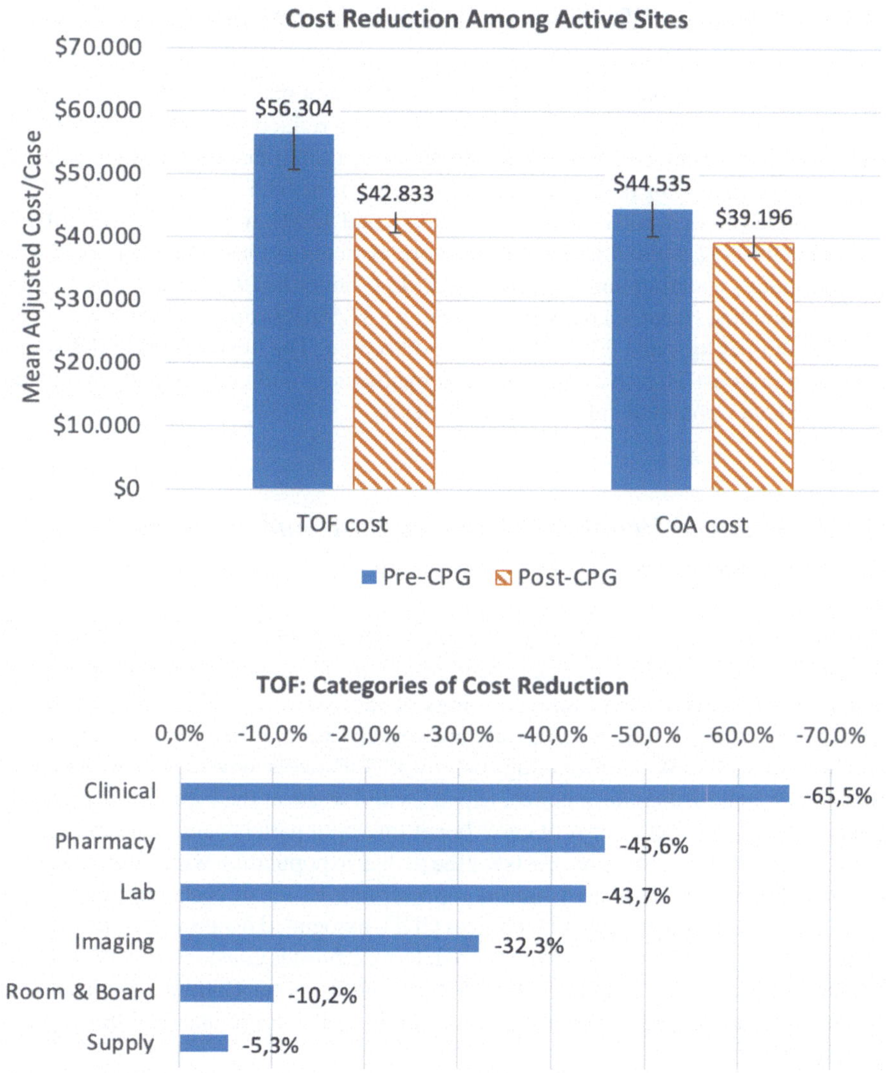

Fig. 12.9 **a** The figure shows the cost reductions for the TOF cohort and the CoA cohort. **b** The graph breaks down cost savings into various categories for cohort TOF

the extraordinarily diverse team of stakeholders magnifies its dissemination potential to other (pediatric) hospitals and to conditions beyond congenital heart disease because intubation is required in many surgical procedures.

12.5.1 Improved Efficiency of Pediatric Heart Care

Since October 2015, our CPG has become routine practice across the active sites. Patients spend less time on mechanical ventilators: a reduction of 78.8% of median intubation time (from 21.2 h to 4.5 h), 37.5% time to first enteral feeds (from 30.7 h to 19.2 h), 25.6% in postoperative ICU LOS (from 68.5 h to 51 h), 16.7% in postoperative LOS (from six days to five days), and 12.9% in postoperative critical care LOS (from 52.9 h to 46.1 h). Furthermore, patients use less medication: 55.7% less continuous sedation and (or) analgesia (43.6 hours to 19.3 hours), 46.3% less dexmedetomidine (from 5.4 per kg to 2.9 per kg), 37.5% less opioid (from 0.8 mg per kg to 0.5 mg per kg), and 33.3% less benzodiazepine. The hospitals have delivered timelier, more cost-effective care with less medication, and consequently with less stress and fewer potential side effects to patients.

12.5.2 Reduced Healthcare Delivery Costs and Reduced Waste

The operations efficiency and reduced medication highlighted above translates to a reduction in the hospital resources required to care for patients. Overall, the early extubation CPG resulted in average savings of approximately 27%, or $13,500 per surgical procedure. It reduced clinical care costs by 65%, pharmacy costs by 46%, laboratory costs by 44%, and imaging costs by 32%. Patients required fewer resources (time, labor, and medication); hence, expenses and waste at the hospitals decreased. It also reduced the overall cost of care (benefiting both patients and insurers).

In the United States, hospital costs for the pediatric population with cardiovascular defects in 2013 were about $6.1 billion, representing 15.1% of costs for all pediatric hospitalizations (Arth et al. 2017). Critical CHD–associated hospitalizations had the highest mean and median cost of the birth defect categories considered ($79,011 and $29,886, respectively), accounted for 26.7% of all costs for CHD hospitalizations, with hypoplastic left heart syndrome, coarctation of the aorta, and TOF having the highest total costs.

Each year, approximately 40,000 babies are born in the United States with a CHD and 10,000 (25%) of these babies need surgery in their first year; many require multiple surgical procedures. Hence, if implemented nationwide, our work would translate to significant procedural savings ($135 million alone for the first surgery of these 10,000 newborns). In addition, there is great potential to magnify the cost savings by adapting the CPG to conditions beyond CHDs, because intubation is required in many surgical procedures.

12.5.3 Improved Quality of Life

Earlier extubation and tube removal improve parental bonding. Earlier resumption of normal feeding and reduction in analgesics usage lessen the risks of complications. Indeed, because analgesics have been associated with cognitive impairment and impaired brain development in children, it is noteworthy that our post-CPG data indicated a reduction of 33–55% in analgesics usage (Amula et al. 2019). Shorter mechanical ventilation time and LOS for patients reduces exposure to critical care therapies and indwelling devices, which subsequently reduces the risk of HAI.

Children with congenital heart diseases face a lifelong risk of health problems such as issues with growth and eating, developmental delays, difficulty with exercise, heart rhythm problems, heart failure, and sudden cardiac arrest, or stroke. Decreased time to oral feeds implies earlier resumption of normal feeding, which lessens the risks. A shorter hospital stay (thus less medication, X-rays, and daily tests) improves overall quality of life and cumulatively could save billions of dollars in subsequent CHD management.

12.5.4 Design of Evidence-Based Practice and Cost-Effective Clinical Trial

The PHN OR-analytic collaborative approach magnifies our capability to advance pediatric heart care practice by offering an effective, flexible way to study adequate numbers of patients with uncommon diseases through a common infrastructure for recruiting, monitoring, and following patients whose conditions will be characterized in a standard fashion.

The framework facilitates the design of a common CPG, and its successful implementation with documented and measurable clinical outcomes. Such a framework permits a flexible clinical transformative environment that can accommodate practice variance while enabling care teams to identify critical system pathways for multiple-site clinical care and process improvement.

The hypothesis testing and dissemination of findings required less than two years, as opposed to conventional randomized trials that often take five years or more. Multiple sites adopted the changes with consistently good outcomes, thus providing high confidence in the clinical findings. Further, this work was accomplished with only a fraction (2%) of the budget of a typical clinical trial. Hence, the collaborative learning framework used herein can serve as a blueprint for other clinical and process-improvement initiatives.

12.5.5 *Broad Applicability*

12.5.5.1 Spillover of Early Extubation Practices

Witte et al. recently assessed early extubation rates for infants undergoing cardiac surgeries not targeted by the CPG to determine whether changes in extubation practices spill over to care of other infants (Witte et al. 2021). The study covers the four active-site hospitals. It focused on both the lower complexity cases (infants undergoing ventricular septal defect repair, atrioventricular septal defect repair, or superior cavopulmonary anastomosis) and the higher complexity cases (arterial switch operation or isolated aortopulmonary shunt). Comparing the aggregated outcome of 12 months pre-CPG practice, and 12 months after CPG study completion, it was found that among infants undergoing lower complexity surgeries, early extubation increased from 18.8% to 30.2%, and hours to initial postoperative extubation decreased. Furthermore, the surgical procedure of ventricular septal defect repair alone was associated with the largest increase from 26 to 47%. The effect differs by surgical subtypes and among sites. No difference was found for infants undergoing higher complexity surgeries. The study shows promising sustainable clinical practice one year after CPG study completion.

12.5.5.2 Beyond Pediatric Heart Care

The OR-analytic collaborative paradigm has been applied successfully to investigate practice variance in performing epidural anesthesia for cesarean sections. Our findings optimized the safe needle dose, identified clinical factors for good outcomes, and pinpointed the changes required, which guided the development of an improved CPG for training and implementation. The improved CPG resulted in a 20% reduction in analgesic usage, and significantly lower side effects (e.g., reduction in hypotension rate from 55 to 31%) (Lee et al. 2018a, b). Furthermore, although almost 50% of the drug combinations used involve fentanyl, our findings showed that fentanyl has little effect on the outcome and should be avoided in epidural anesthesia. The work facilitates delivery improvement and establishes CPG for broad dissemination. Each year, about 4 million babies are born in the United States, and about one-third of the births are delivered via cesarean section. Worldwide, about 10% to 15% of newborns (of 130 million) are delivered via medically necessary cesarean sections. Hence, our generalizable framework has far-reaching global implications.

Coincidentally, these studies reflect the successful reduction of opioid usage through quality and process improvement. Thus, our systems OR-analytic collaborative approach can serve as one viable means for combating the current opioid crisis (U.S. Department of Health and Human Services 2017; National Institute of Drug Abuse 2021).

Acknowledgements The author thanks the collaborators from multiple clinical sites, including Children's Healthcare of Atlanta, Atlanta, Georgia; C. S. Mott Children's Hospital, Ann Arbor,

Michigan; Cardiac Surgery Department, University of Michigan Medical School, Ann Arbor, Michigan; Children's Hospital of Philadelphia, Philadelphia, Pennsylvania; Perelman School of Medicine at the University of Pennsylvania, Philadelphia, Pennsylvania; Texas Children's Hospital, Houston, Texas; Departments of Pediatrics, Baylor College of Medicine, Houston, Texas; Primary Children's Hospital, Salt Lake City, Utah; Pediatric Critical Care Medicine, University of Utah School of Medicine, Salt Lake City, Utah; National Heart, Lung, and Blood Institute, National Institutes of Health, Bethesda, Maryland, and the Pediatric Heart Network Investigators, Bethesda, Maryland. We acknowledge Niquelle Brown, Cory Girard, Jinha Lee, Kevin Yee, TsungLin Wu, and Ruilin Zhou from Georgia Tech for performing some of the time-motion and system process observations. The study was supported by U01 grants from the National Heart, Lung, and Blood Institute, and the National Science Foundation (IIP- 0832390, IIP-1361532). The contents of this work are solely the responsibility of the authors and do not necessarily represent the official views of the National Heart, Lung, and Blood Institute or the National Science Foundation. We acknowledge that the clinical advances and results of this project have been reported elsewhere (Bates et al. 2018; Mahle et al. 2016a, 2016b; McHugh et al. 2019; Wolf et al. 2016; Witte et al. 2021), and we include some excerpts herein for completeness. We thank the editors and reviewers for their critical comments to improve the manuscript.

Appendix: Early Extubation Clinical Practice Guideline

In Fig. 12.10, we show the consensus CPG developed and implemented at the active sites.

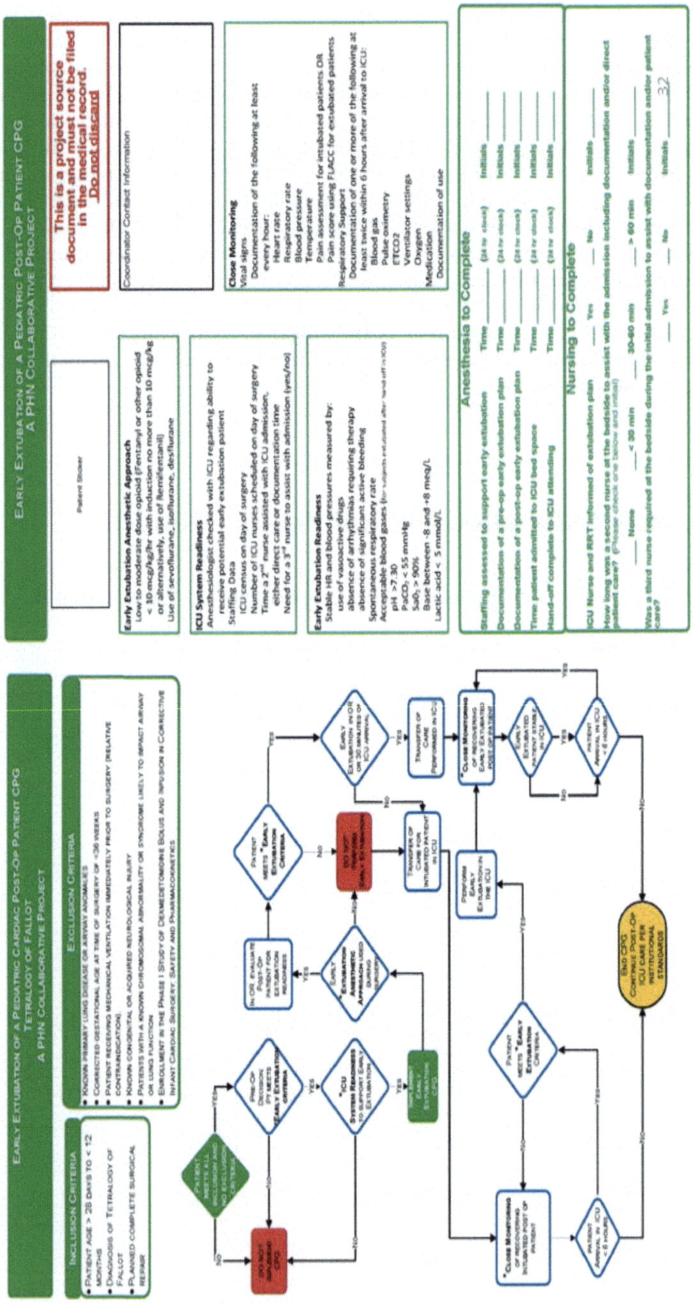

Fig. 12.10 The early extubation clinical practice guidelines facilitate best practice dissemination

References

Anand V, Kwiatkowski DM, Ghanayem NS, Axelrod DM, DiNardo J, Klugman D, Krishnamurthy G et al (2016) Training pathways in pediatric cardiac intensive care: proceedings from the 10th international conference of the pediatric cardiac intensive care society. World J Pediatric Congenital Heart Surgery 7(1):81–88

Anderson JB, Iyer SB, Schidlow DN, Williams RV, Varadarajan K, Horsley M, Lannon C (2011) Variation in interstage weight gain between surgical centers in single ventricle infants: report from the national pediatric cardiology quality improvement collaborative registry. Amer Acad Pediatrics National Conf Exhibition (American Academy of Pediatrics, Itasca, IL)

Arth AC, Tinker SC, Simeone RM, Ailes EC, Cragan JD, Grosse SD (2017) Inpatient hospitalization costs associated with birth defects among persons of all ages—United States, 2013. Centers Disease Control Prevention Morbidity Mortality Weekly Rep 66(2):1–6

Amula V, Vener DF, Pribble CG, Riegger L, Wilson EC, Shekerdemian LS, Ou Z, Presson AP, Witte MK, Nicolson SC (2019) Changes in anesthetic and postoperative sedation-analgesia practice associated with early extubation following infant cardiac surgery: experience from the pediatric heart network collaborative learning study. Pediatr Crit Care Med 10(Oct):931–939

Baker-Smith CM, Neish SR, Klitzner TS, Beekman RH, Kugler JD, Martin GR, Lannon C, Jenkins KJ, Rosenthal GL (2011) Variation in postoperative care following stage I palliation for single-ventricle patients: a report from the joint council on congenital heart disease national quality improvement collaborative. Congenit Heart Dis 6(2):116–127

Bates KE, Mahle WT, Retzloff LB, Donohue JE, Gaies MG, Nicolson SC, Shekerdemian LS et al (2019) Variation in implementation and outcomes of early extubation practices after infant cardiac surgery. Ann Thorac Surg 107(May):1434–1440

Berrington de González A, Mahesh M, Kim KP, Bhargavan M, Lewis R, Mettler F, Land C (2009) Projected cancer risks from computed tomographic scans performed in the United States in 2007. Arch. Internal Medicine 169(22):2071–2077

Bradner S (1991) Benchmarking terminology for network interconnection devices. Request for Comment, Benchmark Working Group, Internet Engineering Task Force; Fremont, CA

Brooks JP, Lee EK (2010) Analysis of the consistency of a mixed integer programming-based multi category constrained discriminant model. Ann Oper Res Data Mining 174(1):147–168

Brooks JP, Lee EK (2014) Solving a multigroup mixed-integer programming-based constrained discrimination model. INFORMS J Comput 26(3):567–585

Brown DW, Connor JA, Pigula FA, Usmani K, Klitzner TS, Beekman RH 3rd, Kugler JD et al (2011) Variation in preoperative and intraoperative care for first-stage palliation of single-ventricle heart disease: a report from the joint council on congenital heart disease national quality improvement collaborative. Congenit Heart Dis 6(2):108–115

Bruffee KA (1993) Collaborative learning: higher education, interdependence, and the authority of knowledge. Johns Hopkins University Press, Baltimore

Camp RC (1989) Benchmarking: the search for industry best practices that lead to superior performance. American Society for Quality Control, Milwaukee

Casparie AF (1996) The ambiguous relationship between practice variation and appropriateness of care: an agenda for further research. Health Policy 35(3):247–265

Chassin MR, Kosecoff J, Park RE, Winslow CM, Kahn KL, Merrick NJ, Keesey J, Fink A, Solomon DH, Brook RH (1987) Does inappropriate use explain geographic variations in the use of health care services? A study of three procedures. JAMA 258(18):2533–2537

Chiu MM (2000) Group problem-solving processes: social interactions and individual actions. J Theory Soc Behav 30(1):26–49

Damoiseaux RA, de Melker RA, Ausems MJ, van Balen FA (1999) Reasons for non-guideline-based antibiotic prescriptions for acute otitis media in The Netherlands. Fam Pract 16(1):50–53

Davis S, Worley S, Mee RB, Harrison AM (2004) Factors associated with early extubation after cardiac surgery in young children. Pediatr Crit Care Med 5(1):63–68

Dillenbourg P (1999) What do you mean by collaborative learning? In: Dillenbourg P (ed) Collaborative learning: cognitive and computational approaches (Emerald Group Publishing Limited, Bingley, UK), 1–19.
Feltus FA, Lee EK, Costello JF, Plass C, Vertino PM (2003) Predicting aberrant CpG island methylation. Proc Natl Acad Sci 100(21):12253–12258
Gallagher RJ, Lee EK, Patterson DA (1997) Constrained discriminant analysis via 0/1 mixed integer programming. Ann Oper Res 74:65–88
Gilboa SM, Salemi JL, Nembhard WN, Fixler DE, Correa A (2010) Mortality resulting from congenital heart disease among children and adults in the United States, 1999 to 2006. Circulation 122(22):2254–2263
Grytten J, Sørensen R (2003) Practice variation and physician-specific effects. J. Health Econom. 22(3):403–418
Hagen MS, Jopling JK, Buchman TG, Lee EK (2013) Priority queuing models for hospital intensive care units and impacts to severe case patients. Proc Amer Medical Informatics Assoc Ann Sympos (AMIA), (AMIA, Bethesda, MD), 841–850
Harris KC, Holowachuk S, Pitfield S, Sanatani S, Froese N, Potts JE, Gandhi SK (2014) Should early extubation be the goal for children after congenital cardiac surgery? J Thoracic Cardiovascular Surgery 148(6):2642–2648
Hu WT, Watts KD, Tailor P, Nguyen TP, Howell JC, Lee RC, Seyfried NT, Gearing M, Hales CM, Levey AI, Lah JJ, Lee EK (2016) CSF complement 3 and factor H are staging biomarkers in Alzheimer's disease. Acta Neuropathologica Comm 4(1):14–24
Jacobs JP, Pasquali SK, Gaynor JW (2011) Invited commentary: the assessment of outcomes and the improvement of quality of the treatment of patients with congenital and pediatric cardiac disease. World J Pediatric Congenital Heart Surgery 2(4):597–602
Kahn KL, Kosecoff J, Chassin MR, Solomon DH, Brook RH (1988) The use and misuse of upper gastrointestinal endoscopy. Ann Internal Medicine 109(8):664–670
Kazmin D, Nakaya HI, Lee EK, Johnson MJ, van der Most R, van den Berg RA, Ballou WR et al (2017) Systems analysis of protective immune responses to RTS, S malaria vaccination in humans. Proc Natl Acad Sci 114(9):2425–2430
Koczor CA, Lee EK, Torres RA, Boyd A, Vega JD, Uppal K, Yuan F, Fields EJ, Samarel AM, Lewis W (2013) Detection of differentially methylated gene promoters in failing and nonfailing human left ventricle myocardium using computational analysis. Physiol Genomics 45(14):597–605
Lee EK (2007) Large-scale optimization-based classification models in medicine and biology. Ann Biomedical Engrg 35(6):1095–1109
Lee EK (2012a) Children's Healthcare of Atlanta cardiac center post-op analysis practice variance report. Report, Children's Healthcare of Atlanta, Atlanta
Lee EK (2012b) C. S. Mott Children's Hospital cardiac center post-op analysis practice variance report. Report, C. S. Mott Children's Hospital, Ann Arbor, MI
Lee EK (2012c) Children's hospital of philadelphia cardiac center post-op analysis practice variance report. Report, Children's Hospital of Philadelphia, Philadelphia
Lee EK (2012d) Texas Children's Hospital cardiac center post-op analysis practice variance report. Report, Texas Children's Hospital, Houston
Lee EK (2013a) Primary Children's Hospital cardiac center post-op analysis practice variance report. Report, Primary Children's Hospital, Salt Lake City
Lee EK (2016) Machine learning for early detection and treatment outcome prediction. In: Kong N, Zhang SF (eds) Decision analytics and optimization in disease prevention and treatment: 367. Wiley Series in Operations Research and Management Science
Lee EK (2017) Innovation in big data analytics: applications of mathematical programming in medicine and healthcare. In: 2017 IEEE international conference big data (Institute of Electrical and Electronics Engineers, Washington, DC), 3586–3595
Lee EK, Gallagher RJ, Patterson D (2003) A linear programming approach to discriminant analysis with a reserved judgment region. INFORMS J Comput 15(1):23–41

Lee EK, Wu TL (2007) Classification and disease prediction via mathematical programming. In: Seref O, Kundakcioglu OE, Pardalos PM (eds) Data mining, systems analysis and optimization in biomedicine (American Institute of Physics, Melville, NY), 1–42

Lee EK, Chen CH, Pietz F, Benecke B (2009) Modeling and optimizing the public health infrastructure for emergency response. Interfaces—The Daniel H. Wagner Prize for Excellence in Operations Research Practice 39(5):476–490

Lee EK, Chen CH, Brown N, Handy J, Desiderio A, Lopez R, Davis B (2012a) Designing guest flow and operations logistics for the dolphin tales. Interfaces 42(5):492–506

Lee EK, Yuan F, Hirsh DA, Mallory MD, Simon HK (2012b) A clinical decision tool for predicting patient care characteristics: Patients returning within 72 hours in the emergency department. Proceedings annual symposium proceedings 2012b American medical informatics association (AMIA, Bethesda, MD), 495–504

Lee EK, Wu TL, Goldstein F, Levey A (2012c) Predictive model for early detection of mild cognitive impairment and Alzheimer's disease. In: Pardalos PM, Coleman TF, Xanthopoulos P (eds) Optimization and data analysis in biomedical informatics. Springer, New York, 83–97

Lee EK, Pietz F, Benecke B, Mason J, Burel G (2013) Advancing public health and medical preparedness with operations research. Interfaces 43(1):79–98

Lee EK, Atallah HY, Wright MD, Post ET, Thomas IV C, Wu DT, Haley LL (2015) Transforming emergency department workflow and patient care. Interfaces—The Franz Edelman Award Achievement in Operations Research 45(1):58–82

Lee EK, Nakaya H, Yuan F, Querec TD, Burel G, Pietz FH, Benecke BA, Pulendran B (2016a) Machine learning for predicting vaccine immunogenicity. Interfaces 46(5):368–390

Lee EK, Wang Y, Hagen MS, Wei X, Davis RA, Egan BM (2016b) Machine learning: multi-site evidence-based best practice discovery. In: Pardalos P, Conca P, Giuffrida G, Nicosia G (eds) Machine learning, optimization, and big data. MOD 2016b. Lecture Notes in Computer Science. Springer, Cham, CH, 1–15

Lee EK, Tian HZ, Lee JH, Wei X, Neeld J, Jr, Smith KD, Kaplan AR (2018a) Investigating a needle-based epidural procedure in obstetric anesthesia. Proceedings annual symposium proceedings. AMIA, Bethesda, MD, 720–729

Lee EK, Tian HZ, Wei X, Lee JH, Smith KD, Neeld J, Jr., Kaplan AR (2018b) Factors influencing epidural anesthesia for cesarean section outcome. Proceedings 2018b IEEE international conference bioinformatics biomedicine. Institute of Electrical and Electronics Engineers, Washington, DC, 1062–1069

Lee EK, Egan B. (2022) A multi-stage multi-group classification model: applications to knowledge discovery for evidence-based patient-centered care. In: Proceedings of the 14th International joint conference on knowledge discovery, knowledge engineering and knowledge management - Volume 1: KDIR, ISBN 978–989–758–614–9, ISSN 2184–3228, pp 95–108

Lee EK, Maheshwary S, Wei X (2022) Facets of conflict hypergraphs. SIAM J. Optim., forthcoming

Mahle WT, Jacobs JP, Jacobs ML, Kim S, Kirshbom PM, Pasquali SK, Austin EH, Kanter KR, Nicolson SC, Hill KD (2016a) Early extubation after repair of tetralogy of Fallot and the Fontan procedure: an analysis of the society of thoracic surgeons congenital heart surgery database. Ann Thoracic Surgery 102(3):850–858

Mahle WT, Nicolson SC, Hollenbeck-Pringle D, Gaies MG, Witte MK, Lee EK, Goldsworthy M et al (2016b) Utilizing a collaborative learning model to promote early extubation following infant heart surgery. Pediatr Crit Care Med 17(10):939–947

Mangione-Smith R, DeCristofaro AH, Setodji CM, Keesey J, Klein DJ, Adams JL, Schuster MA, McGlynn EA (2007) The quality of ambulatory care delivered to children in the United States. New England J Medicine 357(October):1515–1523

McGlynn EA, Asch SM, Adams J, Keesey J, Hicks J, DeCristofaro A, Kerr EA (2003) The quality of health care delivered to adults in the United States. New England J Medicine 348(June):2635–2645

McHugh KE, Mahle WT, Hall M, Scheurer MA, Moga M-A, Triedman JK, Nicolson SC et al (2019) Hospital costs related to early extubation after infant cardiac surgery. Ann Thorac Surg (May) 107(5):1421–1426

Menon SC, McCandless RT, Mack GK, Lambert LM, McFadden M, Williams RV, Minich LL (2013) Clinical outcomes and resource use for infants with hypoplastic left heart syndrome during bidirectional Glenn: summary from the joint council for congenital heart disease national pediatric cardiology quality improvement collaborative registry. Pediatr Cardiol 34(1):143–148

Menze BH, Kelm BM, Masuch R, Himmelreich U, Bachert P, Petrich W, Hamprecht FA (2009) A comparison of random forest and its Gini importance with standard chemometric methods for the feature selection and classification of spectral data. BMC Bioinformatics 10(1):213

Nakaya HI, Wrammert J, Lee EK, Racioppi L, Marie-Kunze S, Haining WN, Means AR et al (2011) Systems biology of vaccination for seasonal influenza in humans. Nat Immunol 12(8):786–795

Nakaya HI, Hagan T, Duraisingham SS, Lee EK, Kwissa M, Rouphael N, Frasca D et al (2015) Systems analysis of immunity to influenza vaccination across multiple years and in diverse populations reveals shared molecular signatures. Immunity 43(6):1186–1198

National Institute of Drug Abuse (2021) Opioid overdose crisis. Accessed December 15, 2021, https://www.drugabuse.gov/drug-topics/opioids/opioid-overdose-crisis

Newburger JW, Sleeper LA, Frommelt PC, Pearson GD, Mahle WT, Chen S, Dunbar-Masterson C et al (2014) Transplant-free survival and interventions at 3 years in the single ventricle reconstruction trial. Circulation 120(20):2013–2020

O'Connor GT, Plume SK, Olmstead EM, Olmstead EM, Morton JR, Maloney CT, Nugent WC, Hernandez F Jr et al (1996) A regional intervention to improve the hospital mortality associated with coronary artery bypass graft surgery. JAMA 275(11):841–846

Ohye RG, Sleeper LA, Mahony L, Newburger JW, Pearson GW, Lu M, Goldberg CS et al (2010) Comparison of shunt types in the Norwood procedure for single-ventricle lesions. New England J Medicine 362(21):1980–1992

Ohye RG, Pearson GD, Lu M, Kaltman J, Caldarone C, Derby C, Dunbar-Masterson C et al (2011) Variation in perioperative management of Norwood surgery in the pediatric heart network's single ventricle reconstruction trial. J Amer College Cardiol 57(14):E428

Ogrinc G, Mooney SE, Estrada C, Foster T, Goldmann D, Hall LW, Huizinga MM et al (2008) The SQUIRE (standards for QUality improvement reporting excellence) guidelines for quality improvement reporting: explanation and elaboration. BMJ Quality Safety 17(Suppl 1):i13–i32

Pasquali SK, Ohye RG, Lu M, Kaltman J, Caldarone CA, Pizarro C, Dunbar-Masterson C et al (2012) Variation in perioperative care across centers for infants undergoing the Norwood procedure. J Thoracic Cardiovascular Surgery 144(4):915–921

Peterson S, Eriksson M, Tibblin G (1997) Practice variation in Swedish primary care. Scandinavian J Primary Health Care 15(2):68–75

Povey B (1997) Book review: benchmarking: a tool for continuous improvement. Technovation 17(4):225–226

Prager R, Armenti FR, Bassett JS, Bell GF, Drake D, Hanson EC, Heiser JC et al (2009) Cardiac surgeons and the quality movement: the Michigan experience. Seminars Thoracic Cardiovascular Surgery 21(1):20–27

Querec TD, Akondy RS, Lee EK, Cao W, Nakaya HI, Teuwen D, Pirani A (2009) Systems biology approach predicts immunogenicity of the yellow fever vaccine in humans. Nat Immunol 10(1):116–125

Schidlow DN, Anderson JB, Klitzner TS, Beekman RH 3rd, Jenkins KJ, Kugler JD, Martin GR et al (2011) Variation in interstage outpatient care after the norwood procedure: a report from the joint council on congenital heart disease national quality improvement collaborative. Congenit Heart Dis 6(2):98–107

Sekimoto M, Imanaka Y, Evans E, Ishizaki T, Hirose M, Hayashida K, Fukui T (2004) Practice variation in perioperative antibiotic use in Japan. Internat J Quality Health Care 16(5):367–373

Shah ND, Montori VM, Krumholz HM, Tu K, Alexander GC, Jackevicius CA (2010) Responding to an FDA warning–geographic variation in the use of rosiglitazone. New England J Medicine 363(22):2081–2084

Song Y, Skinner J, Bynum J, Sutherland J, Wennberg JE, Fisher ES (2010) Regional variations in diagnostic practices. New England J Medicine 363(2):45–53

Stark J, Gallivan S, Lovegrove J, Hamilton JR, Monro JL, Pollock JC, Watterson KG (2000) Mortality rates after surgery for congenital heart defects in children and surgeons' performance. Lancet 355(9208):1004–1007

Thille AW, Richard JC, Brochard L (2013) The decision to extubate in the intensive care unit. Amer J Respiratory Critical Care Medicine 187(12):1294–1302

Tucker AL, Nembhard IM, Edmondson AC (2007) Implementing new practices: an empirical study of organizational learning in hospital intensive care units. Management Sci 53(6):894–907

U.S. Department of Health and Human Services (2017) HHS acting secretary declares public health emergency to address national opioid crisis. Accessed 3 Dec 2018, https://www.hhs.gov/about/news/2017/10/26/hhs-acting-secretary-declares-public-health-emergency-address-national-opioid-crisis.html

Veloski J, Tai S, Evans AS, Nash DB (2005) Clinical vignette-based surveys: a tool for assessing physician practice variation. Amer J Medical Quality 20(3):151–157

Vorhies DW, Morgan NA (2005) Benchmarking marketing capabilities for sustainable competitive advantage. J Marketing 69(1):80–94

Witte MK, Mahle WT, Pasquali SK, Nicolson SC, Shekerdemian LS, Wolf MJ, Zhang W, Donohue JE, Gaies M, Investigators PHN (2021) Spillover of early extubation practices from the pediatric heart network collaborative learning study. Pediatr Crit Care Med 22(2):204–212

Willson DF, Horn SD, Hendley JO, Smout R, Gassaway J (2001) Effect of practice variation on resource utilization in infants hospitalized for viral lower respiratory illness. Pediatrics 108(4):851–855

Wolf MJ, Lee EK, Nicolson SC, Pearson GD, Witte MK, Huckaby J, Gaies M, Shekerdemian LS, Mahle WT (2016) Rationale and methodology of a collaborative learning project in congenital cardiac care. Amer Heart J 174(April):129–137

Wood KA, Angus DC (2001) Reducing variation and standardizing practice in the intensive care unit. Current Opinion Critical Care 7(4):281–283

Dr. Eva K. Lee is Director of the Center for Operations Research in Medicine and Homeland Security and a Distinguished Scholar in Logistics at the Data and Analytics Innovation Institute. The center focuses on biomedicine, public health, and defense, advancing domains from basic science to translational medical research; intelligent, personalized, quality, and cost-effective delivery; medical preparedness; health and homeland security; and protection of critical infrastructures. She is a Subject Matter Expert in information and enterprise systems, risks, and decisions. She also serves as a Senior Data at Accuhealth Technologies. Previously, she served as Virginia C. and Joseph C. Mello Endowed Chair and Professor in the School of Industrial and Systems Engineering at Georgia Tech and Co-director for the NSF-I/UCRC Center for Health Organization Transformation. From 2015 to 2018, she served on the National Biodefense Science Board, a 13-member federal committee that provides advice and guidance to the President of the USA, and Secretary of US Department of Health and Human Services. She was an Expert Scientific Consultant for the Food and Drug Administration Center for Devices and Radiological Health from 2014 to 2018. She has also served as a Senior Health Systems Engineer and Professor in the US Department of Veterans Affairs.

Dr. Lee applies combinatorial optimization, math programming, game theory and machine learning, and parallel computation to biological, medical, health systems, and logistics analyses. Her clinical decision-support systems (DSS) assist in disease diagnosis/prediction, treatment design, drug delivery, treatment and healthcare outcome analysis/prediction, and healthcare operations logistics. In logistics, she tackles operations planning and resource allocation, and her

DSS addresses inventory control, vehicle dispatching, scheduling, transportation, telecom, portfolio investment, disaster emergency treatment response, and facility location/planning. She has received numerous practice excellence awards, including the INFORMS Franz Edelman Award on novel cancer therapeutics, the Daniel Wagner Prize on vaccine immunogenicity prediction, the Pierskalla Award on bioterrorism, emergency response and mass casualty mitigation, and the Caterpillar and Innovative Applications in Analytics Award on machine learning applied to multi-site best practice discovery that involves millions of patients from over 737 clinical sites. She is a Fellow at INFORMS and the American Institute for Medical and Biological Engineering. She has served on NAE/NAS/IOM, NRC, NBSB, DTRA panel committees related to CBRN and WMD incidents, public health and medical preparedness, and healthcare systems innovation. She holds ten patents on medical systems and devices. Her work has been featured in the New York Times, London Times, disaster documentaries, and in other venues.

Index

E. Çetin and H. Özen (eds.), *Healthcare Policy, Innovation and Digitalization*, Accounting, Finance, Sustainability, Governance & Fraud: Theory and Application,
https://doi.org/10.1007/978-981-99-5964-8